Biology and Culture of
Asian Seabass
Lates calcarifer

Biology and Culture of
Asian Seabass
Lates calcarifer

Editor

Dean R. Jerry

Centre for Sustainable
Tropical Fisheries and Aquaculture
School of Marine and Tropical Biology
James Cook University
Townsville, QLD
Australia

CRC Press
Taylor & Francis Group
Boca Raton London New York

CRC Press is an imprint of the
Taylor & Francis Group, an **informa** business

A SCIENCE PUBLISHERS BOOK

CRC Press
Taylor & Francis Group
6000 Broken Sound Parkway NW, Suite 300
Boca Raton, FL 33487-2742

© 2014 Copyright reserved
CRC Press is an imprint of Taylor & Francis Group, an Informa business

No claim to original U.S. Government works

International Standard Book Number: 978-1-4822-0807-8 (Hardback)

This book contains information obtained from authentic and highly regarded sources. Reasonable efforts have been made to publish reliable data and information, but the author and publisher cannot assume responsibility for the validity of all materials or the consequences of their use. The authors and publishers have attempted to trace the copyright holders of all material reproduced in this publication and apologize to copyright holders if permission to publish in this form has not been obtained. If any copyright material has not been acknowledged please write and let us know so we may rectify in any future reprint.

Except as permitted under U.S. Copyright Law, no part of this book may be reprinted, reproduced, transmitted, or utilized in any form by any electronic, mechanical, or other means, now known or hereafter invented, including photocopying, microfilming, and recording, or in any information storage or retrieval system, without written permission from the publishers.

For permission to photocopy or use material electronically from this work, please access www.copyright.com (http://www.copyright.com/) or contact the Copyright Clearance Center, Inc. (CCC), 222 Rosewood Drive, Danvers, MA 01923, 978-750-8400. CCC is a not-for-profit organization that provides licenses and registration for a variety of users. For organizations that have been granted a photocopy license by the CCC, a separate system of payment has been arranged.

Trademark Notice: Product or corporate names may be trademarks or registered trademarks, and are used only for identification and explanation without intent to infringe.

Visit the Taylor & Francis Web site at
http://www.taylorandfrancis.com

CRC Press Web site at
http://www.crcpress.com

Science Publishers Web site at
http://www.scipub.net

Preface

The Asian seabass (*Lates calcarifer*) is an euryhaline fish species within the Family Latidae that is distributed from the Persian Gulf, throughout south-east Asia, India, northern Australia, Papua New Guinea and the western Pacific. Throughout its range the species has many common names including *barramundi* (Australia), *plakapong* (Thailand), *koduva* (Sri Lanka), *kalaanji* (Malaysia), *pandugappa* and *chonak* (India), *apahap* (Philippines) and *siakap* (Indonesia), to name but a few, and is of significant cultural and economic importance, both as an important fishery, as well as being increasingly commercially farmed.

As well as the species' economic importance, due to its relatively unique life-history whereby the species exhibits protandry (sex reverts from male to female as it ages), is catadromous (spawns in marine conditions and then may undergo part of its juvenile development in freshwater), whilst also being an ecologically important keystone predator species, *L. calcarifer* has received substantial scientific attention. Consequently, there has been a plethora of publications generated on the genetics, ecology, physiology, and aquaculture of *L. calcarifer*. However, to date, there has been neither scientific synthesis nor embodiment of information on this species' biology and ecology, as well as its aquaculture exploitation. The aim of this book therefore was to bring together the accumulated knowledge on *L. calcarifer* that is current and in a form that can be easily accessed by biologists, aquaculturists and university students. Of particular emphasis in the book is the science behind aquaculture of this species, as interest in farming of *L. calcarifer* is rapidly increasing with the species now commonly farmed outside its natural range in North America, Europe and southern Australia.

Finally, as a note to readers, a large body of the scientific knowledge related to the biology, ethylogy and ecology of *L.calcarifer* is derived from Australian research and as the book was being edited and compiled it became very obvious that Australian seabass populations may not be indicative of the wider species as whole. There presently are critical knowledge gaps for *L. calcarifer* in south-east Asia and other parts of its distribution. Anecdotal reports exist of populations which are purely marine, are not completely protandrous and that exhibit different morphology and/ or behaviour. Therefore the editor feels that it is critical in the future that more scientific effort is devoted to understanding the species outside of Australia. This is essential not only for the conservation of the species, but also deeper understanding may identify traits and characteristics useful for aquaculture exploitation.

Acknowledgements

I thank all those people who have assisted in making this publication possible. First of all I would like to thank all the chapter contributors for their time and effort and for sharing their substantial knowledge on particular aspects of *L. calcarifer*. I also would like to thank all the external reviewers that took the time to review chapters and provide useful comments. These include Jeremy van der Waal, Igor Pirozzi, Kyall Zenger, Jose Domingos, and Brad Pusey. Also a big thanks to Erica Todd for her tireless editing of chapters and formatting of the bibliographies and to Felix Ayson for his initial help with pulling together the chapters and other valuable contributions.

Dean R. Jerry

Contents

Preface	v
1. **Taxonomy and Distribution of Indo-Pacific *Lates*** *Rohan Pethiyagoda* and *Anthony C. Gill*	1
2. **Early Development and Seed Production of Asian Seabass, *Lates calcarifer*** *Evelyn Grace De Jesus-Ayson, Felix G. Ayson* and *Valentin Thepot*	16
3. **Climate Effects on Recruitment and Catch Effort of *Lates calcarifer*** *Jan-Olaf Meynecke, Julie Robins* and *Jacqueline Balston*	31
4. **Reproductive Biology of the Asian Seabass, *Lates calcarifer*** *Evelyn Grace De Jesus-Ayson* and *Felix G. Ayson*	67
5. ***Lates calcarifer* Wildstocks: Their Biology, Ecology and Fishery** *D.J. Russell*	77
6. **Infectious Diseases of Asian Seabass and Health Management** *Kate S. Hutson*	102
7. **The Genetics of Asian Seabass** *Dean R. Jerry* and *Carolyn Smith-Keune*	137
8. ***Lates calcarifer* Nutrition and Feeding Practices** *Brett Glencross, Nick Wade* and *Katherine Morton*	178
9. **Post-Harvest Quality in Farmed *Lates calcarifer*** *Alexander G. Carton* and *Ben Jones*	229
10. **Farming of Barramundi/Asian Seabass: An Australian Industry Perspective** *Paul Harrison, Chris Calogeras* and *Marty Phillips*	258

11. **Nursery and Grow-out Culture of Asian Seabass,** 273
    ***Lates calcarifer*, in Selected Countries in Southeast Asia**
    Felix G. Ayson, Ketut Sugama, Renu Yashiro and
    Evelyn Grace de Jesus-Ayson

12. **Muscle Proteins in Asian Seabass—Parvalbumin's Role** 293
    as a Physiological Protein and Fish Allergen
    Michael Sharp and Andreas L. Lopata

Index 307

Color Plate Section 311

1

Taxonomy and Distribution of Indo-Pacific *Lates*

Rohan Pethiyagoda[1,]* and Anthony C. Gill[2]

1.1 Introduction

Lates calcarifer is among the most important food fishes in tropical Australasia and Asian countries bordering the Indian Ocean (Yingthavorn 1951, Rabanal and Soesanto 1982, Pender and Griffin 1996, Rimmer and Russell 1998). This species has been recorded to attain a body weight in excess of 44 kg and reach a total length of more than 135 cm in Australia (IGFA 2011). *Lates calcarifer* has long been a species of cultural and economic significance (e.g., it figures in Australian aboriginal rock art dating 15,000–8,000 ybp: Chaloupka 1997) and is an important component of recreational and commercial fisheries and aquaculture throughout its entire geographical distribution. The existence of as many as 75 local names (Mathew 2009, FAO 2011) in 14 of the countries within its range (Fig. 1.1) is additional evidence of its popularity as a species of commercial significance.

Although two excellent investigations of the higher-level relationships of *Lates* exist (Greenwood 1976, Otero 2004), a revision of the species-level taxonomy of the Indo-Pacific members of the genus has remained a desideratum (Pethiyagoda and Gill 2012). This chapter reviews the

---

[1]Australian Museum, 6 College Street, Sydney NSW 2010, Australia.
Email: rohan.pett@austmus.gov.au
[2]Macleay Museum, University of Sydney, Sydney NSW 2006, Australia.
Email: anthony.c.gill@sydney.edu.au
*Corresponding author

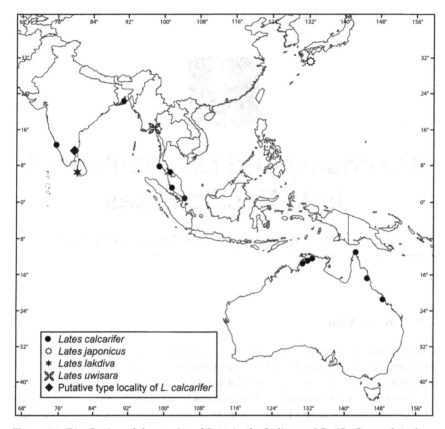

Figure 1.1. Distribution of the species of *Lates* in the Indian and Pacific Oceans based on specimens examined by Pethiyagoda and Gill (2012). *Lates* have been recorded from rivers, estuaries and coastal seas in a longitudinal range extending approximately from the eastern end of the Persian Gulf to Japan, in the approximate latitudinal range ± 25°.

current state of taxonomic knowledge of *Lates* in the countries bordering the Indian and (western) Pacific Oceans, provides aids to the identification of the known species, and remarks on conservation issues that relate to taxonomic knowledge.

1.2 Family Latidae

The family Latidae as presently understood (Mooi and Gill 1995, Otero 2004) includes three extant genera, *Psammoperca*, *Hypopterus* and *Lates*. *Psammoperca* and *Hypopterus* each include only a single species, *P. waigiensis* (Waigeo Seaperch) and *H. macropterus* (Spikey Bass), respectively. They are easily distinguished from *Lates* by having the lower edge of the preopercle smooth (vs. serrated in *Lates*), the posterior edge of the maxilla falling under (vs. distinctly behind) the eye, and the tip of the lower jaw level with that of the upper jaw (vs. lower jaw produced in *Lates*) (Figs. 1.2, 1.3). Adult *Psammoperca* and *Hypopterus* are also much smaller than *Lates*, rarely exceeding total lengths of 40 cm and 14 cm, respectively, whereas *L. calcarifer* may reach 135 cm. *Psammoperca waigiensis* is a coastal-marine fish that never enter rivers, being restricted to tropical regions of the western Indian Ocean, approximately from India to Japan. *Hypopterus macropterus* is also a coastal marine fish, occurring in sand-sea grass habitat along the central coast of Western Australia (Allen et al. 2006).

Figure 1.2 **A)** *Psammoperca waigiensis*, AMS IA.1456, 175 mm SL, Queensland, Australia; and **B)** *Hypopterus macropterus*, WAM P.30162-008, 90 mm SL, Shark Bay, Western Australia. Arrows indicate characters that distinguish the genus from *Lates* (see Fig. 1.3). 1, nostrils set widely apart; 2, lower jaw not in advance of upper jaw; 3, maxilla falls under the eye, not behind it. [Figure 1.2B courtesy Barry Hutchins, Western Australian Museum.]

Color image of this figure appears in the color plate section at the end of the book.

The genus *Lates*, as presently understood, comprises 15 extant species, 11 of which are restricted to fresh and brackish waters in tropical Africa, principally the Nile basin (Otero 2004, Pethiyagoda and Gill 2012). A number of species are known only from fossils dating to the Lower Miocene, suggestive of a much wider former distribution in rivers draining into the Mediterranean. With the severing of the Tethys Sea during the Mid-Miocene (Harzhauser and Piller 2007) the Mediterranean and Indian Ocean populations of *Lates* were isolated and embarked on separate evolutionary trajectories.

The Latidae were recognised as a distinct family by Mooi and Gill (1995). Previously, Greenwood (1976) had considered them as a subfamily of the Centropomidae, a family in which he also included the New World subfamily Centropominae—the Snooks (genus *Centropomus*). Greenwood diagnosed the family on the basis of two purportedly specialised characters: pored lateral-line scales extending to the posterior margin of the caudal fin; and the neural spine of the second vertebra markedly expanded in an anteroposterior direction. Mooi and Gill, however, noted that the first character also occurs in various other perch-like fishes and may be a primitive feature rather than a specialised one, and that the second character is found in *Centropomus* but not in Greenwood's Latinae. They further noted that Greenwood's two subfamilies differed in how the epaxial (dorsal body) musculature is attached to dorsal-fin supports. Although Mooi and Gill examined two of the extant latid genera (*Lates* and *Psammoperca*), subsequent examination of *Hypopterus* confirms that it too has a typical latid arrangement of the epaxial musculature (A. Gill, pers. obs.). Otero (2004) provided additional evidence for recognising the Latidae as a distinct family, in which she also included the fossil genus *Eolates*; she did not, however, examine *Hypopterus*.

Springer and Johnson (2004) noted that *Centropomus* and *Lates* differ little in their dorsal gill arches. In an appendix to that paper, Springer and Orrell's (2004) phylogenetic analysis of gill-arch characters of 147 spiny-finned fish families retrieved the Centropomidae and Latidae as sister taxa, but without support from uniquely shared specialisations. In contrast, other morphological studies have not supported a close relationship between the two families. Most recently, Friedman's (2012) analysis suggested that latids are more closely related to flatfishes (Pleuronectiformes) than to centropomids. Molecular analyses have similarly yielded mixed results, with some analyses indicating that latids and centropomids are each other's closest relatives (e.g., Li et al. 2011, Near et al. 2012) and others suggesting a more distant relationship (e.g., Smith and Craig 2007, Near et al. 2013). Despite marginal statistical support, Li et al. (2011) argued for inclusion of latids in the Centropomidae, even though continued recognition of separate families is not at odds with either their results or the various results of other studies. It is noteworthy that no recent morphological or molecular phylogenetic analyses have included *Hypopterus*.

1.3 Diversity within *Lates calcarifer*

Until Katayama and Taki (1984), Indo-Pacific *Lates* were considered to comprise a single species, *L. calcarifer* (Bloch 1790), in the synonymy of which several other names are placed: *L. heptadactylus* (Lacepède 1802) (locality unknown), *Lates vacti* (Hamilton 1822) (Ganges River, India), *L. nobilis* (Cuvier 1828, in Cuvier and Valenciennes 1828) (from Puducherry, Tamil Nadu, India), *L. cavifrons* (Alleyne and Macleay 1877) (from "somewhere in Torres Straits or the coast of New Guinea") and *L. darwiniensis* (Macleay 1878) (from Darwin, Northern Territory, Australia).

The subsequent delineation of the *Lates* species *L. japonicus* by Katayama and Taki (1984), however, suggested that greater diversity lies within Indo-Pacific *Lates* than hitherto suspected. Nevertheless, a comprehensive taxonomic treatment has not yet been possible for a number of reasons, principally the difficulty of obtaining series of preserved adult specimens of both sexes from across the range. Indo-Pacific *Lates* are protandrous hermaphrodites, males changing sex at a total length of about 60 cm (Dunstan 1958, 1962); specimens of this size are rarely present in museum collections. As a result, the taxonomy of these fishes has been based almost entirely on juveniles and males, reducing opportunities to assess heterochronic characters that may be differentially expressed only in the females of the various species.

Several other studies too, had suggested substantial variation among the various populations of '*Lates calcarifer*' and called for a taxonomic reassessment. Dunstan (1958, 1962) noted variation in colour and morphology in New Guinean and Australian populations of *L. calcarifer*, but these were not sufficiently unambiguous as to allow putatively distinct populations to be taxonomically differentiated (see also Grey 1987). Katayama et al. (1977) and Katayama and Taki (1984) showed the Australian and Japanese populations of *Lates* to be distinct and made available the name *L. japonicus* for the latter. However, they were unable to shed light on the true identity of *L. calcarifer sensu stricto* because the name-bearing type was thought to be lost, and because the type locality, "Japan", had been in doubt since Cuvier (in Cuvier and Valenciennes 1828) argued that it was more likely Java. Specimens said to be types have since been found (Paepke 1999), providing an opportunity to address these questions.

Advances in technology have allowed also for several other approaches to the problem, all of which have shown that there is considerable genetic variation between populations of the fishes currently assigned to *L. calcarifer*. Based on an electrophoretic analysis, Salini and Shaklee (1988) and Keenan and Salini (1990) showed that as many as 14 widely-spaced locations in Western Australia and the Northern Territory harbour genetically distinct populations of *Lates*. More recently mitochondrial DNA (Chenoweth et al. 1998) and microsatellite analyses (Jerry and Smith-Keune 2013—Chapter 7) have demonstrated the existence of an even larger number of genetically distinct stocks throughout the species' range in Australia and Southeast Asia.

Ward et al. (2008), in a study of the mitochondrial cytochrome *c* oxidase I (CO1) gene, showed the Australian and Myanmarese populations of *L. calcarifer* to differ by a Kimura 2-parameter distance of *ca* 9.5% and a cytochrome-*b* distance of 11.3%. They correctly mooted the existence of a cryptic species of *Lates* in Myanmar. Following on that study, Yue et al. (2009) genotyped populations of *L. calcarifer* from Southeast Asia and Australia and demonstrated that the Australian and Southeast Asian stocks showed substantial divergence (Jerry and Smith-Keune 2013—Chapter 7). Because of the different methods involved, it is not possible reliably to combine the results of all these authors to elucidate the relationships of the various populations studied. Nevertheless, these analyses suggest that *L. calcarifer* across its currently accepted distribution likely comprised more than a single species, and a thorough morphological examination of samples from Australia and Asia was clearly needed. This was supplied in part by Pethiyagoda and Gill (2012), who showed that in addition to *L. calcarifer* and *L. japonicus*, at least two other Indian-Ocean species exist: *L. lakdiva* and *L. uwisara*. These are briefly characterized below.

1.4 The species of Indo-Pacific *Lates*

The four species of Indo-Pacific *Lates* currently recognized are distinguished from each other as follows.

1.4.1 Lates calcarifer

Holocentrus calcarifer Bloch 1790: 100, pl. 244 (Japan).

Holocentrus heptadactylus Lacepède 1802: 344, 389 (no type locality).

Coius vacti Hamilton 1822: 86, 369, pl. 16 (Ganges River, Bengal, India).

Lates nobilis Cuvier, in Cuvier & Valenciennes 1828: 96, pl. 13 (Pondicherry, India).

Pseudolates cavifrons Alleyne and Macleay 1877: 262, pl. 3 (Torres Strait or coast of New Guinea).

Lates darwiniensis Macleay 1878: 345 (Darwin, Australia.)

Lates calcarifer (Fig. 1.3) is distinguished from *L. lakdiva* by its greater body depth (28.9–34.6% standard length (SL), vs. 26.6–27.6% SL); possessing 6 (vs. 5) rows of scales between the base of the third dorsal-fin spine and lateral line; 16–26 (vs. 31–34) serrae on the posterior edge of the preoperculum; and having the third dorsal-fin spine 2.1–2.8 (vs. 3.0–3.5) times the length of the second. It differs from *L. japonicus* in possessing 6 (vs. 7 or 8) rows of scales between the base of the third dorsal-fin spine and lateral line; 52–56 (vs. 58–63) lateral-line scales; and having the third anal-fin spine longer (vs. shorter) than the second. It differs from *L. uwisara* in possessing 6 (vs. 7) scales between the base of the third dorsal-fin spine and lateral line; having an eye diameter greater than (vs. less than) depth of maxilla; and having the length of the pelvic spine greater than that of dorsal-fin spine 5.

Distribution: Estuaries and coastal seas from south-western India to north-eastern Australia, within the approximate latitudes ± 25°, possibly extending westwards to coastal regions of the Persian Gulf and eastwards to the South China Sea and Philippine Sea.

Figure 1.3. *Lates calcarifer*, AMS I.4005-001, 183 mm SL, Northern Territory, Australia. *Color image of this figure appears in the color plate section at the end of the book.*

1.4.2 Lates japonicus

Lates japonicus Katayama and Taki 1984 (Shikoku and Kyushu Islands, Japan).

Lates japonicus (Fig. 1.4) is distinguished from *L. calcarifer* in possessing 7 or 8 (vs. 6) rows of scales between the base of the third dorsal-fin spine and lateral line; and 58–63 (vs. 52–56) lateral-line scales; and having the third anal-fin spine shorter (vs. longer) than the second. It differs from *L. lakdiva* by having the third anal-fin spine shorter (vs. longer) than the second; possessing 7 or 8 (vs. 5) rows of scales between the base of the third

Figure 1.4. *Lates japonicus*, AMS I.25742-001, 132 mm SL, Kyushu, Japan.
Color image of this figure appears in the color plate section at the end of the book.

dorsal-fin spine and the lateral line; a greater body depth (30.5–33.7% SL, vs. 26.6–27.6% SL); and 58–63 (vs. 47–52) lateral-line scales. It is distinguished from *L. uwisara* by having the third anal-fin spine shorter (vs. longer) than the second; a longer caudal peduncle (18.2–20.1% SL, vs. 15.8–17.6% SL) and a shorter preanal length (67.5–72.3% SL, vs. 72.9–75.6% SL).

Distribution: Estuaries and coastal seas off south-eastern Shikoku and Kyushu Islands, Japan.

1.4.3 Lates lakdiva

Lates lakdiva Pethiyagoda and Gill 2012 (western Sri Lanka).

Lates lakdiva (Fig. 1.5) is distinguished from *L. calcarifer* by its shallower body depth (26.6–27.6% SL, vs. 28.9–34.6% SL); possessing 5 (vs. 6) rows of scales

Figure 1.5. *Lates lakdiva*, AMS I.37516-001, 220 mm SL, western Sri Lanka.
Color image of this figure appears in the color plate section at the end of the book.

between the base of the third dorsal-fin spine and lateral line; 31–34 (vs. 16–26) serrae on the posterior edge of the preoperculum; and having the third dorsal-fin spine 3.0–3.5 (vs. 2.1–2.8) times the length of the second. It differs from *L. japonicus* by having the third anal-fin spine longer (vs. shorter) than the second; and possessing 5 (vs. 7 or 8) rows of scales between the base of the third dorsal-fin spine and lateral line, a shallower body depth (26.6–27.6% SL, vs. 30.5–33.7% SL), and 47–52 (vs. 58–63) lateral-line scales. *Lates lakdiva* differs from *L. uwisara* by having the maxilla depth less than (vs. greater than) the eye diameter, a lesser body depth (26.6–27.6% SL, vs. 28.4–32.0% SL), a shorter dorsal-fin base (41.1–42.0% SL, vs. 43.9–45.0% SL), and by possessing 5 (vs. 7) rows of scales between the base of the third dorsal-fin spine and lateral line.

Distribution: Estuaries and coastal seas off south-western Sri Lanka.

1.4.4 Lates uwisara

Lates uwisara Pethiyagoda and Gill 2012 (estuaries between Yangon and Sittang, Myanmar).

Lates uwisara (Fig. 1.6) is distinguished from *L. calcarifer* by possessing 7 (vs. 6) scales between the base of the third dorsal-fin spine and lateral line; having an eye diameter less than (vs. greater than) depth of maxilla; and having the length of pelvic spine less than or equal to that of dorsal spine 5. It differs from *L. lakdiva* by having the maxilla depth greater than (vs. less than) the eye diameter, a greater body depth (28.4–32.0% SL, vs. 26.6–27.6% SL), a longer dorsal-fin base (43.9–45.0% SL, vs. 41.1–42.0% SL), and by possessing 7 (vs. 5) rows of scales between the base of the third dorsal-fin spine and lateral line. It is distinguished from *L. japonicus* by having the third anal-fin spine shorter (vs. longer) than the second; a shorter caudal

Figure 1.6. *Lates uwisara*, ANFC H.6316-10, 353 mm SL, Myanmar.
Color image of this figure appears in the color plate section at the end of the book.

peduncle (15.8–17.6 % SL, vs. 18.2–20.1% SL) and a greater preanal length (72.9–75.6% SL, vs. 67.5–72.3% SL).

Distribution: Estuaries between Yangon and Sittang, Myanmar. The species appears to have been translocated for aquaculture before it was recognized as distinct from L. *calcarifer*, and is now known also from French Polynesia (Ward et al. 2008, FAO 2011).

1.5 A key to the species of Indo-Pacific *Lates*

1. Third anal-fin spine longer than the second   2
   - Third anal-fin spine shorter than the second   *Lates japonicus*
2. Scales between base of 3rd dorsal spine and lateral line, 5; body depth 26.6–27.6% SL   *Lates lakdiva*
   - Scales between base of 3rd dorsal spine and lateral line, 6 or 7; body depth 28.4–34.6% SL   3
3. Scales between base of 3rd dorsal spine and lateral line, 7; eye diameter 4.4–4.7% SL;
   lateral-line scales 56–59; eye diameter less than depth of maxilla
        Lates uwisara
   - Scales between base of 3rd dorsal spine and lateral line, 6; eye diameter 4.7–6.9% SL;
   lateral-line scales 52–56; eye diameter greater than depth of maxilla
        Lates calcarifer

1.6 Type Locality of *Lates calcarifer*

Given that two new species have already been recognized from within the range of L. *calcarifer*, and due also to the substantial genetic divergence that exists between several populations tentatively identified as this species, it is likely that L. *calcarifer* will be further split in the future. An important consideration then, is to establish the identity of the 'real' L. *calcarifer*.

This, however, is difficult. Bloch (1790) gave the type locality of this species as Japan. This was contested soon afterwards, however, by Cuvier (in Cuvier and Valenciennes 1828), who observed that many of the specimens Bloch thought to be from Japan were in fact from Java. This problem seemed only of passing interest so long as all the populations of Indo-Pacific *Lates* were assigned to a single species, L. *calcarifer*. When Katayama and Taki (1984) came to describe L. *japonicus*, however, the problem of the type locality of L. *calcarifer* had to be addressed. By then Bloch's type specimen/s at the Museum für Naturkunde, Berlin, had been misplaced and the authors had to rely entirely on Bloch's brief (1790) original description and somewhat inaccurate illustration (Fig. 1.7), and Cuvier's (in Cuvier and Valeciennes 1828) conjecture that the type locality of L. *calcarifer* was not Japan.

Paepke (1999) announced the discovery of two specimens thought to be Bloch's types and designated the larger of these (290 mm SL; Fig. 1.8) as lectotype of *Holocentrus calcarifer*. Pethiyagoda and Gill (2012) expressed doubt that either of these specimens were in fact those upon which Bloch based his original description. Nevertheless, in his own handwritten catalogue, Bloch referred to the specimen of *L. calcarifer* depicted on Plate 244 of Bloch (1790) as "aus Tranquebar malaiisch Atukottalei" ("from Tranquebar, in Malay [language] Atukottalei"). "Atukottalei" was apparently a misspelling of Pattukottai, a town in Tamil Nadu, India, about 90 km from Tranquebar (now Tharangambadi). Given that Bloch (1790) was clearly in error in mentioning Japan as the type locality, it appears reasonable now to conclude that the type locality of *L. calcarifer* is in fact Pattukottai.

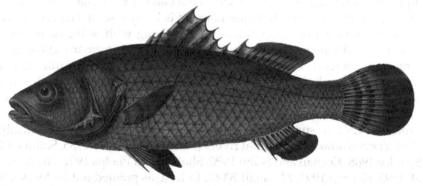

Figure 1.7. Iconotype of *Holocentrus calcarifer* Bloch 1790; pl. 244, laterally inverted.

Color image of this figure appears in the color plate section at the end of the book.

Figure 1.8. *Holocentrus calcarifer* Bloch 1790, lectotype, ZMB 13652, 273 mm ["290 mm": Paepke 1999] SL; right skin, photograph laterally inverted. Photo: courtesy of Jörg Freyhof.

This offers an opportunity to establish the natural population of *Lates* at Pattukottai as the 'true' *L. calcarifer*, with future molecular analysis possibly paving the way to confirming or rejecting the validity of one or both of Bloch's putative type specimens in Berlin, thereby allowing for the identity of *L. calcarifer* to be established with certainty. This prospect may, however, have already been confounded by the introduction from elsewhere of *Lates* for aquaculture or fisheries enhancement in the Pattukottai area.

1.7 Conservation and Genetics

Lates calcarifer has been successfully bred in captivity since 1973 (Barnabe 1995), with widespread introductions for fisheries enhancement and aquaculture having taken place across Asia and Australasia in the course of the past four decades. In many places it is likely also that escapes have occurred from captive populations, hybridizing with indigenous—and potentially distinct—populations of *Lates* (see Lintermans 2004, and references therein). With *Lates* being commonly cultured or introduced in countries such as India, Malaysia, Sri Lanka and Thailand, such hybridization has probably occurred already on a more or less wide scale. This is particularly worrying in the light of several studies that have produced data to suggest that wild populations of *Lates* are genetically distinct even among individual rivers (Shaklee and Salini 1985, Salini and Shaklee 1988, Keenan and Salini 1990, Shaklee and Phelps 1991, Shaklee et al. 1993, Keenan 1994, Marshall 2005). Indeed, as pointed out by Marshall (2005), "Despite the barramundi's catadromous life history, and ability to disperse through marine waters, the present genetic structure indicates a division principally among river drainages. From a population [genetics] viewpoint the species can be regarded as [a] freshwater, rather than marine [fish]."

The recent description of two new species of seabass—*L. lakdiva* and *L. uwisara*—in countries bordering the Bay of Bengal (Pethiyagoda and Gill 2012) suggests that others await discovery in countries bordering the Persian Gulf, Arabian Sea and South China Sea, from which *Lates calcarifer* has been reported, but as yet not subjected to taxonomic scrutiny. In some of these countries, translocations may make it difficult to work out the identity of the indigenous population. Thus, in addition to regulating translocations and aquaculture practices, the preservation of museum specimens becomes important, with museum specimens that predate the era of introductions acquiring added value. The data of Pethiyagoda and Gill (2012) suggest that the external morphology of *L. calcarifer* is remarkably conserved across its range despite substantial genetic variation, an example of morphologically static cladogenesis (Bickford et al. 2006). They also suggest that *L. calcarifer* is a widely distributed species within whose range pockets of distinct species of *Lates* occur. Further species almost certainly exist within the range of *L. calcarifer* as here understood, and are likely to be uncovered by future exploration. Until then conservative translocation policies restricting movement of '*Lates calcarifer*' throughout the western and south east Asian region of its distribution need to be developed.

1.8 Acknowledgements

We thank Barry Hutchins (Western Australian Museum, Perth) and Jörg Freyhof (Leibniz-Institute of Freshwater Ecology and Inland Fisheries, Berlin) for permission to reproduce the photographs Figs. 1.2 and 1.8, respectively; Maurice Kottelat (Cornol, Switzerland) and Barry Russell (Museums and Art Galleries of the Northern Territory, Darwin) for comments that helped improve the paper (Pethiyagoda and Gill 2012) upon which this chapter is substantially based; and Mark McGrouther and Amanda Hay (Australian Museum, Sydney) for facilities to conduct this research and for many kindnesses, especially in sourcing material.

References

Allen, G.R., C.F. Allen and D.F. Hoese. 2006. Latidae. *In:* P.L. Beesley and A. Wells (eds.). Zoological Catalogue of Australia (vol. 35). ABRS and CSIRO Publishing, Canberra, pp. 966–968.

Alleyne, H.G. and W. Macleay. 1877. The ichthyology of the Chevert expedition. Proc. Linn. Soc. N.S.W. 1 (3-4): 261–281, 321–359, pls. 3–9, 10–17.

Barnabe, G. 1995. The Sea Bass. *In:* C.E. Nash and A.J. Novotny (eds.). Production of aquatic animals: Fishes. World Animal Science C8, Elsevier Science B.V., Amsterdam, pp. 269–287.

Bickford, D., D.J. Lohman, N.S. Sodhi, P.K.L. Ng, R. Meier, K. Winker, K.K. Ingram and I. Das. 2006. Cryptic species as a window on diversity and conservation. Trends Ecol. Evol. 22: 148–155.

Bloch, M.E. 1790. Naturgeschichte der ausländischen Fische. Morino, Berlin.

Chaloupka, G. 1997. Journey in Time: the world's longest continuing art tradition. The 50,000-year story of the Australian aboriginal rock art of Arnhem Land. Reed Books, Sydney.

Chenoweth, S.F., J.M. Hughes, C.P. Keenan and S. Lavery. 1998. When oceans meet: a teleost shows secondary intergradation at an Indian-Pacific interface. Proc. R. Soc. Lond., B. 265: 415–420.

Cuvier, G. and A. Valenciennes. 1828. Histoire naturelle des poissons. Tome second. Livre Troisième. Des poissons de la famille des perches, ou des percoïdes. Levrault, Paris.

Dunstan, D.J. 1958. The barramundi, *Lates calcarifer* (Bloch) in Queensland waters. CSIRO Division of Fisheries and Oceanography Technical Paper 5: 1–22.

Dunstan, D.J. 1962. The barramundi in New Guinea waters. Papua New Guinea Agric. J. 15: 23–31.

[FAO] United Nations Food and Agriculture Organization. 2011. *Lates calcarifer*. Cultured Aquatic Species Information Programme, Fisheries and Aquaculture Department. Available from http://www.fao.org/fishery/culturedspecies/Lates_calcarifer/en (accessed 28 November 2012).

Friedman, M. 2012. Osteology of †*Heteronectes chaneti* (Acanthomorpha, Pleuronectiformes), an Eocene stem flatfish, with a discussion of flatfish sister-group relationships. J. Vertebr. Paleontol. 32: 735–756.
Greenwood, P.H. 1976. A review of the family Centropomidae (Pisces, Perciformes). Bull. Br. Mus. Nat. Hist. (Zool.) 29: 1–81.
Grey, D.L. 1987. An overview of *Lates calcarifer* in Australia and Asia. *In:* J.W. Copland and D.L. Grey (eds.). Management of wild and cultured sea bass/barramundi (*Lates calcarifer*). Australian Centre for International Agricultural Research, Canberra, pp. 15–21.
Harzhauser, M. and W.E. Piller. 2007. Benchmark data of a changing sea—palaeogeography, palaeobiogeography and events in the Central Paratethys during the Miocene. Palaeogeogr. Palaeoclimatol. Palaeoecol. 253: 8–31.
Hamilton, F. 1822. An account of the fishes found in the river Ganges and its branches. Edinburgh and London.
[IGFA] International Game Fishing Association. 2011. Available from http://www.igfa.org/records/Fish-Records.aspx?Fish=BarramundiandLC=ATR (accessed 28 November 2012).
Jerry, D.R. and C. Smith-Keune. 2013. The Genetics of Asian Sea Bass *Lates calcarifer*. *In:* D.R. Jerry (ed.). Biology and Culture of Asian Sea Bass *Lates calcarifer*. CRC Press.
Katayama, M., T. Abe and T. Nguyen. 1977. Notes on some Japanese and Australian fishes of the family Centropomidae. Bull. Tokai Reg. Fish. Res. Lab. 90: 45–55.
Katayama, M. and M. Taki. 1984. *Lates japonicus*, a new centropomid fish from Japan. Jpn. J. Ichthyol. 30: 361–367.
Keenan, C.P. 1994. Recent evolution of population structure in Australian barramundi, *Lates calcarifer* (Bloch): an example of isolation by distance in one dimension. Aust. J. Mar. Freshwater Res. 45: 1123–1148.
Keenan, C.P. and J. Salini. 1990. The genetic implications of mixing barramundi stocks in Australia. Proc. Aust. Soc. Fish Biol. 8: 145–150.
Lacepède, B.G.E. 1802. Histoire naturelle des poisons (vol. 4.). Plassan, Paris.
Li, C., B.-R. Ricardo, W.L. Smith and G. Orti. 2011. Monophyly and interrelationships of snook and barramundi (Centropomidae sensu Greenwood) and five new markers for fish phylogenetics. Mol. Phylogenet. Evol. 60: 463–471.
Lintermans, M. 2004. Human-assisted dispersal of alien freshwater fish in Australia. N. Z. J. Mar. Freshwater Res. 38: 481–501.
Macleay, W. 1878. The fishes of Port Darwin. Proc. Linn. Soc. N. S. W. 2: 344–367, Pls. 7–9.
Marshall, C.R.E. 2005. Evolutionary genetics of barramundi (*Lates calcarifer*) in the Australian region. Ph.D. Thesis, Murdoch University, Perth, Western Australia.
Mathew, G. 2009. Taxonomy, identification and biology of Seabass (*Lates calcarifer*). Cage Culture of Seabass. Proc. Cent. Mar. Fish. Res. Inst. (India), pp. 38–43.
Mooi, R.D. and A.C. Gill. 1995. Association of epaxial musculature with dorsal-fin pterygiophores in acanthomorph fishes, and its phylogenetic significance. Bull. Nat. Hist. Mus. Lond. (Zool.). 61: 121–137.
Near, T., M. Sandel, K. L. Kuhn, P. J. Unmack, P. C. Wainwright and W. L. Smith. 2012. Nuclear gene-inferred phylogenies resolve the relationships of the enigmatic pygmy sunfishes, Elassoma (Teleostei: Percomorpha). Mol. Phylogenet. Evol. 63: 388–395.
Near, T.J., A. Dornburg, R.I. Eytan, B.P. Keckb, W.L. Smith, K.L. Kuhn, J.A. Moore, S.A. Price, F.T. Burbrink, M. Friedman and P.C. Wainwright. 2013. Phylogeny and tempo of diversification in the superradiation of spiny-rayed fishes. Proc. Natl. Acad. Sci. USA., www.pnas.org/cgi/doi/10.1073/pnas.1304661110.
Otero, O. 2004. Anatomy, systematics and phylogeny of both recent and fossil latid fishes (Teleostei, Perciformes, Latidae). Zool. J. Linn. Soc. 141: 81–133.
Paepke, H.-J. 1999. Bloch's fish collection in the Museum für Naturkunde der Humboldt-Universität zu Berlin: an illustrated catalog and historical account.*Theses Zoologicae*, 32. A.R.G. Gantner, Ruggel.

Pender, P.J. and R.K. Griffin. 1996. Habitat history of barramundi *Lates calcarifer* in a north Australian river system based on barium and strontium levels in scales. Trans. Am. Fish. Soc. 125: 679–689.

Pethiyagoda, R. and A.C. Gill. 2012. Description of two new species of sea bass (Teleostei: Latidae: *Lates*) from Myanmar and Sri Lanka. Zootaxa 3314: 1–16.

Rabanal, H.R. and V. Soesanto. 1982. Report of the training course on seabass spawning and larval rearing—Songkhla, Thailand—1 to 20 June 1982. FAO Corporate Document Repository, SCS/GEN/82/39, 120 pp. Available from http://www.fao.org/docrep/field/003/Q8694E/Q8694E00.htm (accessed 27 November 2012).

Rimmer, M.A. and D.J. Russell. 1998. Survival of stocked barramundi, *Lates calcarifer* (Bloch), in a coastal river system in far northern Queensland, Australia. Bull. Mar. Sci. 62: 325–336.

Salini, J.P. and J.B. Shaklee. 1988. Genetic structure of barramundi (*Lates calcarifer*) stocks from northern Australia. Aust. J. Mar. Freshwater Res. 39: 317–329.

Shaklee, J.B. and S.R. Phelps. 1991. Analysis of fish stock structure and mixed-stock fisheries by the electrophoretic characterization of allelic isozymes. *In*: D.H. Whitmore (ed.). Electrophoretic and isoelectric focusing techniques in fisheries management. CRC Press, Boca Raton, Florida, pp. 173–196.

Shaklee, J.B. and J.P. Salini. 1985. Genetic variation and population subdivision in Australian barramundi, *Lates calcarifer* (Bloch). Aust. J. Mar. Freshwater Res. 36: 203–218.

Shaklee, J.B., J.P. Salini and R.N. Garrett. 1993. Electrophoretic characterization of multiple genetic stocks of barramundi (*Lates calcarifer*) in Queensland, Australia. Trans. Am. Fish. Soc. 122: 685–701.

Smith, W.L. and M.T. Craig. 2007. Casting the percomorph net widely: The importance of broad taxonomic sampling in the search for the placement of serranid and percid fishes. Copeia 2007: 35–55.

Springer, V.G. and G.D. Johnson. 2004. Study of the dorsal gill-arch musculature of teleostome fishes, with special reference to the Actinopterygii. Bull. Biol. Soc. Wash. 11: 1–260.

Springer, V.G. and T.M. Orrell. 2004. Phylogenetic analysis of the families of acanthomorph fishes based on dorsal gill-arch muscles and skeleton. Bull. Biol. Soc. Wash. 11: 237–254.

Ward, R.D., B.H. Holmes and G.K. Yearsley. 2008. DNA barcoding reveals a likely second species of Asian sea bass (barramundi) (*Lates calcarifer*). J. Fish Biol. 72: 458–463.

Yingthavorn, P. 1951. Notes on Pla-Kapong (*Lates calcarifer* Bloch) culturing in Thailand. [FAO] United Nations Food and Agriculture Organization Fisheries Biology Technical Paper 20: 1–6.

Yue, G.H., Z.Y. Zhu, L.C. Lo, C.M. Wang, G. Lin, F. Feng, H.Y. Pang, J. Li, P. Gong, H.M. Liu, J. Tan, H. Lim and L. Orban. 2009. Genetic variation and population structure of Asian seabass (*Lates calcarifer*) in the Asia-Pacific region. Aquaculture 293: 22–28.

2

Early Development and Seed Production of Asian Seabass, *Lates calcarifer*

Evelyn Grace De Jesus-Ayson,[1,]* *Felix G. Ayson*[1] and *Valentin Thepot*[2]

2.1 Introduction

The Asian seabass *Lates calcarifer* (Bloch) is widely distributed in the Indo-West Pacific. Because of its high commercial value, it is widely cultured in most of Asia and Australia. However, seed supply from the wild is not abundant; hence the aquaculture of seabass is based on mass seed production in hatcheries. Seabass spawn naturally in captivity (Toledo et al. 1991). Alternatively, they can be induced to spawn by hormonal or environmental manipulations (Kungvankij 1987, Garcia 1989a, b). The development of methods for broodstock management for reliable reproduction of seabass under captive conditions paved the way for studies on the environmental, physiological and nutritional requirements of the developing larvae and the development of protocols for seed production in the hatchery (e.g., Parazo et al. 1998, Schipp et al. 2007).

---

[1]Aquaculture Department, Southeast Asian Fisheries Development Center, Tigbauan, Iloilo 5021.
[2]Center for Sustainable Tropical Fisheries and Aquaculture, James Cook University, Townsville, Queensland, Australia.
*Corresponding author

Survival of marine fish larvae is dependent on the interplay of various environmental factors (e.g., temperature and water quality, as well as availability of good quality and nutritionally adequate food supply) with a suite of species-specific characteristics including egg and larval size, amount of energy reserves (volume of yolk and oil globule, if present), utilization rates of energy stores (resorption rates of yolk and oil globule), metabolic demand, initiation of feeding and feeding success, growth rates, swimming and feeding behavior. In general, size provides an advantage (i.e., large eggs give rise to large larvae with large yolk reserves which will sustain larval development and allow the larvae sufficient time to initiate feeding before the onset of irreversible starvation (May 1974)).

This Chapter outlines the characteristics of *L. calcarifer* eggs and larvae, the changes during embryonic and larval development, advances in seed production and at the same time highlights the relative ease in its mass production.

2.2 Embryonic Development

Seabass spawn naturally in tanks and in floating net cages. Spawning is highly predictable and occurs 3–4 days before or after the quarter moon phase, usually at night (Toledo et al. 1991). The fertilized egg is pelagic and measures around 0.80 mm in diameter.

Embryonic development lasts about 14–16 h at 27–30°C (Table 2.1, Kungvankij 1987, Maneewongsa and Tattanon 1982a, b). Cell division starts within 30 min after fertilization. The developing embryo attains blastula, gastrula and neurula stages around 5.5, 6.5–7 and 8.5–9.5 h after

Table 2.1. Timing of embryonic development stages in *L. calcarifer* at 27–28°C.

Developmental Stage	Time after fertilization
2-cell stage	35–40 min
4-cell stage	45–55 min
8-cell stage	1–1 h and 10 min
16-cell stage	1 h and 30 min
32-cell stage	1 h and 50 min to 2 h and 15 min
64-cell stage	2 h and 20 min to 2 h and 45 min
128-cell stage	2 h and 55 min to 3 h
Blastula	5.5 h
Gastrula	6.5–7 h
Neurula	8.5–9 h
Early embryo with eye vesicle formed	11 h and 20 min to 11 h and 50 min
Heart and twitching observed	15.5 h
Hatching	17.5–18 h

fertilization, respectively. The embryo with the eye vesicle already formed is observed between 11–12 h post fertilization. Twitching of the embryo is observed beginning around 15 h and hatching occurs between 17.5–18 h after fertilization. Tables 2.2 and 2.3 list summary data on *L. calcarifer* egg and larval characteristics and timing of critical developmental stages, respectively, whilst Figs. 2.1 and 2.2 illustrate the key developmental stages from egg to juvenile metamorphosis.

Newly-hatched seabass larvae measure ~1.40 mm total length. The endogenous nutrients of seabass larvae are the yolk and the oil globule. At hatching, the yolk of seabass larvae measure 0.0801 ± 0.0007 µl and the oil globule 0.0058 ± 0.0006 µl in volume (Bagarinao 1986). About 87% of the yolk is resorbed in the first 16 h after hatching and is usually completely resorbed

Table 2.2. Characteristics of *L. calcarifer* eggs and larvae (from Bagarinao 1986).

Characteristics	Description
Egg type	Pelagic
Egg diameter	0.80 mm
Egg volume	0.2681 µl
Length at hatching	1.72 mm
Yolk volume at hatching	0.0859 µl
Maximal larval SL attained on the yolk reserves	2.45 mm
Yolk volume remaining when maximal larval SL attained	0.0072 µl
Larval SL at onset of feeding	2.3 mm
Mouth width at opening	250 µm

Table 2.3. Timing of events during early development of *L. calcarifer* larvae (from Bagarinao 1986, Kohno et al. 1986).

Events	Time (hr from hatching)
Eye lens and auditory vesicles developed	0 h
Neuromasts first appear on head and body	16 h
First eye pigment	24 h
Pectoral fin buds appear	29 h
Total eye pigmentation	32 h
Opening of the mouth (mouth width at opening is 250 µm)	32 h
Earliest availability of food	32 h
Earliest food incidence in the gut	50 h
Complete resorption of energy reserves	120 h
Irreversible starvation	60 h
100% death of unfed larvae	144 h
Time available to initiate feeding	88 h
Time taken to initiate feeding	18 h

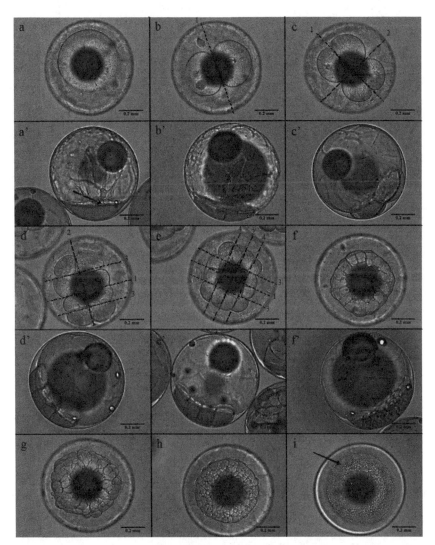

Figure 2.1. Development of *Lates calcarifer* polar and meridian view (') from zygote to germinal ring. Cytoplasmic movements start 5 min post-fertilization and stay visible until the second cleavage (a', b', c'). During the zygote period, 1-cell stage (a), the cytoplasm flows to the animal pole and segregates the blastoderm from the yolk (arrow) (a'). The first cleavage separates the blastoderm into two undifferentiated blastomeres of equal volume (b). The second cleavage occurs on a plane (2) perpendicular to the first plane (1) (c). This alternation of dividing planes will stay obvious until the 16-cell stage (e). The 8-cell and 16-cell stages are the result of two synchronous cleavages (d, e). From the 32-cell stage onward the cell estimate is based on the cell size rather than count. The 128-cell stage marks the onset of the blastula period (h). At the germinal ring stage the blastoderm folds at its margin and spreads over the yolk as epiboly starts (i) Photo: V. Thepot.

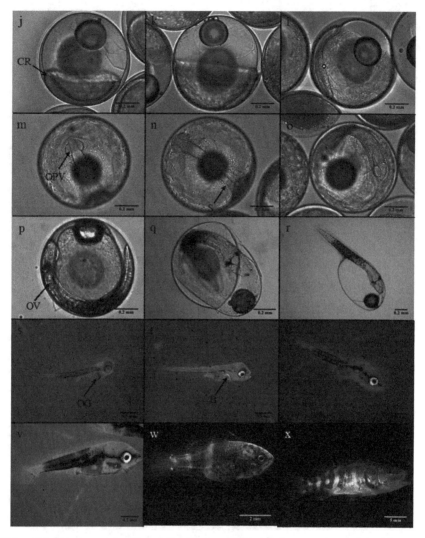

Figure 2.2. Development of *Lates calcarifer* from 30% epiboly to metamorphosed larvae. The epiboly process, during which the cranial region (CR) is identifiable, consists of the envelopment of the yolk by the blastodermal cells to form the yolk sac (j, 30%; k, 50%; l, 85%). Soon after 100% epiboly is reached the optic vesicles (OPV) are in their ending position (m). The first somites to form are anteriorly located and develop rapidly to form muscle segments (n). Prior to hatch the two largest otoliths are clearly visible within the otic vesicle (OV) (p). At 1 day post hatch (1 DPH) the larvae has absorbed the totality of the yolk sac (s) and the oil globule (OG) is absorbed between 2 and 3 DPH, also coinciding with the swim bladder inflation (t). Flexion starts between 5 and 6 DPH (u); final fin shape is acquired by 16 DPH (w). At 23 DPH the metamorphosed larvae is fully pigmented although patterns are still of a juvenile fish (x). Photo: V. Thepot

by 71 h post-hatch (PH). The oil globule is scarcely resorbed while the yolk is still available. Oil globule resorption commences at about the same time as when the yolk is completely resorbed and totally disappears at 120–145 h PH. Based on these observations, the pattern of utilization of the endogenous source of energy is as follows, (1) rapid yolk resorption from hatching to 16 h PH, (2) slow yolk resorption from 17 to 60–70 h PH, and (3) oil globule resorption from 60–70 h to 120–145 h PH (Kohno et al. 1986).

Protein and lipids are the main energy reserves of fish larvae (Heming and Buddington 1988). In *L. calcarifer* larvae, lipid content decreases by about half from hatching to 9 days PH, where 62% of the lipid consumed were neutral lipids, while protein decreased by only 4.3% during the same period. This indicates that lipids are the major energy source of the developing larvae (Southgate et al. 1994). The neutral lipid triglycerides make up the bulk of yolk lipids in fish eggs (Terner 1979). This is in agreement with the observation that the yolk of seabass is almost completely resorbed at 70 h after hatching.

L. calcarifer larvae grow to 2.45 ± 0.04 mm standard length (SL) in about 24 h, by which time yolk had been reduced to about 5% and the oil globule to about 60%. Growth also plateaus on Days 2 to 5. Unfed larvae measure 2.36 ± 0.02 mm SL on Days 3 to 6, whereas fed larvae measure 2.57 ± 0.04 mm SL on Day 5. Larval growth phase could be described into (a) rapid early growth from hatching to about 15 h after hatching (dependent on yolk), (b) morphological differentiation and slow growth dependent on energy from remaining yolk reserves until about 50 h (2 days) after hatching, (c) slow growth with initiation of feeding and swimming activities supported by energy from oil globule and exogenous food to about 110 h (4–5 days) PH, (d) accelerated growth and effective feeding and swimming fueled by energy from remaining oil globule and exogenous feeding up to 120–140 h (5–6 days) after hatching and (e) fast growth with energy derived solely from exogenous food (Kohno et al. 1986).

Seabass larvae hatch with non-functional mouth and eyes, although the eye lens and auditory vesicles are already developed (Kohno et al. 1986). Neuromasts first appear on the head and body at 16 h and increase in number at 23 h after hatching (Kohno et al. 1986). Eye pigmentation starts at 24 h and is completed (total pigmentation) at 32 h PH. At the same time, the mouth opens at 32 h after hatching with mouth width opening of 250 µm. The earliest food incidence in the gut is observed at 50 h after hatching. The energy reserves (yolk and oil globule) are completely resorbed at 120 h PH. The time available to initiate feeding is 88 h after hatching, however, seabass larvae start feeding early when food is available with the time taken to initiate feeding at just 18 h (Bagarinao 1986).

Kohno et al. (1986) also studied the behavioral characteristics of seabass larvae. At hatching, the larvae are floating with head directed upward. At

10 h post hatching, two pairs of neuromasts appear on the head and body and the larvae shows positive response to pipetting stimulus; at 23 h post hatching, the body neuromasts increased to four pairs and the larvae become highly sensitive. The pectoral fin buds appeared at 29 h and at 47 h PH, the larvae drifted with the water current. The pectoral fins are movable and larvae form loose patches (partial grouping) in the tank at 71 h, while at 96 h PH the larvae form compact patches (dense crowding) in the tank and are able to maintain their position against the water current. At 121 h PH, the larvae exhibits positive phototaxis and use the pectoral fins to maintain position in the water column (free positioning of body).

2.3 Osteological Development of Feeding Apparatus

The development of bony structures important for feeding like the neurocranium, jaws, suspensorium, hyoid, branchial arches, opercular bones and fins in seabass are described by Kohno et al. (1996). The bony maxilla appeared at 9.5 h after mouth opening. The Meckel's cartilage (MC) is the first jaw element to appear (~15.5 h after mouth opening) with its anterior tip located midway between the eyes and initial ossification occurring roughly at 58 h after mouth opening. Premaxillary teeth (PT) also appeared at this time. Dentary teeth (DT) were first evident at 81 h and the number of PT and DT reached 8–19 and 3–5, respectively, at 156.5 h after mouth opening.

For the suspensorium, the symplectic–hyomandibular cartilage (SHC) appears first at 7.5 h after mouth opening followed by the cartilaginous quadrate. A small foramen develops first on the upper part of the SHC at 35.5 h and initial ossification is first evident on the lower part of SHC at 58 h after mouth opening. The cartilaginous palatine which was first visible at 68.5 h is contiguous with the quadrate forming a wide, triangular cartilaginous plate called the palatoquadrate cartilage at 104.5 h after mouth opening. At this time, the hyomandibular starts ossifying at the upper part of the SHC as well.

In the hyoid arch, the ceratohyal-epihyal cartilage (CEC) appears at 15.5 h, while the hypohyal and interhyal cartilages appear at 9.5 h after mouth opening. At 58 h, the ceratohyal starts to ossify and a single branchiostegal ray is formed, whereas the interhyal starts to ossify at 156.5 h after mouth opening and 5-6 branchiostegal rays are already evident.

In the branchial arch, the lower arch elements, the cartilaginous basibranchial and two ceratobranchials (1–2) form at 7.5 h, ceratobranchials 3–4 at 9.5 h and ceratobranchial 5 at 68.5 h after mouth opening. In the upper arch, the first to appear is the cartilaginous pharyngobranchial 3 bearing 2 teeth at 58 h, while pharyngobranchial 2 appeared at 81 h after mouth opening. Four cartilaginous epibranchials were observed at 68.5 h after mouth opening. The number of upper pharyngeal teeth increased to 6–7 at 156.5 h after mouth opening.

As to the opercular bones, both the opercle and preopercle appeared at 58 h while the interopercle and the subopercle came much later at 81 and 104.5 h after mouth opening, respectively.

For the neurocranium, the cartilaginous trabercula appears at 7.5 h, while the cartilaginous ethmoid, auditory capsule and paracordal form at 9.5 h after mouth opening. The occipital process becomes evident at 44.5 h followed by the supraorbital bar at 58 h after mouth opening. Contiguous with the supraorbital bar, the cartilaginous epiphysial tectum and ectethmoid bar appears at 104.5 h after mouth opening.

For the fins, the bony cleithum appears first at 7.5 h and the coracoscapular cartilage at 21.5 h after mouth opening. The bony supracleithum appears at 105.5 h and the three cartilaginous hypurals in the caudal fin are first evident at 104.5 h after mouth opening.

Within 10 h after mouth opening, the larva is already equipped with the fundamental bony elements forming the oral cavity such as the trabecular roof, some elements of the lower branchial arch and all of the hyoid arch forming the floor, and the quadrate and symplectic-hyomandibular cartilages making up the sides.

The development of the structures involved in feeding can be divided into three phases: a) the first phase can be characterized by the presence of the basic oral cavity elements that prepares the larvae for a "sucking" mode of feeding and which occurs from initial mouth opening to 50–60 h afterwards; b) the second phase can be characterized by the acquisition of "new" elements and their functional development, and the initial ossification of the earlier feeding structures which allows the larvae to acquire "grasping" capabilities in addition to "sucking" which takes place from between 50–60 to 100–110 h after mouth opening; c) the third phase in which development of the feeding elements are completed and ossified which brings about an improved feeding ability using both sucking and grasping movements to capture and ingest food organisms from 110 h after mouth opening.

2.4 Development of the Digestive System

Minjoyo et al. 2003 studied the development of the digestive tract and the ontogeny of some digestive enzymes in larval to juvenile seabass. The digestive tract of the newly-hatched seabass larvae was a simple tube that became differentiated into esophagus, stomach and intestine in the 2-day old larvae. This indicated that larvae can be provided food before the yolk sac is completely resorbed. The first critical period in the larval rearing of seabass is at day 3, when the yolk sac is resorbed. Taste buds are present in the epithelium of the entire buccal cavity and pharynx, while alkaline phosphatase and esterase are localized in the intestine (Minjoyo et al. 2003). These changes correlated with endogenous to zooplanktivorous feeding. α-amylase activity is also present in newly-hatched larvae and the activity increased at first feeding. Moreover, the activity of the enzyme is up-regulated by cortisol and thyroid hormone treatment (Ma et al. 2004). Incidentally, both cortisol and thyroid hormones, presumably of maternal origin, are present in fertilized seabass eggs and although their levels decrease after hatching as the yolk is resorbed, the levels again start to increase one day after hatching (Nugegoda et al. 1994, Sampath-Kumar et al. 1995).

At 5 days post-hatching, large supranuclear cells in the epithelium of the posterior intestine and rectum, and esterase were present in the esophagus. Sac-like cells lined the epithelium of the buccal cavity and pharynx of 10-day old seabass larvae. The early development of mucosal folds in the esophagus at day 5 may help provide peristaltic movement for bulky food. It may also be related to the development of osmoregulatory ability in seabass larvae.

At the beginning of metamorphosis (20 days post-hatching), pharyngeal teeth were developed in the roof and dorsal surface of the tongue. Gastric glands and goblet cells were formed in the stomach and intestine, respectively. Pyloric caeca were first observed. Alkaline phosphatase, esterase and aminopeptidase were localized in the brush borders of the epithelial cells of the pyloric caeca. Aminopeptidase and lipase were observed in the brush border of the epithelial cells. In 30-day old larvae, amylase was observed in both columnar cells and gastric glands (Minjoyo et al. 2003). Protease and amylase were present in the pyloric caeca and intestine. The pH of the anterior gut changed from alkaline (pH 7.7 in 8-day old larvae) to acidic (stomach pH 3.7) in 22 day-old larvae where pepsin-like activity became well-established (Walford and Lam 1993). Development in various regions of the digestive tract at this stage was correlated with the change of feeding habit from zooplanktivorous to carnivorous. The second critical period occurs at metamorphosis when the seabass larvae assume the typical adult form.

Cannibalism was reported in 25 to 30 day old larvae. By this time, the entire region of the digestive tract is well developed. The presence of the pharyngeal teeth in the roof and dorsal surface of the tongue at the beginning of metamorphosis may restrain strong moving prey when seabass assumes a carnivorous feeding habit.

2.5 Seed Production in the Hatchery

Lates calcarifer is widely cultured throughout Asia, however, fingerling supply from the wild is not reliable, hence aquaculture of this species is based primarily on hatchery-produced fry (FAO/SCSP 1982). Larval rearing of seabass requires the provision of live food in the form of rotifers and brine shimp (*Artemia*) in increasing sizes (newly-hatched nauplii, sub-adult or adult biomass) (Parazo et al. 1998). At 28°C, seabass larvae begin to feed 50 h after hatching, but it is advisable to introduce the feed at an earlier time. Larvae should be weaned to each new feed type by gradually increasing the amount of the new feed type while reducing that of the preceding food type over a 3 day period. This is done to allow the larvae to recognize and accept the new food type. If weaning is undertaken properly, feed wastage and mortalities due to starvation will be minimized.

Rotifers are given as first food for the larvae. These are added into the rearing tank on the second day of larval rearing at a density of 15–20 individuals/ml. The density is maintained by daily addition of rotifers until day 12. *Chlorella* is also added at a density of $1-3 \times 10^5$ cells/ml to maintain water quality and as food for the rotifers. The larvae are weaned to *Artemia* from day 12–14 given at densities from 0.5 to 2 individuals/ml. This is increased to 5–10 individuals/ml from day 15–23. As the larvae grow bigger, they ingest larger food particles and are fed sub-adult or adult *Artemia* biomass at 1/individual/ml or higher. Feeding the larvae with highly-unsaturated fatty acids (HUFA)-enriched rotifers and brine shimp improves larval vigor and survival (Rimmer and Reed 1989, Dhert et al. 1990).

Brine shimp are expensive, hence other potential live food organisms were evaluated as partial or complete replacement for this food item. Among those found to be suitable are the freshwater cladoceran *Moina macropora* (Fermin 1991, Fermin and Bolivar 1994) and the brackishwater cladoceran *Diaphanosoma celebensis* (de la Pena et al. 1998, de la Pena 2006). Copepods and mysids have also been tested in view of studies that have shown that these can be used as replacement for *Artemia* in the larval rearing of groupers (Toledo et al. 1999, Eusebio et al. 2010). Some hatcheries feed the older larvae with trash fish. In such cases, only fresh trash fish should be used (Parazo et al. 1998). The head, entrails and bones are removed and the flesh chopped into fine bits that are easily swallowed by the larvae. It is advised to remove the head because trash fish are potential carriers of nodavirus (de la Pena et al. 2011, Hutson 2013—Chapter 6). Feeding is conducted slowly to minimize wastage as trash fish also pollute the rearing water if not removed from the system quickly.

Aside from other live food organisms, there have also been efforts towards the development of suitable artificial feeds for seabass larvae since their use can considerably simplify hatchery operations and reduce tank requirements for natural food culture and thus investment. Commercially-prepared protein-walled microcapsules were tested as food for seabass larvae, however, while these were readily ingested, the larvae were not able to digest the protein wall of the microcapsules resulting in mass mortalities (Walford et al. 1991). However, the protein membrane of the microcapsules was observed to be broken down in the larval gut with some assimilation when the microcapsules were fed to the larvae together with rotifers (Walford et al. 1991). Feeding seabass larvae solely with microbound diets containing fish roe and using gelatin or carrageenan as binder resulted in total mortality on day 5, although the larvae actively fed on the diets indicating that they were unable to digest the artificial feeds (Southgate and Lee 1993). Currently, artificial feeds are used in the larval rearing of seabass in combination with live food. Artificial feeds may be given starting on day 8 or day 15.

The presence of "shooters", or fast growing fish in the population are observed during the later part of the hatchery phase (Parazo et al. 1991). Shooters are cannibalistic and will prey on their smaller siblings. Usually, cannibalistic fish swallow their prey whole. Since the maximum size of prey that a cannibal may ingest is approximately 60–67% or 2/3 of its length, fish with a length difference of 33% or more should be separated from the larger sibs during size grading. Sorting of the stock every 3–4 days is usually necessary to minimize mortalities due to cannibalism.

2.6 Deformities

Like many other cultured fish species, a range of skeletal deformities manifest during early development of hatchery-bred seabass (Fraser and de Nys 2005). Left unchecked, these deformities would lead to poor performance of the stocks, as deformed fry usually exhibit slow growth rates due to impaired feeding capability—especially if the deformities involve the jaws. Once deformed fish reach marketable size, the fish also looks unattractive to the buyer because of its "strange" shape and appearance.

In *L. calcarifer* common deformities manifest in the jaw and the operculum. The jaw deformities include shortened upper jaw, shortened lower jaw, twisted lower jaw and pinched jaw affecting both the upper and lower jaws. All jaw deformities are usually first observed at around 18 days after hatching. Deformity in the operculum is also observed, but these manifest later than the jaw deformities at around 36 days after hatching. The opercular bone and a portion of the subopercula are affected and the operculum is folded inward exposing a portion of the gill surface. Associated with this deformity are curled branchiostegal rays which result in the shortening of the branchiostegal membrane (Fraser and de Nys 2005). The incidence of deformities can be minimized by Vitamin C supplementation, i.e., incorporating Vitamin C in the enrichment for rotifers and brine shrimp. Vitamin C is important in the production of collagen, an important component for bone formation.

2.7 Diseases during the Hatchery Phase

Early stages of larval development are susceptible to diseases of varying etiology (Chong and Chao 1986, Leong and Wong 1986, Munday et al. 1992). Preventive and prophylactic measures can be performed effectively with basic information on the onset of innate immune function of the fish. The thymus, lymphoid kidney (pronephos) and spleen were first observed from histological sections at 2 days after hatching (Azad et al. 2009). The thymus is a bi-lobed organ, situated dorso-posteriorly in the oro-pharyngeal cavity, in the angle between the opercular bone and the head bone. The pronephos was observed with undifferentiated stem cells, though the excretory cells were already present at hatching. Gut associated lymphoid tissue (GALT) was observed at 5 days post hatching. Recently, viral nervous necrosis caused by nodavirus infection is a major problem causing heavy mortalities in the hatchery of seabass (Munday et al. 1992, Maeno et al. 2004, Azad et al. 2005, Hutson 2013—Chapter 6).

2.8 Concluding Notes

Seabass eggs are considered "medium-sized" and the larvae possess average yolk volume at hatching. Although about 90% of the yolk is used up during the first 24 h from hatching, the oil globule persists for a long time (about 5 days) giving the larvae sufficient time to initiate feeding. At the same time, although larvae exhibit slow growth during the first 5 days after an initial growth spurt during the first 15 h after hatching, this is compensated by accelerated development of organs and organ systems needed by the developing larvae to recognize and ingest food organisms as well as digest and assimilate nutrients from its food items. Hence, seabass larvae are able to initiate feeding at 50 h after hatching, long before the energy reserves are completely used up. These characteristics may explain, at least in part, the relative ease of producing seabass fry in the hatchery.

References

Azad, I.S., M.S. Sheknar, A.R. Thirunavukkarasu, M. Poomima, M. Kailasam, J.J.J. Rajan, S.A. Ali, M. Abelar and P. Ravichandran. 2005. Nodavirus infection causes mortalities in hatchery produced larvae of *Lates calcarifer*: First report for India. Dis. Aquat. Org. 63: 113–118.

Azad, I.S., A.R. Thirunavukkarasu, M. Kailasam, R. Subburaj and J.J.S. Rajan. 2009. Ontogeny of lymphoid organs in the Asian seabass (*Lates calcarifer*, Bloch). Asian Fish. Sci. 22: 901–913.

Bagarinao, T. 1986. Yolk resorption, onset of feeding and survival potential of larvae of thee tropical marine fish species reared in the hatchery. Mar. Biol. 91: 449–459.

Chong, Y.C. and T.M. Chao. 1986. Common diseases of marine foodfish. Fisheries Handbook No. 2, Primary Production Department, Ministry of National Development, Republic of Singapore, pp. 9–22.

de la Pena, M.R. 2006. Use of juvenile instar *Diaphanosoma celebensis* (Stingelin) in hatchery rearing of Asian seabass *Lates calcarifer* (Bloch). Isr. J. Aquacult.—Bamidgeh 53: 128–138.

de la Pena, M.R., A.C. Fermin and D.P. Lojera. 1998. Partial replacement of *Artemia* sp. by the brackishwater cladoceran, *Diaphanosoma celebensis* (Stingelin), in the larval rearing of seabass *Lates calcarifer* (Bloch). Isr. J. Aquacult.—Bamidgeh 50: 25–32.

de la Pena, L.D., V.S. Suarnava, G.C. Capulos and M.N.M. Satos. 2011. Prevalence of viral nervous necrosis (VNN) virus in wild caught and trash fish in the Philippines. Bull. Eur. Assoc. Fish Pathol. 31: 129–138.

Dhert, P., P. Lavens, M. Duray and P. Sorgeloos. 1990. Improved larval survival at metamorphosis of Asian seabass (*Lates calcarifer*) using ω3-HUFA-enriched live food. Aquaculture 90: 63–74.

Eusebio, P.S., R.M. Coloso and R.S.J. Gapasin. 2010. Nutritional evaluation of mysids *Mesopodopsis orientalis* (Crustacea: Mysida) as live food for grouper, *Epinephelus fuscoguttatus* larvae. Aquaculture 306: 286–294.

[FAO/SCSP] United Nations Food and Agriculture Organization / South China Sea Fisheries Development and Coordinating Programme. 1982. Report of training course on seabass spawning and larval rearing. Manila, Philippines. 120 pp.

Fermin, A.C. 1991. Freshwater cladoceran *Moina macropora* (Strauss) as an alternative live food for seabass *Lates calcarifer* (Bloch) fry. J. Appl. Ichthyol. 7: 8–14.

Fermin, A.C. and M.E.C. Bolivar. 1994. Feeding live or frozen *Moina macropora* (Strauss) to Asian seabass *Lates calcarifer* (Bloch) larvae. Isr. J. Aquacult.—Bamidgeh 46: 132–139.

Fraser, M.R. and R. de Nys. 2005. The morphology and occurrence of jaw and operculum deformities in cultured barramundi (*Lates calcarifer*) larvae. Aquaculture 250: 496–503.

Garcia, L.M.B. 1989a. Dose-dependent spawning response of mature female seabass, *Lates calcarifer* (Bloch), to pelleted luteinizing hormone-releasing hormone analogue (LHH-a). Aquaculture 77: 85–96.

Garcia, L.M.B. 1989b. Spawning response of mature female seabass, *Lates calcarifer* (Bloch) to a single injection of luteinizing hormone-releasing hormone analogue and 17 α-methyltestosterone. J. Appl. Ichthyol. 5: 155–184.

Heming, T.A. and R.K. Buddington. 1988. Yolk resorption in embryonic and larval fishes. *In*: W.S. Hoar, D.J. Randall and J.R. Brett (eds.). Fish Physiology (vol. XI, part A). Academic Press, London, pp. 407–446.

Hutson, K. 2013. Infectious diseases of Asian seabass and health management, *In*: D.R. Jerry (ed.). Biology and Culture of Asian Seabass, *Lates calcarifer*. CRC Press.

Kohno, H., S. Hara and Y. Taki. 1986. Early larval development of the seabass *Lates calcarifer* with emphasis on the transition of energy sources. Bull. Jap. Soc. Sci. Fish. 52: 1719–1725.

Kohno, H., R. Ordonio-Aguilar, A. Ohno and Y. Taki. 1996. Osteological development of the feeding apparatus in early stage larvae of the seabass, *Lates calcarifer*. Ichthyol. Res. 43: 1–9.
Kungvankij, P. 1987. Induction of spawning of seabass (*Lates calcarifer*) by hormone injection and environmental manipulation. *In*: J.W. Copland and D.L. Grey (eds.). Management of Wild and Cultured Seabass/Barramundi (*Lates calcarifer*). Australian Centre for International Agricultural Research, Canberra, pp. 120–122.
Leong, T.S. and J.Y.S. Wong. 1986. Parasite fauna of seabass *Lates calcarifer* Bloch from Thailand and from floating cage culture in Penang, Malaysia. *In*: J.L. Maclean, L.B. Dizon and L.V. Hosillos (eds.). Proceedings of the First Asian Fisheries Forum. Asian Fisheries Society, Manila, Philippines, pp. 251–254.
Ma, P., P. Sivaloganathan, K.P. Reddy, W.K. Chan and T.J. Lam. 2004. Hormonal influence on amylase gene expression during seabass (*Lates calcarifer*) larval development. Gen. Comp. Endocrinol. 138: 14–19.
Maeno, Y., L.D. de la Pena and E.R. Cruz-Lacierda. 2004. Mass mortalities associated with viral nervous necrosis in hatchery reared seabass, *Lates calcarifer* in the Philippines. JARQ 38: 69–73.
Maneewongsa, S. and T. Tattanon. 1982a. Nature of eggs, larvae and juveniles of the seabass. *In:* Report of training course on seabass spawning and larval rearing. SCS/GEN/82/39. South China Sea Fisheries Development and Coordinating Programme, Manila, Philippines.
Maneewongsa, S. and T. Tattanon. 1982b Growth of seabass larvae and juveniles. *In:* Report of Training course on seabass spawning and larval rearing. SCS/GEN/82/39. South China Sea Fisheries Development and Coordinating Programme, Manila, Philippines
May, R.C. 1974. Larval mortality in marine fishes and the critical period concept. pp. 3–19. *In:* J.H.S. Blaxter (ed.). The early life history of fish, Springer Verlag, New York.
Minjoyo, H. 1990. Histochemical Studies on the Early Stages of Development of the Digestive Tract of SeabAss, *Lates calcarifer* (Bloch). University of the Philippines. Diliman, Quezon City, Philippines 56 pp.
Minjoyo, H., J.D. Tan-Fermin and J.M. Macaranas. 2003. Localization of enzymes in the digestive tract during the larval to early juvenile stages of seabass (*Lates calcarifer* Bloch). Indonesian Fish. Res. J. 9: 46–53.
Munday, B.L., J.S. Langdon, A. Hyatt and J.D. Humphey. 1992. Mass mortality associated with a viral-induced vacuolating encephalopathy of larval and juvenile barramundi, *Lates calcarifer* Bloch. Aquaculture 103: 197–211.
Nugegoda, D., J. Walford and T.J. Lam. 1994. Thyroid hormones in the early development of seabass (*Lates calcarifer*) larvae. J. Aquacult. Trop. 9: 279–292.
Parazo, M., E. Avila and D. Reyes. 1991. Size and weight dependent cannibalism in hatchery-bred seabass (*Lates calcarifer* Bloch). J. Appl. Ichthyol. 7: 1–7.
Parazo, M.M., L.M.B. Garcia, F.G. Ayson, A.C. Fermin, J.M.E. Almendras, D.M. Reyes, E.M. Avila and J.D. Toledo. 1998. Seabass hatchery operations. Southeast Asian Fisheries Development Center, Aquaculture Department, Tigbauan, Iloilo, Philippines 42 pp.
Rimmer, M.A. and A. Reed. 1989. Effects of nutritional enhancement of live food organisms on growth and survival of barramundi/seabass *Lates calcarifer* (Bloch) larvae. Adv. Trop. Aquacult. AQUACOP IFREMER Actes de Colloque 9: 611–623.
Sampath-Kumar, R., R.E. Byers, A.D. Munro and T.J. Lam. 1995. Profile of cortisol during ontogeny of the Asian seabass, *Lates calcarifer*. Aquaculture 132: 349–359.
Schipp, G., J. Bosmans and J. Humphey. 2007. Northern Territory barramundi farming handbook. Department of Primary Industry, Fisheries and Mines, Darwin Aquaculture Center, Darwin 80 pp.
Southgate, P.C. and P.S. Lee. 1993. Notes in the use of microbound artificial diets for larval rearing of seabass (*Lates calcarifer*). Asian Fish. Sci. 6: 245–247.
Southgate, P.C., P.S. Lee and M.A. Rimmer. 1994. Growth and biochemical composition of cultured seabass (*Lates calcarifer*) larvae. Asian Fish. Sci. 7: 241–247.

Terner, C. 1979. Metabolism and energy conversion during early development. *In*: W.S. Hoar, D.J. Randall and J.R. Brett (eds.). Fish Physiology (vol. III). Academic Press, London, pp. 261–278.

Toledo, J.D., C.L. Marte and A.R. Castillo. 1991. Spontaneous maturation and spawning of seabass *Lates calcarifer* in floating net cages. J. Appl. Ichthoyl. 7: 217–222.

Toledo, J.D., M.S. Golez, M. Doi and A. Ohno. 1999. Use of copepod nauplii during early feeding stages of grouper *Epinephelus coioides*. Fish. Sci. 65: 390–399.

Walford, J. and T.J. Lam. 1993. Development of digestive tract and proteolytic enzyme activity in seabass (*Lates calcarifer*) larvae and juveniles. Aquaculture 109: 187–205.

Walford, J., T.M. Lim and T.J. Lam. 1991. Replacing live food with microencapsulated diets in the rearing of seabass (*Lates calcarifer*) larvae: Do the larvae ingest and digest protein–membrane microcapsules? Aquaculture 92: 225–235.

3

Climate Effects on Recruitment and Catch Effort of *Lates calcarifer*

Jan-Olaf Meynecke,[1,]* Julie Robins[2] and Jacqueline Balston[3]

3.1 Introduction

Worldwide coastal fisheries are under pressure. Changes to fisheries populations are expected to accelerate with climate shifts as a result of global warming, in particular for fisheries species that rely on access to coastal wetlands for reproduction like *L. calcarifer*. Understanding how climate variability interacts with *L. calcarifer* catch is helpful when considering the wider implications of climate change on coastal fisheries.

Lates calcarifer are widely distributed throughout the coastal and littoral waters of the Indo- West Pacific from Iran to Australia (Davis 1982, Pillay 1993) including China, Taiwan and Papua New Guinea. In Australia, the

---

[1]Australian Rivers Institute, Griffith University, Gold Coast QLD 4222.
Email: j.meynecke@griffith.edu.au
[2]Sustainable Fisheries Unit, Agri-Science Queensland, Queensland Department of Agriculture, Fisheries and Forestry, Ecosciences Precinct, Dutton Park QLD 4102.
Email: julie.robins@daff.qld.gov.au
[3]Barbara Hardy Institute, University South Australia, Jacqueline Balston & Associates, P.O. Box 315, Bridgewater, SA 5155.
Email: jacqueline.balston@jbalston.com
*Corresponding author

species occurs as far south as the Noosa River (latitude 26°30′ S) on the east coast and the Ashburton River (latitude 22°30′ S) on the northwestern coast (Schipp 1996).

Lates calcarifer supports important commercial wild-capture fisheries and is a major component of the recreational fishery in Queensland, the Northern Territory and Western Australia (Coleman 2004, Pusey et al. 2004). Wild stocks of *L. calcarifer* are also under pressure from habitat alteration and destruction in parts of northern Australia as a result of mining and port development in large river systems (Keenan and Salini 1989, Hamilton and Gehrke 2005). Highly intensive shore-based grow out aquaculture systems (freshwater or saltwater) combined with a year-round supply of hatchery produced fish is currently practiced in a number of Australian states (estimated aquaculture value in 2006 was $45 million) (Australian Barramundi Farmers Association 2012).

3.2 General Biology of *L. calcarifer* Relevant for Impacts of Climate Factors

To identify the potential impacts future climate changes will have on *L. calcarifer* it is important to understand the species biology and ecology. *Lates calcarifer* has a complex and spatially variable life history, and displays non-obligatory catadromy, i.e., migration from freshwater to saltwater to spawn. The proportion of the population that migrates to permanent freshwater habitats for a portion of their juvenile to adult life-stage varies between catchments and between years within catchments (Pender and Griffin 1996, Milton et al. 2008, Halliday et al. 2012). *Lates calcarifer* are euryhaline, and are capable of inhabiting a wide variety of habitats, including rivers, billabong, swamps, estuaries and rocky coastal habitats.

Lates calcarifer are protandrous hermaphrodites and in Australia initially mature as males at two to five years of age and then change sex to female at between five and seven years of age (Moore 1979, Davis 1982). Movement of adults to spawning areas is thought to be triggered by the seasonal increases in water temperature (Grey 1987). High salinity (32 to 38 ppt) appears to be the main requirement of spawning grounds (Davis 1987), a finding confirmed by aquaculture studies (Garcia 1989). Gametogenesis in *L. calcarifer* is initiated by seasonal increases in water temperature and photoperiod (Russell 1990). In Australia, *L. calcarifer* have an extended and variable spawning season, the timing and duration of which varies depending on water temperatures and lunar and/or tidal cycles (Dunstan 1959, Davis 1985, Williams 2002). Russell and Garrett (1985) and later Pusey et al. (2004) suggested that the completion of the major part of the spawning season prior to the onset of the monsoonal wet season (and heightened river flows) is probably a strategy for eggs and larvae to avoid low-salinity water. The timing also allows juveniles to take advantage of the vast aquatic habitats that result from wet season rains, as suggested by Davis (1985) and confirmed by Jardine et al. (2011). The timing of monsoonal rains in northern Australia varies between years, particularly towards the higher latitudes of *L. calcarifer's* distribution in Australia. An extended spawning season by *L. calcarifer* in northern Australia also provides several within-year cohorts of post-larval *L. calcarifer* that can take advantage of the temporally variable monsoon rains and associated aquatic habitats.

The larvae of *L. calcarifer* hatch in near shore waters and are passively delivered by tidal action to supra-littoral swamps near a river mouth or in floodplains further upstream (Pusey et al. 2004). By the end of the Australian tropical monsoon season (around February to March) 0+ age class juveniles and post-larvae recruits depart tidal creek habitats (stimulus unknown) and occupy nursery swamps, where yearlings remain until the mid-dry season. Active movement occurs by a variable proportion of the juvenile *L. calcarifer* population in the 1+ age class, and comprises mostly immature males that move upstream and colonise a range of freshwater habitats where access allows (Pusey et al. 2004). Adult *L. calcarifer* migrate downstream to estuarine spawning grounds (Griffin 1987). In the Gulf of Carpentaria, this change occurs when fish reach about 68 to 90 cm total length, while on the east coast of Australia the change occurs when fish reach about 85 to 100 cm total length (Dunstan 1959, Davis 1982). There is limited movement of *L. calcarifer* between major river systems and, as such the stock, is generally confined within river systems and along near shore habitat (Davis 1985, Russell and Garrett 1985).

3.3 River Flow, Temperature and *Lates calcarifer*

There are many studies that demonstrate strong covariance between *L. calcarifer* commercial marine and estuarine fisheries catch data and natural variations in freshwater flows (Loneragan and Bunn 1999, Quiñones and Montes 2001, Lloret et al. 2004), often with time lags that equal the approximate age at which species enters the fishery (Staunton-Smith et al. 2004). Significant positive correlations between river flow and various aspects of the *L. calcarifer* fishery in northern Australia are known; for example catch rates are higher during above average river flow (Sawynok 1998, Staunton-Smith et al. 2004, Meynecke et al. 2006, Robins et al. 2006, Balston 2009). However, field studies are yet to establish the actual casual mechanisms between environmental factors and catch rates (Robins et al. 2005). Significant positive lagged correlations between catch and freshwater flows have suggested the importance of flows for the survival of fish during their first year of life. Year-class strength, a measure of recruitment success, was variable for *L. calcarifer* of the Fitzroy River estuary in Queensland and was significantly and positively correlated to flows in spring and summer (Staunton-Smith et al. 2004, Halliday et al. 2011). Sawynok and Platten (2011) reported positive relationships between catch rates of recreationally caught 0+ age class barramundi with rainfall and flow variables in the Fitzroy River (central Queensland), with January monsoonal rain having the highest correlation value with catch (i.e., r = 0.56 p<0.01). Changes in catch in the Fitzroy River between 1945 and 2002 showed notable 15 to 20-year cycles in the data that coincided with the Pacific Ocean Dipole (POD) (Halliday and Robins 2007), a climate system that affects rainfall and hence freshwater flow (MacDonald and Case 2005).

Seasonal pulses of freshwater proximate to reproductive periods may strongly influence catchability by stimulating spawning migrations in estuarine-dependent fish (Gillson et al. 2009). For instance, increased freshwater flow into Princess Charlotte Bay of north eastern Australia resulted in increased catchability of *L. calcarifer* based on commercial fishing records over a 14 year period. Balston (2009) speculated that this occurs as a result of mature males migrating towards estuarine spawning grounds at the beginning of the wet season, although such mechanisms have not yet been substantiated. However, this correlation between freshwater flow and catch rates demonstrate the importance of river flow as a key driver of catch, particularly in relation to the "flood-pulse" concept (Bunn and Arthington 2002). The "flood-pulse" concept defines annual floods as the most productive and important time where dynamic interactions between water and land take place—for *L. calcarifer* this time means more available nursery area. Freshwater flow is also critical for population scale responses, such as the timing, intensity and direction of fish migration, most likely in interaction with other key environmental variables such as water temperature and light (Jonsson 1991).

In tropical climates, the direct effects of air temperature may not have an obvious influence on *L. calcarifer*. However, seasonal variations in air temperature in northern Australia (e.g., Normanton, Gulf of Carpentaria) can range from a monthly minimum average of 15.7°C in the dry season, to a monthly maximum average of 37°C in the wet season (BoM 2012). The use of air temperature as a proxy for water temperature has shown that warmer temperatures may positively influence growth rates and survival of juvenile *L. calcarifer*. Indeed, Xiao (2000) demonstrated that a seasonally varying von Bertalanffy growth model provided a better fit to research growth data for *L. calcarifer* in the Northern Territory, than did a non-seasonally varying growth model. Xiao's (2000) study therefore provides strong support towards seasonally varying temperature playing a major role in *L. calcarifer* growth rate regulation, possibly as a consequence of the wider increased biological productivity within ecosystems when water temperatures increase (Liston et al. 1992). Besides these temperatures effects on *L. calcarifer* growth, a relationship between annual and monthly sea surface temperature (SST) and *L. calcarifer* catches has also been observed, but correlations are weaker than with rainfall (Meynecke and Lee 2011). It is also likely that other climate variables including evaporation and wind affect the successful recruitment of the species and translate to increased catches (Balston 2009), however, studies correlating the effect of these particular environmental variables on recruitment and catch rates are presently lacking.

Despite the previous research outlined above, knowledge gaps still remain with respect to the role and degree climate variability has on influencing wild-caught populations of *L. calcarifer*. In particular, there is no synthesis of the effect wider aspects of climate variability have on *L. calcarifer* population dynamics. Consequently, we combined life-history knowledge with correlative studies of climate-catch and climate-recruitment data to better understand the effects of climate on *L. calcarifer*. Four Australian case studies as presented below were used to illustrate relationships. Two of the case studies explored relationships between climate variables and catch data, one over a large spatial scale covering Queensland and parts of Northern Territory in Australia (> 1000 km) and one at a defined regional scale (i.e., Princess Charlotte Bay, northeast Queensland). One case study modeled recruitment indices for five regions, geographically spread along the central east coast of Queensland, the Gulf of Carpentaria and the Northern Territory, whilst the final case study investigated a quantitative population dynamics model for *L. calcarifer* that included climate effects on growth and recruitment and explores population responses under outputs of an A1FI climate change scenario (IPCC 2000).

3.3.1 Case studies

3.3.1.1 Case Study 1—Influence of climate variables on L. calcarifer catch in Queensland, Australia—a large scale comparison

In the first case study, we focus on the east and north coast of Australia (i.e., Queensland; Fig. 3.1) for which there is reasonably good catch data (compared to other countries) to investigate the influence of climate variables on *L. calcarifer* catch rates. The Queensland coastline has a wide range of mean monthly and annual air temperatures with annual average water temperatures varying between 23 and 31°C. Between 80% and 90% of annual coastal rainfall occurs in the monsoonal wet season (November –April) and annual average rainfall ranges from 700 mm in the Gulf Plains Bioregion to 1500 mm in the Wet Tropics Bioregion (BoM 2012).

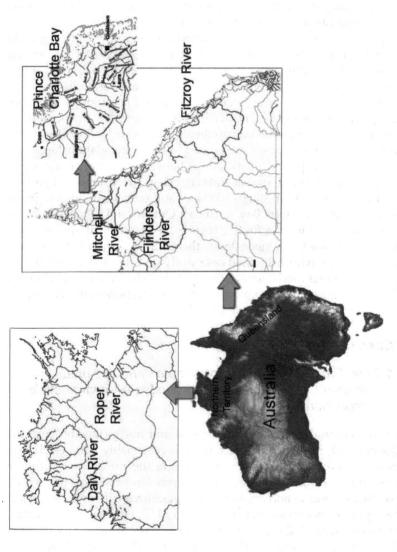

Figure 3.1. Location of study sites. The catchment areas are indicated by bold lines.

Methods and data used in Case Study 1. For this study, we used coastal catch data from the commercial *L. calcarifer* fishery to assess regional differences in the relationship between climate variables and fisheries catch rates. Eight Queensland coastal regions (Lower Carpentaria (LC), North Peninsula (NP), South Peninsula (SP), Barron (B), Herbert (H), East Central Coast (ECC), Port Curtis (PC), Moreton (M)) were selected to evaluate the influence of climate variables on *L. calcarifer* catch rates in accordance with the Bureau of Meteorology rainfall districts (BoM 2004) (Fig. 3.1). Data on catch and effort (number of fishing days) for commercial *L. calcarifer* between 1988 and 2004 were provided by Fisheries Queensland (Department of Agriculture, Fisheries and Forestry) (Fig. 3.2). Effort was calculated for the monthly net fishery for *L. calcarifer* (i.e., excluding line fishery data from the analyses)

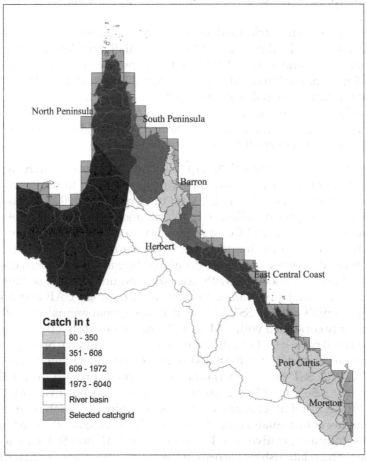

Figure 3.2. Distribution of *Lates calcarifer* catches along eight coastal regions in Queensland, Australia.

to infer an index of abundance. The original catch data were recorded in 30 nautical mile grids (=1,668 km^2). Catch was adjusted for effort using residuals from the regression of log-transformed catch and effort, and the catch adjusted for effort (CAE) was calculated for each year, season and month. Monthly mean and maximum sea surface temperature (SST) data points with a four km^2 resolution (obtained from http://podaac.jpl.nasa.gov/DATA_PRODUCT/SST, NASA JPL Physical Oceanography) based on daily means were selected for a 50-km buffer zone along the coastline of Queensland. Monthly Southern Oscillation Index (SOI) values and interpolated mean and maximum rainfall data for the eight coastal regions were obtained from the Bureau of Meteorology. Terrestrial rainfall data derived from local weather stations for the years 1981–2008 was used and then interpolated for selected coastal region at a resolution of 5km grid cell size.

Analyses were performed using Pearson correlations for seasonal configuration of the data set for SST and rainfall data. We tested annual (calendar year) and seasonal CAE data against five variables: SST max, rainfall max, annual/monthly rainfall, rainfall wet season and SOI. Non-metric multidimensional scaling (nMDS) was used to visually represent the similarity of the coastal regions on the basis of temperature, rainfall and catch in kg/day. Data was standardised by sample and resemblance measure was Euclidean distance.

Results from Case Study 1. The commercial *L. calcarifer* catch rates in Queensland between 1988 and 2004 vary from south to north with the highest commercial catches in the Gulf of Carpentaria (2000 to 6000 t for the observation period) followed by North and South Peninsula of Cape York and the East Central Coast region (600 to 2000 t for the observation period). The highest catch rates occur in northern Queensland and reflect the natural abundance of *L. calcarifer* along the east and northern coast of tropical Australia associated with extensive and undisturbed wetlands.

Pearson correlation values for annual *L. calcarifer* CAE and annual average rainfall and/or SST indicated that regional annual rainfall was significantly correlated with CAE (e.g., North Peninsula ($r = 0.62$, $p < 0.01$), Barron ($r = 0.54$, $p < 0.01$), Gulf ($r = 0.49$, $p < 0.05$) regions). No significant correlations were detected for the Moreton region and a higher correlation for CAE and SST than with rainfall was shown for the Central Coast region ($r = 0.36$; Fig. 3.3). All eight regions showed positive correlation between CAE and at least one of the selected climate variables, with the correlations with rainfall being the highest. In seven coastal regions there was a significant positive correlation between CAE and SST max for a 2 year lag. The relationships correspond with the life cycle of *L. calcarifer* in such way that more rainfall provides more habitat, and higher temperatures result in better growth rates and hence catch weights (Table 3.1).

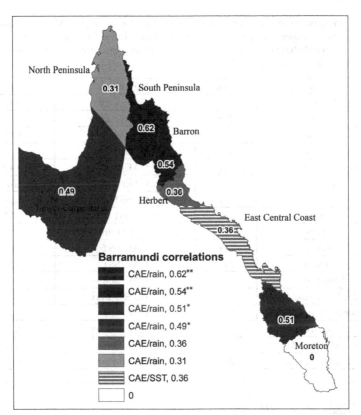

Figure 3.3. Pearson correlation values for *L. calcarifer* CAE and annual average rainfall for each region between 1988–2004 (* < 0.05; ** < 0.01).

An MDS plot of annual rainfall and SST max for each region and the correlation with CAE showed three distinct groups: Southern Queensland, the Central Coast and northern Queensland (Fig. 3.4). This is in accordance with genetically different stocks that have been identified for PC/ECC, Barron/Herbert, SP/NP and the Gulf region (Shaklee et al. 1993).

3.3.1.2 Case Study 2—Short-term climate impacts on L. calcarifer catches in northern Queensland

This case study concentrates on the South Peninsula of Cape York, Queensland, Australia in Princess Charlotte Bay (Fig. 3.1). The first case study identified significant correlation values between *L. calcarifer* CAE and regional rainfall and SST values in this coastal region.

Methods and data used in Case Study 2. Rainfall for the study area is highly variable (400 to 2000 mm per annum; BoM 2005). July and August are

Table 3.1. Pearson coefficient of correlation (*r*) values for seasonal CAE and relationships with sea surface temperature (SST), Southern Oscillation Index (SOI) and rainfall (rain). Wet and dry season data has been used instead of annual time configuration. N=35 for no lag and n=31 for 2 yr lag. $p < 0.05$*; $p < 0.01$**.

Region	Lag	Parameter	r
Barron	2yr	CAE/rain	0.41*
Barron	2yr	CAE/SST max	0.44**
Barron	no	CAE/rain	0.42**
East Coast Central	no	CAE/SST	0.38*
East Coast Central	2yr	CAE/SST	0.40*
Gulf	no	CAE/rain	0.64**
Gulf	no	CAE/SST max	0.65**
Gulf	2yr	CAE/SST max	0.58*
Herbert	no	CAE/rain	0.47**
Herbert	2yr	CAE/SST	0.40*
Moreton	2yr	CAE/SST	0.39*
Moreton	no	CAE/SOI	0.45*
North Peninsula	no	CAE/SST max	0.41**
North Peninsula	2yr	CAE/SST	0.40*
Port Curtis	no	CAE/rain max	0.36*
South Peninsula	no	CAE/rain max	0.34*
South Peninsula	2yr	CAE/rain max	0.25

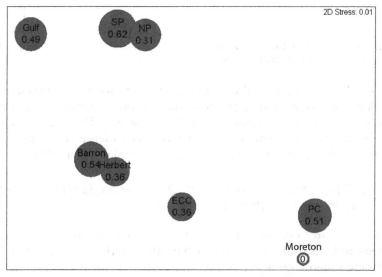

Figure 3.4. MDS based on rain, SST and total catch with overlay of annual r values for *L. calcarifer* catch and rainfall. Size of the circles represents the correlation value.

the driest months (4 mm each on average) and February the wettest (292 mm on average) (Clewett et al. 2003). Air temperature in the study area ranges from a minimum of about 13°C in July up to a maximum of 35°C in December (BoM 2005).

Monthly catch and effort data were extracted for commercial logbook grid squares for the years 1989 to 2002 for the northern edge of Princess Charlotte Bay to Cape Flattery from the Fisheries Queensland database. Total catch for each financial year (1 July to 30 June) was calculated to include the individuals spawned in a single year. Fishing effort was recorded as the number of netting days and calculated as kg/day. Catch was adjusted for effort using residuals from the regression of catch and effort for each year.

Because of a scarcity of actual rainfall and temperature recordings in the study area, the Bureau of Meteorology SILO interpolated climate data surface at a 0.05 degree resolution was used to provide data for these two parameters (http://www.bom.gov.au/silo). Interpolated maximum and minimum monthly average air temperature and monthly total pan evaporation data were extracted for Lakefield station (14°57'S, 144°12'E) and total monthly rainfall data was averaged over five grid squares within the study area. Monthly freshwater-flow data were collected from the Queensland Department of Environment and Resource Management stream gauge website database (www.nrw.qld.gov.au/watershed/) for eight gauged rivers flowing into Princess Charlotte Bay and modeled for the years 1988 onwards. Monthly averaged SST data were extracted for a 1°point (−14°S, 114°E) from satellite derived temperature (obtained from http://podaac.jpl.nasa.gov/DATA_PRODUCT/SST, NASA JPL Physical Oceanography). Madden–Julian Oscillation (MJO) indices were taken from the BoM Research Centre website and include longitudinal position of the centre of the oscillation (phase) and the number of days in each phase (http://www.bom.gov.au/bmrc/clfor/cfstaff/matw/maproom/RMM/index.htm). The variable used in the analysis was a summation of the number of days for phases one, four and six from 1 November to 30 April (northern wet season).

Seasonal variables were calculated as defined by Vance et al. (1998): pre-wet (October–December); wet (January–March); early dry (April–June); and dry (July–September). Variables that were not normally distributed prior to statistical analysis were transformed and checked for normality using histograms and the Shapiro Wilk test (Shapiro et al. 1968). A lag analysis to determine the amount of variability in *L. calcarifer* catch explained by climate was undertaken by correlating Princess Charlotte Bay *L. calcarifer* catch rates adjusted for effort (CAE) against each climate parameter for lags of up to 5 years. A linear regression analysis was used to test the relationship between each selected climate variable and commercial *L. calcarifer* CAE for lags of up to three years. To reduce the risk of spurious relationships that did not have a likely causal link to *L. calcarifer* catch, results of the regression analysis were viewed in context of a life-cycle model.

Results from Case Study 2. Rainfall in the Princess Charlotte Bay area shows an increasing trend over the past 100 years. The selected time period for this case study spans approximately 80% of the variability in rainfall for the region, with at least one year in the highest decile (wettest 10% of years) and one in the second lowest decile (driest 20% of years).

A number of scatter plots at different time lags showed significant linear relationships between climate and CAE. Other plots showed a non-linear relationship between climate and CAE. A Pearson correlation matrix of CAE versus each climate variable and each climate index was generated for lags of up to five years. Concurrent correlations (no lag) identified seven significant correlations: maximum temperature December (negative), pre-wet season (October–December) rainfall (positive), wet season (January–March) rainfall (positive), pre-wet season freshwater flow (positive), wet season freshwater flow (positive), annual evaporation (negative) and average pre-wet season SOI (positive). At the one-year lag there were only two significant correlations: annual evaporation (negative) and MJO Phase 6 (positive). For a lag of two years, there were a number of significant correlations: dry season (July–September) rainfall (positive), wet season rainfall (positive), wet season freshwater flow (positive), annual evaporation (negative) and average wet season SST (positive). Three years prior to catch minimum temperature July was the only variable significantly correlated with CAE (negative). Four years prior to catch there were significant correlations between CAE and dry season rainfall (positive), average pre-wet season SST (positive) and MJO Phase 4 (positive). At the five year lag again average pre-wet season SST (positive) and MJO Phase 4 (positive) were both significant as well as minimum temperature July (negative) (Tables 3.2a, 3.2b). The results support the findings made on a larger spatial scale (Case Study 1) that wetter conditions provide good connectivity between the marine spawning environments and create extensive nursery habitats for young of the year enhancing overall recruitment. Increased overall juvenile recruitment success, coupled with wetter conditions that result in connectivity with the marine environment for two plus year cohorts, and warmer SSTs for increased growth rate, all result in higher barramundi CAE.

Table 3.2a. Pearson coefficient of correlation (r) between Princess Charlotte Bay *L. calcarifer* catch adjusted for effort (1989/90–2001/02) and climate variables (zero–five year lag). Highlighted correlations significant at the $p < 0.05*$ level.

Climate Variable	Zero lag	1 year lag	2 year lag	3 year lag	4 year lag	5 year lag
Maximum air temperature Dec (°C)	–0.55*	–0.25	–0.02	0.06	0.21	0.11
Rainfall Jul–Sept (mm)	–0.01	0.02	0.77*	0.46	0.48	0.28
Rainfall Oct–Dec (mm)	0.56*	0.30	0.38	0.15	–0.06	–0.31
Rainfall Jan–Mar (mm)	0.56*	0.31	0.62*	0.12	0.26	–0.05
Rainfall Apr–Jun (mm)	0.37	0.40	–0.02	0.14	0.08	–0.19
Flow Jul–Sep (Ml)	0.36	0.41	0.33	0.18	–0.34	–0.11
Flow Oct–Dec (Ml)	0.71*	0.37	0.29	–0.02	–0.13	–0.09
Flow Jan–Mar (Ml)	0.52*	0.35	0.76*	0.36	0.43	0.00
Flow Apr–Jun (Ml)	0.33	0.33	0.13	0.27	0.08	–0.26
Evaporation Annual (mm)	–0.73*	–0.48	–0.62*	–0.34	–0.15	0.38
Average Jan–Mar SST (°C)	0.32	0.17	0.58*	0.03	0.29	0.33

Table 3.2b. Pearson coefficient of correlation (r) between Princess Charlotte Bay *L. calcarifer* catch adjusted for effort (1989/90–2001/02) and short-term climate indices (zero–five year lag). Highlighted correlations significant at the $p < 0.05*$ level.

Climate Index	Zero lag	1 year lag	2 year lag	3 year lag	4 year lag	5 year lag
MJO Phase 1	–0.55	0.00	–0.26	–0.08	0.17	–0.16
MJO Phase 4	0.04	–0.39	–0.18	0.05	0.54	0.67*
MJO Phase 6	0.38	0.50	0.16	0.21	–0.09	0.06
Average Jul–Sept SOI	0.47	0.12	0.29	0.13	0.06	–0.47
Average Oct–Dec SOI	0.62*	0.19	0.29	0.14	0.14	–0.20
Average Jan–Mar SOI	0.47	–0.18	0.10	0.19	0.37	0.34
Average Apr–Jun SOI	0.41	0.26	–0.04	0.15	0.09	–0.06

3.3.1.3 Case Study 3—Recruitment indices for L. calcarifer in northern Australia

Methods and data used in Case Study 3. There are two documented methods of calculating recruitment indices for *L. calcarifer* in northern Australia. The first method relies on sampling young-of-the-year *L. calcarifer* in selected nursery habitats (e.g., Griffin 1991, Sawynok and Platten 2011). This method is a direct measure of young-of-the-year recruitment, but is time and labour intensive, requires adequate sampling of sites within a catchment to ensure that the results are representative of the whole population rather than a reflection of dynamics within select habitat types, and should include robust standardisation of the effort used during sampling. Another way of estimating recruitment indices for *L. calcarifer* in northern Australia uses the age-structure of adult *L. calcarifer* sampled from the estuarine *L. calcarifer* population to calculate variation in year-class strength using catch-curve regression, i.e., regression of the natural log of the number of fish observed in each year-class against age. Deviation from the expected abundance of each year-class, given its age, is assumed to indicate variation in recruitment. Large positive deviations are indicative of strong year-classes and large negative deviations are indicative of weak year-classes. It is an application of the method developed by Maceina (1997) for estimating recruitment variation in North American fishes and has been applied in Australia to golden perch (*Macquaria ambigua*) (Roberts et al. 2008), king threadfin (*Polydactylus macrochir*) (Halliday et al. 2008, 2012) and black bream (*Acanthopagrus butcheri*) (Jenkins et al. 2010). The main advantage of this method is that several years of recruitment can be hind cast from a minimum of three years sampling (Staunton-Smith et al. 2004). The method does rely on the ability to confidently estimate the absolute age of individual fish so that a birth-year can be assigned, but this is not generally a problem with *L. calcarifer* in northern Australia due to well pronounced otolithannuali (Fig. 3.5) (Stuart and McKillup 2002, McDougall 2004, Halliday et al. 2012).

Recruitment indices for *L. calcarifer* populations are available for five geographically wide-spread catchments in northern Australia (for details see Staunton-Smith et al. 2004, Halliday et al. 2011, 2012). These are: the Fitzroy, Mitchell and Flinders Rivers in Queensland and the Roper and Daly Rivers in the Northern Territory (Fig. 3.1). These rivers differ in their geographic location, catchment type, climatic influences and flow regimes (Table 3.3). Therefore, whilst each of these regions is known as a *L. calcarifer* 'hot spot' and supports high levels of commercial and recreational catch, it is very likely that *L. calcarifer* populations in each region are responding to slightly different climatic influences, geographic factors and their interactions.

The age-structure of adult *L. calcarifer* in the estuarine sections from these rivers were sampled from commercial fisheries catches in all but the Mitchell River, where samples were obtained from indigenous and recreational fishers. It should be noted that the Fitzroy study was conducted at a different time to that of the remainder of the regions (i.e., 2000 to 2005 *cf.* 2007 to 2010). Age-structures were derived from visual assessment of *L. calcarifer* otoliths that were individually blocked in resin, sectioned at 300 μm and viewed under a microscope using reflected light.

Lates calcarifer has variable catch size limits in northern Australia, with a minimum legal size of 550 mm total length (TL) in the Northern Territory; 580 mm TL on the Queensland east coast and 600 mm TL in the Queensland

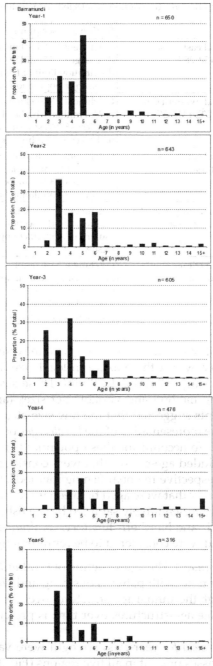

Figure 3.5. Proportional age-structure for *Lates calcarifer* sampled from the Fitzroy River estuary for five consecutive years.

Table 3.3. Location and regional description of selected river systems.

Region	Location	Regional description
Fitzroy	Queensland east coast ~23°23'S, 150°28'E	Catchment 142,537 km^2; mean annual discharge of 5.2 million ML; near the southern distribution limit of *L. calcarifer*; ~ 50% estuarine *L. calcarifer* population had accessed freshwater
Mitchell	Queensland, western Cape York~15°12'S, 141°35'E	Catchment 71,000 km^2; mean annual discharge of 11.3 million ML; headwaters in the Wet Tropics; ~55% estuarine *L. calcarifer* population had accessed freshwater
Flinders	Queensland south east Gulf of Carpentaria	Catchment 109,400km^2; 3220km^2 of wetlands, of which 13% are estuarine; mean annual discharge of 3.86 million ML; ~ 40% estuarine *L. calcarifer* population had accessed freshwater
Roper	Northern Territory, western Gulf of Carpentaria	Catchment 81,800 km^2; mean annual discharge (of the non-tidal section) is 3289 GL; ~40% of estuarine *L. calcarifer* population had accessed freshwater
Daly	Northern Territory, south west of Darwin	Catchment 54,400 km^2; mean annual discharge of 8653 GL; perennial flow with dry season flow sustained by groundwater inflow; 1590 km^2 of floodplain; >80% of estuarine *L. calcarifer* population had accessed freshwater

section of the Gulf of Carpentaria. There is no maximum legal size in the Northern Territory, whilst the maximum legal size in Queensland (east coast and Gulf of Carpentaria) is 1200 mm. These size limits, along with the type of fishing method and variable net sizes (and thus selectivity) between catchments suggested that the raw size-structure of commercial catches may not be representative of the *L. calcarifer* population within any single catchment. To account for these potential biases, the catch-curve analyses only included age-classes that were considered to be effectively sampled by the respective regional fishing methods. On the Queensland east coast, *L. calcarifer* that were aged between three and 11 years-old were considered to be representatively sampled by the fishery. However, in the Gulf of Carpentaria and Northern Territory, *L. calcarifer* aged between two and nine-years-old were considered to be representatively sampled.

Results from Case Study 3. The age-structure of *L. calcarifer* in all five regions was variable over time, with patterns of strong and weak age-classes that progressed through the annual age-structures over the sampling years. An example of variable age-structures over time is presented for the Fitzroy region in Fig. 3.6.

Indices of *L. calcarifer* year-class strength were variable, with strong and weak year-classes present in all five regions (Fig. 3.6). For *L. calcarifer*, the standardized residuals from the catch-curve regression greater than 0.5

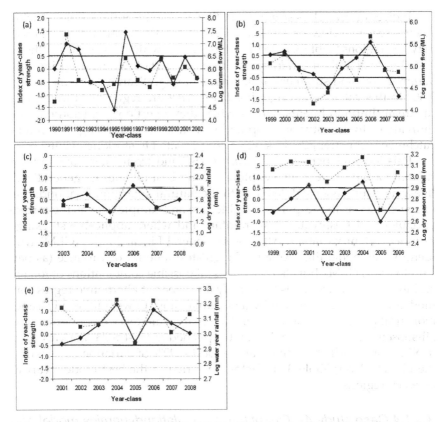

Figure 3.6. Indices of year-class strength (solid line) for *Lates calcarifer* sampled from five regions in northern Australia against seasonal rainfall or river flow. (a) Fitzroy River, Queensland east coast; (b) Mitchell River, Queensland, Gulf of Carpentaria; (c) Flinders River, Queensland, Gulf of Carpentaria; (d) Roper River, Northern Territory, Gulf of Carpentaria; and (e) Daly River, Northern Territory, "Top End". Solid line = Year-Class Strength; Dashed line = flow or rainfall.

are considered indicative of strong recruitment, while those less than −0.5 are considered indicative of weak recruitment. Residuals between −0.5 and 0.5 are considered to be neither strong nor weak and may indicate average recruitment for a region (Staunton-Smith et al. 2004).

In the Fitzroy River, strong year-classes were apparent in 1991, 1992, 1996 and weak year-classes were apparent in 1995 and 2000. In the Mitchell River, strong-year classes occurred in 1999, 2000 and 2006 and weak year-classes occurred in 2003 and 2008. In the Flinders River, a strong year-class occurred in 2006 and a weak year-class occurred in 2005. In the Roper River, strong year-classes occurred in 2001 and 2004 and weak year-classes occurred in 1999, 2002 and 2005. In the Daly River, strong year-classes occurred in 2004 and 2006, but no weak year-classes were detected.

Indices of year-class strength (YCS) were significantly correlated with several seasonal flow or rainfall variables, but the relationships differed between regions (Table 3.4). In the Fitzroy River, which is near the southern distributional limit of *L. calcarifer* on the east coast of Australia, spring and summer flow and summer coastal rain were significantly correlated with YSC. In this catchment, fingerling *L. calcarifer* are regularly stocked into upstream impoundments, which when they overflow, contribute adult individuals to the estuarine population. It was not surprising that in this catchment year strength class was also correlated to stocking events.

Relationships between *L. calcarifer* YSC and freshwater flow and/or rainfall variables were also explored using best-all subsets generalized linear modeling (Genstat 2008), with the natural logarithm of abundance of age-classes as the response variable, age and sample year as the forced variables and freshwater flow, rainfall, and where applicable stocking, as independent variables. All-subsets General Linear Model (GLM) will identify the model that explains the greatest amount of variance (as per step-forward GLM), but also calculates all possible combinations of forced and independent variables to identify a number of alternative regression models that can be evaluated by their explanatory power (adjusted r^2) and biological plausibility. Here, several alternative models were identified by all-subsets GLM in each region that explained at least > 78% of variation in YCS (for the Mitchell River) and at best < 90% of variation in YCS (for the Fitzroy River; Table 3.5). Factors included in the best models varied between regions.

3.3.1.4 Case study 4—Quantitative populationdynamics model of *L. calcarifer* for the Fitzroy River region considering the A1FI climate change scenario

One way to explore the possible effects of climate change on *L. calcarifer* is to model the population, incorporating known climate effects, and to then simulate what might happen under predicted global climate change. Despite *L. calcarifer*'s prominence as an important fishery species in Australia and southeast Asia, population modeling to date has been relatively simplistic as a consequence of data limitations (Welch et al. 2002, Campbell and O'Neil 2007), or highly complex (i.e., SHASSAM), but incomplete simulations (Grace et al. 2008), and none of these models included regional climate factors. Tanimoto et al. (2012) is the first population model for *L. calcarifer* that is monthly age- and length-structured, calibrated to three different fishery data sets collected between 1945 and 2005, and incorporates a climate-related factor (observed river discharge) and its impact on variable annual recruitment and variable seasonally-adjusted growth rates. Tanimoto et al. (2012) explored the response of the modeled population to river

Table 3.4 Correlation coefficients (r) between indices of barramundi year-class strength (YCS) and seasonal freshwater flow and rainfall variable for five regions in northern Australia.

Season	Qld East Coast		Gulf of Carpentaria			NT Top End
	Fitzroy River	Mitchell River	Flinders River	Roper River	Daly River
Water-year flow	0.55***	0.19		0.64**/0.73***	
Water-year rain	0.67***	0.06/-0.07	0.42	-	0.59***
Spawning (Sep to Feb) flow	0.61***	-	-	-	-
Spawning (Sep to Feb) rain	0.63***	-	-	-	-
Wet (Nov to Apr) flow	-	0.16	-	-	-
Wet (Nov to Apr) rain	-	-0.12	0.42	0.72***	0.51
Dry (May to Oct) flow	-	0.57**	-	-	-
Dry (May to Oct) rain	-	0.15	0.84***	0.53*	0.58***
Spring (Sep to Nov) flow	0.66***	0.42*	-	-	-
Spring (Sep to Nov) rain	0.17	0.16	0.65*	0.27	-0.25
Summer (Dec to Feb) flow	0.60***	0.26	-	-	-
Summer (Dec to Feb) rain	0.50**	-0.35	-0.48	0.59**	0.34
Autumn (Mar to May) flow	0.16	-0.00	-	-	-
Autumn (Mar to May) rain	0.09	0.33	0.80**	0.46*	0.49**
Winter (Jun to Aug) flow	-0.01	0.51**	-	-	-
Winter (Jun to Aug) rain	-0.24	0.03	-0.46	0.45	0.02
Upstream stocking fingerlings	0.37*	n/a	n/a	n/a	n/a

*=P<0.05, **= P<0.01, ***, =P<0.001

Table 3.5 Percent variance accounted for by significant best sub-sets regression models for *Lates calcarifer* from five regions in northern Australia.

Region	Significant best all sub-sets regression models[A]	Percent variance account for (adjR²)
Fitzroy River	Age, sample year, summer flow, stocking autumn coastal rain	90.4
	Age, sample year, summer flow stocking	88.2
	Age, sample year, spring flow, summer flow	86.3
	Age, sample year, spring flow summer coastal rain	85.9
Mitchell River	Age, sample year, dry-season flow, autumn flow	77.7
	Age, sample year, summer rain (-ve), autumn rain, winter rain	75.8
	Age, sample year, water-year rain, summer rain (-ve)	75.3
	Age, sample year, summer rain (-ve), autumn rain, spring rain	74.8
	Age, sample year, summer rain (-ve), autumn rain, spring flow	74.8
	Age, sample year, summer rain (-ve), wet season rain	72.7
	Age, sample year, dry season flow, summer rain (-ve)	72.3
	Age, sample year, summer rain (-ve), autumn rain	67.0
	Age, sample year, dry season flow	64.9
Flinders River	Age, sample year, dry season rain	94.9
	Age, sample year, autumn rain	93.6
	Age, sample year, summer rain (-ve)	92.2
Roper River	Age, sample year, wet season rainfall	84.6
	Age, sample year, summer rainfall, autumn rainfall	84.2
	Age, sample year, water-year rainfall	83.8
Daly River	Age, sample year, autumn rainfall, wet season rainfall	82.9
	Age, sample year, dry season rainfall	81.8
	Age, sample year, autumn rainfall, summer rainfall	81.8
	Age, sample year, water-year rainfall	80.9

[A]Factors in the multiple regression were positively related to age-class abundance unless otherwise indicated. Age and sample year were forced variables.

discharge sequences that were modified to reflect: (i) total abstraction of all licensed water; and (ii) projected regional climate change derived from downscaled outputs of SRES emissions scenario A1FI.

Methods. The modeling involved two stages: (i) constructing parameter posterior distributions for a monthly age- and length-structured population model using Markov Chain Monte Carlo sampling (MCMC with Metropolis–Hastings algorithm) calibrated to data for the Fitzroy River region between 1945 and 2005; and (ii) running likely parameter values and uncertainty in simulations through the population model (i.e., the projection stage). Detailed methods are provided in Tanimoto et al. (2012) and its accessory publication. The dynamics of the population were tracked monthly by applying recruitment, growth and mortality functions. Expected recruitment (R_y) was based on the spawning stock size of the previous year ($\hat{R}_y$ derived from a Beverton-Holt recruitment function), adjusted by the effect of anomalies in total summer flow (i.e., December to February) and total spring flow (i.e., September to November). Threshold values were based upon the quantitative results of Staunton-Smith et al. (2004) and Halliday et al. (2011). Recruitment was further modified to account for environmental fluctuations to recruitment (i.e., recruitment error) that were additional to and independent of flow. In the second stage (i.e., the projection stage), the model was set so that variation in recruitment was approximately 50% due to river flow (Halliday et al. 2011) and 50% to random error.

Tanimoto et al. (2012) considered five scenarios for river flow to the Fitzroy River estuary. The first scenario, referred to as '*Status Quo*', was the observed historic flow to the estuary and included actual levels of upstream water abstraction (Robins et al. 2005). The other four scenarios were hypothetical and were supplied by the Queensland Department of Science, Information Technology, Innovation and the Arts (DSITIA). The four hypothetical scenarios all assumed total abstraction of all licensed water (including latent licenses), plus the following levels of projected climate change: (i) no change (Latent + Nil Climate Change; LCCNil); (ii) 10th percentile projected climate change (Latent + Climate Change Wet; LCCWet based on GCMMIUBECHO G); (iii) the 50th percentile of projected climate change (Latent + Climate Change Median; LCCMed based on GCMIAPFGOALSG1.0); and (iv) the 90th percentile of projected climate change (Latent + Climate Change Dry; LCCDry based on GCMUKMOHADGem1). Flow scenarios that included climate change were based on the historical flow time-series modified by the parameter change percentages to rainfall and potential evaporation in the central Queensland region under SRES emissions scenario A1FI for the 2050 projection year (DERM 2009). Emissions under the A1FI scenario most closely follow current trends and assume a high-reliance on fossil fuels. Of the 23 GCMs adopted for use by CSIRO and BoM (2007), the

11 GCMs most appropriate for predicting rainfall change in Queensland were selected by the Queensland Climate Change Centre of Excellence (QCCCE). Projected rainfall, temperature and evaporation climate change factors were converted to the A1FI emission scenario and 2050 projection year using the methodology detailed in CSIRO and BoM (2007, p. 36–46). In order to indicate the range of uncertainty in climate change projections, DSITIA generated climate change adjusted data based on three of the 11 suggested models. The three GCMs (10th, 50th and 90th percentile) were selected based on rainfall ranking as described in Voogt et al. (2009). The method of generating the climate change adjusted rainfall, evaporation and inflow is also explained in Voogt et al. (2009). The river flow percentiles are based on exceedences taking into account the 11 GCMs selected by QCCCE, i.e., 10th percentile case is the GCM where the flow is exceeded 10% of the time (i.e., the wetter case), while the 90th percentile case is the GCM where the flow is exceeded 90% of the time (i.e., the drier case).

The effects of changed flow on the *L. calcarifer* population of the Fitzroy River were evaluated by comparing relative differences in equilibrium maximum sustainable yield (MSY) between scenarios, as well as the following indicators averaged over the last 20 years of the simulations (i.e., 2030 to 2050): exploitable biomass, spawning stock size (number of eggs x 10^6), annual catch, mean catch-at-age and mean catch-at-length. The effect of flow scenario on these indicators was tested using a Residual Maximum Likelihood Model (REML) (Genstat 2008), with scenario as a fixed effect and replicate as a random effect. Predicted means for each fishery indicator from the REML analysis were tested for significant differences between scenarios using the criteria of means having differences greater than twice the average standard error for each term (i.e., estimated LSD).

Results. Values for the fishery indicators are presented as percent change compared to the *Status Quo* scenario, to emphasise changes as a consequence of altered river flow. Mean values for exploitable biomass, spawning stock size (SpSS), maximum sustainable yield (MSY), annual catch and fish length were greatest for the *Status Quo* scenario and reduced for all other scenarios, with the differences significant at $p<0.001$ (Fig. 3.7). Mean values of catch-at-age were the reverse of this pattern ($p<0.001$), with the mean age of fish (caught) youngest in the *Status Quo* scenario and oldest in the LCCDry scenario. Increases in the mean age of fish (caught) in the LCCDry scenario was a consequence of fish growing more slowly under hypothetical scenarios of reduced flow and therefore taking longer to reach minimum legal size.

Reductions in mean exploitable biomass compared to the *Status Quo* scenario ranged from 4% for the LCCWet scenario up to 25% for the LCCDry scenario. Based on the results for the LCCNil scenario, about 10% of the

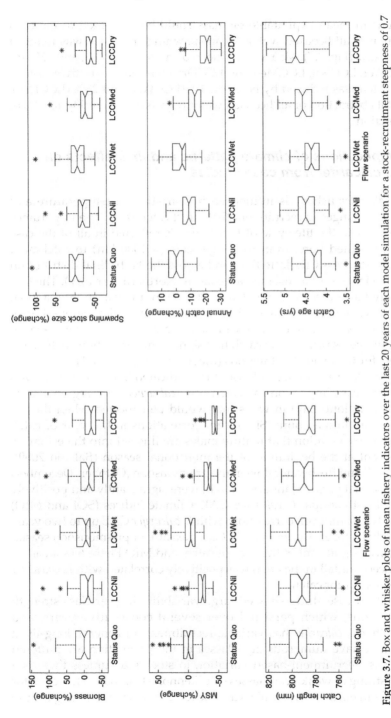

Figure 3.7. Box and whisker plots of mean fishery indicators over the last 20 years of each model simulation for a stock-recruitment steepness of 0.7 under alternate flow scenarios, where Status Quo = historic observed flow to the estuary; LCCNil = hypothetical Latent + No Climate Change; LCCWet = hypothetical Latent + Climate Change Wet, LCCMed = hypothetical Latent + Climate Change Median, LCCDry = hypothetical Latent + Climate Change Dry. Values presented as percent change relative to the Status Quo scenario. The boxes represent the 25th, median and 75th percentiles. The error bars indicate 1.5 times the inter-quartile range, and (*) represent outlier values (n=50).

reductions in mean exploitable biomass could be attributed to the total abstraction of all licensed water. Mean spawning stock size was reduced by 5% to 42% (Fig. 3.7). MSY was reduced by a 4%, 23%, 33%, and 45% for the LCCWet, LCCNil, LCCMed and LCCDry scenarios respectively. Mean annual catch was reduced by between 4% (LCCWet) and 23% (LCCDry), with total abstraction of all licensed water (LCCNil) accounting for 9% of the reduction.

3.3.2 Implications of climate patterns and their effects on L. calcarifer from case studies

The *L. calcarifer* fishery is influenced by variation in temperature and rainfall via changes in recruitment that affect catch rates. The importance of freshwater in the life cycle of *L. calcarifer* is reflected in all of the case studies presented here, from the large scale (> 1000 km) to catchment scale (e.g., Princess Charlotte Bay) and across the latitudinal distribution of this species (i.e., Queensland east coast to Northern Territory). The first case study has indicated that between 30–40% of the variation in annual *L. calcarifer* CAE may be explained by rainfall for particular regions. For all regions, wet season rainfall was the most relevant variable, and wet season rainfall two years prior to the catch had significant correlation coefficients (r values) for the majority of regions (Meynecke and Lee 2011).

The results from the second case study indicated that climate variables had a significant impact on the catch of *L. calcarifer* on a regional scale (Princess Charlotte Bay) in ways that would be expected when the life cycle is taken into account. Significant correlations in the year of catch support the observation that mature males are flushed into the estuarine environment at the beginning of the monsoonal season (Balston 2009). Both rainfall and freshwater flow in the pre-monsoon (October–December) and monsoonal period (January–March) were significantly and positively correlated with annual *L. calcarifer* CAE. Climate indices (SOI and MJO) returned two significant correlations with *L. calcarifer* CAE up to two years prior catch: October–December SOI (as a measure of pre-monsoon season rainfall) was significant in the year of catch, and MJO Phase 6 (a measure of enhanced rainfall in the area) was positively correlated with *L. calcarifer* CAE the year prior catch.

The third case study showed large variability in year-class strength across regions, which persisted over several consecutive years, and indicate that *L. calcarifer* has variable recruitment. In general, long-lived species that have numerous age-classes in the population are buffered against vast recruitment-based variation in stock size, unless there is a series of strong or weak year-classes (McGlennon et al. 2000). However, for *L. calcarifer* in northern Australia, variable recruitment was a strong feature across all regions studied, and was of such a magnitude that it influenced adult stock sizes and the associated fisheries catches (i.e., CAE's). The Fitzroy River region had the greatest extremes of variation in year-class strength. The Fitzroy River catchment has a much greater inter-year variation in rainfall and subsequent flood patterns than the other regions in the Gulf of Carpentaria and Northern Territory. It is not uncommon for the Fitzroy region to have several wet years (with good *L. calcarifer* recruitment) followed by several dry/drought years (with poor *L. calcarifer* recruitment). At the other extreme, the Daly River in the Northern Territory had no year-classes that were classified as weak, but had two that were classified as strong. The Daly River is a perennially flowing river that consistently experiences a monsoon and high rainfall season every year. In addition, otolith microchemistry found that >85% of estuarine caught *L. calcarifer* in the Daly River had accessed freshwater habitats for a period greater than three months (Halliday et al. 2012), compared to the Fitzroy River, where about 50% of fish examined had accessed freshwater habitats before capture (Milton et al. 2008). Therefore, a consistent monsoonal season plus consistent access to upstream freshwater habitats are possible reasons why *L. calcarifer* recruitment in the Daly River was never observed to be weak.

The fourth case study, an environmentally responsive population dynamics model, calibrated to historic data for *L. calcarifer*, indicated that reduced river flows projected to occur under future climate change (Hughes 2003) would result in significantly reduced barramundi production. The simulation suggested that under a worst case scenario (i.e., LCCDry—full use of all water entitlements and a dry climate change) that exploitable biomass of barramundi would be reduced (on average) by 24% from the current estimate, with a resultant reduction in (average) MSY by 45% and (average) catch of 20%. However, the effects of altered river discharge on barramundi fishery production were variable and non linear. This suggests that caution should be used in extrapolating observed river-flow catch relationships in trying to predict how a species may respond to changed climate conditions.

3.4 Role of Climate Variables in Relation to the *L. calcarifer* Life Cycle

In general, our studies showed a clear pattern throughout the regions and different spatial scales with high wet season rainfall and freshwater flows, high dry season rainfall, warm wet season SSTs and low annual/wet season evaporation were all significant in the same year or two years prior to catch. Each of these variables describes climatic conditions that would result in extensive and conducive nursery habitats for young-of-year *L. calcarifer* and a likely increase in survival. High rainfall during the spawning and early life-cycle stages of *L. calcarifer* (wet season and early dry season) would be expected to increase both the (i) extent of wetland habitat available for fingerlings and juveniles and (ii) access to upstream habitats for maturing males, as suggested by Davis (1985) and confirmed by Jardine et al. (2011). The same year correlations identified between rainfall, flow, SST and *L. calcarifer* catches are speculated to be explained by increased catchability as a result of downstream migration of mature individuals into estuarine areas. This is where the majority of commercial and recreational fishing activity is located and increased food availability leading to faster growth rates and earlier entry to the size-restricted fishery (Robins et al. 2005).

The positive correlation between *L. calcarifer* CAE and the Southern Oscillation Index (SOI) can be explained in similar ways. Positive values of the SOI are generally associated with a La Niña and above-average rainfall across the study area, while negative values of the SOI are associated with El Niño conditions and below-average rainfall. The Madden–Julian Oscillation (MJO) is another index for monsoonal activity. Pulses of the MJO are associated with increased convection and modulation of monsoonal westerly's and often result in increased rainfall and 'active bursts' of a few days in the north Australian monsoon followed by a strong stabilizing and drying influence after passing (Hendon and Liebmann 1990).

Higher temperatures (air and sea) also enhance primary production and increase growth rates as well as fish activity—all factors contributing to higher *L. calcarifer* catch rates and a likely explanation for the correlations between SST and CAE (e.g., Lower Gulf $r = 0.65$, $p < 0.01$ for seasonal CAE and maximum SST). Warm SSTs are also associated with increased egg and larvae survival (Barlow et al. 1995), so a positive relationship between SSTs and subsequent CAE was expected. Warm water temperatures are expected to improve the survival of young fish, although beyond a certain threshold, very high water temperature in shallow wetlands may be detrimental. Cold winter water temperatures below 15°C have been shown to reduce the survival of even adult *L. calcarifer* (Agcopra et al. 2005), so the relationship between CAE and minimum temperature in July was as expected negative in the Princess Charlotte Bay study ($r = -0.62$, $p < 0.05$).

Overall, the sign of the relationships between climate variables and *L. calcarifer* catch examined here were consistent although the strength of the relationships described varied with the geographical region. Contributing factors to this variability in relationship strength may include the variable and complex interactions between climatic variables observed throughout Queensland and the differences in the stock structure of different populations of fish, that have been confirmed by genetic analyses (Marshall 2005) and are likely caused by isolation and possibly driven by climatic differences between regions. For example, the high temperatures in the Gulf of Carpentaria require better adaptation to extreme temperatures than on the more moderate east coast of Australia.

Some considerations need to be made when undertaking correlation analyses. The analyses were undertaken without adjusting either the fisheries or climate data for autocorrelation (Robins et al. 2005) as this can increase the risk of a Type II error (by removing an authentic significant relationship between the variables), or bias results if the source of autocorrelation is due to covariance (Pyper and Peterman 1998). In addition, it must be noted that correlative relationships may change over time if the causal mechanism is not embodied in the model (Solow 2002). This limitation of correlative and time series modeling must be compensated for through the use of sound theoretical and life cycle modeling that explains the causal links, and eliminates irrelevant variables from the analysis (Robins et al. 2005).

3.5 Impacts of Climate Change on *L. calcarifer* Stocks

Warming water and air temperatures and the greater climate extremes projected for Australia as a result of global warming (Hughes 2003) will likely alter primary productivity of aquatic ecosystems, regional currents and water quality, and may cause a change in fish migration, abundance, growth and survival (Frye 1983, Kapetsky 2000). Such changes have already been reported, e.g., in the Tejo River estuary in Portugal (Europe's largest estuary) where rising water temperature has resulted in a dramatic change of fish species assemblages (Cabral et al. 2007). In northern Australia, where *L. calcarifer* are distributed, altered rainfall and river flows are suggested to be important environmental drivers of potential change (Hobday et al. 2008) in the short to medium term (i.e., up to 2050), whilst in southeastern Australia changing temperature will be important. Other parameters projected to change include sea surface temperature, sea level, ocean currents, ocean chemistry (including pH) and the frequency of extreme climate events (Hobday and Lough 2011). Although there are projections as to how these parameters may change at global scales (e.g., oceans are now about 0.1 pH unit lower than they were prior to the industrial era (30% more acid) (Allison et al. 2009), it is very uncertain how parameters may change in estuaries, which experience a lot more variability than relatively stable oceans (Gillanders et al. 2011, Hobday and Lough 2011).

Vulnerability of *L. calcarifer* to climate change is ultimately linked to impacts on nursery grounds. For example, years with high rainfall and large areas of nursery habitat will see recruitment peaks whereas periods of dry and/or cooler periods will likely have less recruitment success. The following factors associated with climate change are also likely to impact on the life cycle of *L. calcarifer*: sea level rise, changed rainfall patterns, severe weather events, water acidification and increased temperature (Poloczanska et al. 2007) and are further discussed below.

3.5.1 Sea level rise

In general, wetland and mangrove ecosystems are at high risk from sea level rise and flooding due to the low elevation of such environments (Crimp and Balston 2003). Urbanisation and farming will limit the inland migration of wetland and mangrove habitats in the future, and therefore reduce their availability to aquatic organisms (Woodroffe 1995). Manson et al. (2005) showed that links between mangrove area and coastal fisheries production could be detected for *L. calcarifer* and other species at a broad regional scale (1000s of kilometers) on the east coast of Queensland, Australia. Her, mangrove area and/or scale of perimeter, as well as the area of shallow water, correlated with *L. calcarifer* catch and explained 37% of the variation in the dataset. Thus mangrove/wetland area available is a significant driver influencing the population dynamics of *L. calcarifer*. However, the loss of *L. calcarifer* nursery habitat due to sea level rise will have an unknown effect on population dynamics of the species, as being euryhaline *L. calcarifer* has the ability to thrive across a very wide range of environmental conditions, from freshwater to hypersaline conditions; an adaptability that allows them to potentially utilize and exploit a diversity of habitats. In addition, *L. calcarifer* stocks along Queensland's east coast are distinct as each tends to be confined to a particular climatic region close to nursery habitat and available freshwater flow (Marshall 2005). Different stocks appear to use nursery grounds differently and this may be helpful for adaptation to climate change induced SLR.

3.5.2 Rainfall and severe weather events

For most *L. calcarifer* stocks, access to freshwater habitats enhances survival and recruitment. *Lates calcarifer* stocks are predominately found in areas that have large slow flowing rivers that flow at least part of the year (Dunstan 1959). Future reductions in freshwater flow as a result of climate change may well reduce the distribution, recruitment success and catchability of the species. Changes to rainfall, and hence freshwater flows, are likely to alter the extent of nursery habitats and connectivity between freshwater and estuarine areas for mature males returning to spawn. Increased rainfall variability in the region (IPCC 2007) is likely to increase the fluctuation of recruitment size, making the species more vulnerable to overfishing and natural mortality (e.g., from diseases) during extended drought conditions. Sawynok and Platten (2011) suggested that the increase in the duration between large flood events may impact on strong recruitment years for *L. calcarifer*. Currently, strong recruitment years are a feature of several regional stocks of *L. calcarifer* (Halliday et al. 2012) and appear to drive the productivity of associated fisheries for several years. Sawynok and Platten (2011) then suggest that if the length of time between large recruitment events exceed eight years, then there may be issues with the sex ratio of the spawning population with "uncertain consequences" due to the fact that *L. calcarifer* is a sequential protandrous hermatophrodite and if the period of poor recruitment is prolonged then the number of reproductive males in the population will decrease.

3.5.3 Acidification

Lates calcarifer uses a variety of habitats throughout its life history and is more tolerant of a wide range of water acidity than many temperate fishes and has been recorded in several northern Australian locations where pH ranged from ~4 up to 9.12 units (Pusey et al. 2004). Recorded pH in the Fitzroy River estuary (a primary habitat of *L. calcarifer*) can vary from between 8.6 to 6.8 (J.B. Robins unpublished data January 2003 to May 2005). Therefore, it is likely that juveniles and adults of *L. calcarifer*, which occur in waters between pH 4 and up to pH 9.12, will be relatively insensitive to changes in water acidification as a consequence of CO_2 dissolving into the world's oceans. The severity of pH reduction (e.g., pH 3.5) as a result of acid-sulfate soil acidification is likely to be much greater than that expected under climate-change scenarios (0.5 change in pH) (Gillanders et al. 2011). However, less is known about the sensitivity of *L. calcarifer* to pH at critical stages, such as eggs and larvae. De (1971) reported that larval *L. calcarifer* have a narrow range of pH for hatching success, i.e., 7.4 to 7.6 units. Changes outside these bounds may cause an increased mortality for juvenile *L. calcarifer* and larvae e.g. otolith development and an impairment of olfactory discrimination that may reduce successful settlement of larvae, in estuarine waters (Munday et al. 2009). More research is required to determine how tolerant the early life stages of *L. calcarifer* are to variable or altered pH.

In addition, indirect effects of ocean acidification on estuarine fish are likely to occur through the effects on prey with calcified structures (e.g., crustaceans, molluscs). As a major component of the *L. calcarifer* diet is crustacean based potential impacts on early life history and reproductive stages of marine calcifiers may affect *L. calcarifer* population size by reducing food availability and increasing competition, with flow-on effects to food-web interactions (Gillanders and Kingsford 2002, Pörtner 2008, Gillanders et al. 2011).

3.5.4 Temperature

Increased air and water temperatures (IPCC 2007, CSIRO and BoM 2007) will change ocean, estuary, stream and nursery habitats and affect egg hatch, juvenile development and adult growth rates and maturation. Air temperature is projected to increase in northern Australia, but the amount of change is variable between regions (DERM 2009). In the Fitzroy region, average summer temperature (currently 26.9°C) is projected to increase by 2.0°C by 2050 and by 3.2°C by 2070 under high emissions scenarios (DERM 2009). *Lates calcarifer* is likely to be able to cope with water temperature changes in the medium term, as this species has a wide thermal optimum for growth and protein metabolism, i.e., 27 to 33°C (Katersky and Carter 2007) and a critical thermal maximum of 44.5°C (Rajaguru 2002). An increase in SSTs southward along the coast would be expected to extend the current range of *L. calcarifer* southward along the Australian coast. The current average sea surface temperature that defines the existing southerly extent of the species (approximately 24.6°C) could be expected to have moved poleward to Tweed Heads on the east coast, and Carnarvon on the west coast of Australia, by the year 2030 if current trends in global warming continue. Increased water temperatures, in particular in shallow waters, can cause aquatic weeds such as *Salvinia molesta*, water hyacinth *Eichornia crassipes* and Pistia to form rafts that completely block light, heat and oxygen flux in *L. calcarifer* nurseries. These rafts can result in complete overgrowth of small wetlands and channels in the space of several seasons. Electrofishing surveys in the Herbert River floodplain lagoons that were affected by weeds have found differences in the proportions of 0+ and 1+ year class *L. calcarifer* juveniles (Burrows 2004). In some wetlands where weeds have overgrown lagoons, juveniles died within the year when flow ceased. The dieback is a result of anoxia from decomposing weeds.

3.6 Conclusion

Changes in climate will have an effect on the dynamics of rivers, wetlands, coasts and estuaries and thus have an effect on *L. calcarifer* fisheries. The results from our research have shown that climate variability over different temporal and spatial scales has significant and at times non-linear impacts on *L. calcarifer* in northern Australia. Important key climate variables (freshwater flow and water temperature) were identified as drivers of *L. calcarifer* catch and we also identified potential impacts from climate change to *L. calcarifer*. Simulations of the effects of altered river flows projected to occur with global climate change (assuming high emissions) indicated that *L. calcarifer* populations would be expected to have reduced productivity, by the degree to which flow is reduced.

There is a need to identify the likely changes in biotic distributions and production processes in response to climate change to understand the ways in which the fishing industry could adapt to accommodate climate driven changes in catch. Further studies are needed to investigate regional differences and how regional management plans for coastal fisheries can be implemented to account for increased variability of fish stocks as a result of climate change. Better information on the location of spawning grounds, juvenile growth rates, genetic diversity and quantification of threshold climate events (e.g., the predictions of low temperature events reduce the risk of stock losses in aquaculture) for *L. calcarifer* are needed to understand the response of *L. calcarifer* to climate change. There is the opportunity to improve the sustainable management of the fishery by incorporating measures of climate into both catch models and adaptive management strategies.

The available knowledge on possible impacts of climate change on Asian populations of *L. calcarifer* is sparse. Due to rapidly increasing human populations, in particular in low lying coastal areas, immediate risks for wild *L. calcarifer* populations are more likely associated with impacts on habitat such as pollution and removal of wetlands. However, given the paucity of knowledge on Asian populations of *L. calcarifer* there is a need to undertake similar analyses as we have performed in the current chapter to identify the key environmental drivers of recruitment and likely impacts of climate change into the future on these stocks of fish.

3.7 Acknowledgment

The authors like to thank the various governmental agencies that provided data for the analyses including the Bureau of Meteorology; the Queensland Department of Agriculture, Fisheries and Forestry; the Queensland Department of Environment and Resource Management; the Queensland Department of Science Information Technology Innovation and the Arts; and the Northern Territory Department of Resources.

References

Agcopra, C., J.M. Balston, R. Bowater, L.J. Rodgers and A.A.J. Williams. 2005. Predictive system for aquaculture ponds. Queensland Aquaculture News 3 pp.

Allison, I., N.L. Bindoff, R.A. Bindschadler, P.M. Cox, N. de Noblet, M.H. England, J.E. Francis, N. Gruber, A.M. Haywood, D.J. Karoly, G. Kaser, C. Le Quéré, T.M. Lenton, M.E. Mann, B.I. McNeil, A.J. Pitman, S. Rahmstorf, E. Rignot, H.J. Schellnhuber, S.H. Schneider, S.C. Sherwood, R.C.J. Somerville, K. Steffen, E.J. Steig, M. Visbeck and A.J. Weaver. 2009. Updating the World on the Latest Climate Science. The University of New South Wales Climate Change Research Centre (CCRC), Sydney 60 pp.

Australian Barramundi Farmers Association. 2012. Barramundi Aquaculture (vol. 2012).

Balston, J. 2009. Short-term climate variability and the commercial barramundi (*Lates calcarifer*) fishery of north-east Queensland, Australia. Mar. Freshwater Res. 60: 912–923.

Barlow, C.G., M.G. Pearce, L.J. Rodgers and P. Clayton. 1995. Effects of photoperiod on growth, survival and feeding periodicity of larval and juvenile barramundi *Lates calcarifer* (Bloch). Aquaculture 138: 159–168.

[BoM] Australian Bureau of Meteorology. 2004. Australian Rainfall Districts. Commonwealth of Australia, Canberra.

[BoM] Australian Bureau of Meteorology. 2005. SILO (vol. 2005). Commonwealth of Australia, Canberra.

[BoM] Australian Bureau of Meteorology. 2007. Climate Change in Australia: Technical report. CSIRO, Melbourne.

[BoM] Australian Bureau of Meteorology. 2012. Climatology Normanton. Available from http://www.bom.gov.au/.

Bunn, S.E. and A.H. Arthington. 2002. Basic principles and ecological consequences of altered flow regimes for aquatic biodiversity. Environ. Manage. 30: 492–507.

Burrows, D.W. 2004. Translocated fishes in streams of the Wet Tropics Region, North Queensland: Distribution and potential impact. Cooperative Research Centre for Tropical Rainforest Ecology and Management, Rainforest CRC, Cairns. 83. pp.

Cabral, H.N., R. Vasconcelos, C. Vinagre, S. Frana, V. Fonseca, A. Maia, P. Reis-Santos, M. Lopes, M. Ruano, J. Campos, V. Freitas, P.T. Santos and M.J. Costa. 2007. Relative importance of estuarine flatfish nurseries along the Portuguese coast. J. Sea Res. 57: 209–217.

Campbell, A.B. and M.F. O'Neil. 2007. Assessment of the barramundi fishery in Queensland, 1989–2007. Queensland Department of Primary Industries and Fisheries, Brisbane, Queensland. 27 pp.

Clewett, J.F., N.N. Clarkson, D.A. George, S.H. Ooi, D.T. Owens, I.J. Partridge and G.B. Simpson. 2003. Rainman Stream Flow version 4.3: a comprehensive climate and streamflow analysis package on CD to assess seasonal forecasts and manage climate risk. Queensland Department of Primary Industries, Brisbane, Australia.

Coleman, A.P.M. 2004. The National Recreational Fishing Survey: The Northern Territory. Northern Territory Department of Business, Industry and Resource Development, Darwin.

Crimp, S. and J.M. Balston. 2003. A study to determine the scope and focus of an integrated assessment of climate change impacts and options for adaptation in the Cairns and Great Barrier Reef region. Australian Greenhouse Office, Canberra. 117 pp.

[CSIRO and BoM] Commonwealth Scientific and Industrial Research Organisation and Australian Bureau of Meteorology. 2007. Climate Change in Australia: Technical report. CSIRO, Melbourne. 148pp.

Davis, T.L. 1982. Maturity and sexuality in barramundi, *Lates calcarifer* (Bloch), in the Northern Territory and south-eastern Gulf of Carpentaria. Aust. J. Mar. Freshwater Res. 33: 529–545.

Davis, T.L.O. 1985. Seasonal changes in gonad maturity, and abundance of larval and early juveniles of barramundi, *Lates calcarifer* (Bloch), in Van Diemen Gulf and the Gulf of Carpentaria. Aust. J. Mar. Freshwater Res. 36: 177–190.

Davis, T.L.O. 1987. Biology of wildstock *Lates calcarifer* in northern Australia. In: J.W. Copland and D.L. Grey (eds.). ACIAR International Workshop on the Management of wild and cultured sea bass barramundi (*Lates calcarifer*). Australian Centre for International Agricultural Research, Canberra, pp. 22–29.

De, G.K. 1971. On the biology of post-larval and juvenile stages of *Lates calcarifer* Bloch. J. Indian Fish. Assoc. 1: 51–64.

(DERM) Department of Environment and Resource Management. 2009. ClimateQ: toward a greener Queensland. Queensland Government, Brisbane. 424 pp.

Dunstan, D.J. 1959. The barramundi *Lates calcarifer* (Bloch) in Queensland waters. CSIRO, Melbourne.

Frye, R. 1983. Climatic change and fisheries management. Nat. Resour. J. 23: 77–96.

Garcia, L.M.B. 1989. Dose-dependent spawning response of mature female sea bass, *Lates calcarifer*(Bloch), to pelleted luteinizing hormone-releasing hormone analogue (LHRHa). Aquaculture 77: 85–96.
Genstat. 2008. Genstat for Windows, Release 11.1, Eleventh Edition. VSN International Ltd., Oxford.
Gillanders, B.M. and M.J. Kingsford. 2002. Impact of changes in flow of freshwater on estuarine and open coastal habitats and the associated organisms.Oceanogr. Mar. Biol. 40: 233–309.
Gillanders, B.M., T.S. Elsdon, I.A. Halliday, G. Jenkins, J.B. Robins andF.J. Valesini. 2011. Potential effects of climate change on Australian estuaries and fish utilising estuaries: a review. Mar. Freshwater Res. 62: 1115–1131.
Gillson, J., J. Scandol and I. Suthers. 2009. Estuarine gillnet fishery catch rates decline during drought in eastern Australia. Fish. Res. 99: 26–37.
Grace, B., A. Handley and H. Bajha. 2008. Managing, monitoring, maintaining and modelling barramundi.Proceedings of the National Barramundi Workshop, 6–8 July 2005, Darwin, Northern Territory and Overview of the barramundi modelling workshop, 27 February – 3 March 2006. Northern Territory Government, Darwin.
Grey, D.L. 1987. An overview of *Lates calcarifer* in Australia and Asia. *In*: J.W. Copland and D.L. Grey (eds.). ACIAR International Workshop on the Management of wild and cultured sea bass/barramundi (*Lates calcarifer*). Australian Centre for International Agricultural Research, Canberra, pp. 15–21.
Griffin, R.K. 1987. Life history, distribution, and seasonal migration of barramundi in the Daly River, Northern Territory, Australia. *In*: M.J. Dadswell, R.J. Klauda, C.M. Moffitt, R.L. Saunders, R.A. Rulifson and J.E. Cooper (eds.). American Fisheries Society Symposium Series(vol. 1).Bethesda, MD, pp. 358–363.
Griffin, R.K. 1991. Barramundi population assessment by closed area depletion methods. pp. 34. *In*:18th Annual Conference of the Australian Society for Fish Biology(vol. 21). Australian Society for Fish Biology, Hobart.
Halliday, I. and J. Robins. 2007. Environmental flows for sub-tropical estuaries: understanding the needs of estuaries for sustainable fisheries production and assessing the impacts of water regulation. Final Report to the Fisheries Research and Development Corporation for Project No. 2001/022 and Coastal Zone Project FH3/AF. Queensland Department of Primary Industries and Fisheries, Brisbane. 212pp.
Halliday, I.A., J.B. Robins, D.G. Mayer, J. Staunton-Smith andM.J. Sellin. 2008. Effects of freshwater flow on the year-class strength of a non-diadromous estuarine finfish, king threadfin (*Polydactylusmacrochir*), in a dry-tropical estuary. Mar. Freshwater Res. 59: 157–164.
Halliday, I.A., J.B. Robins, D.G. Mayer, J. Staunton-Smith and M.J. Sellin. 2011. Freshwater flows affect the year-class strength of barramundi *Lates calcarifer* in the Fitzroy river estuary, central Queensland. Proc. R. Soc. Queensl. 116: 1–11.
Halliday, I.A., T.Saunders, M. Sellin, Q. Allsop, J.B. Robins, M. McLennan and P. Kurnoth. 2012. Flow impacts on estuarine finfish fisheries in the Gulf of Carpentaria. Department for Agriculture, Forestry and Fisheries, Brisbane. 66pp.
Hamilton, S.K. and P.C. Gehrke. 2005. Australia's tropical river systems: current scientific understanding and critical knowledge gaps for sustainable management. Mar. Freshwater Res. 57: 619–633.
Hendon, H. andB. Liebmann. 1990. The Intraseasonal (30–50 day) Oscillation of the Australian summer monsoon. J. Atmos. Sci. 47: 2909–2923.
Hobday, A.J. and J.M. Lough. 2011. Projected climate change in Australian marine and freshwater environments. Mar. Freshwater Res. 62: 1000–1014.
Hobday, A.J., E.S. Poloczanska and R.J. Matear. 2008. Implications of Climate Change for Australian Fisheries and Aquaculture: a Preliminary Assessment. Department of Climate Change, Canberra.

Hughes, L. 2003. Climate change and Australia: Trends, projections and impacts. Aust. Ecol. 28: 423–443.
[IPCC] Intergovernmental Panel on Climate Change. 2000. In: N. Nakićenović and R. Swart (eds.). Special Report on Emissions Scenarios. Working Group III of the Intergovernmental Panel on Climate Change Cambridge University Press, Cambridge.
[IPCC] Intergovernmental Panel on Climate Change. 2007. Climate Change: Impacts, adaptation and vulnerability. Contribution of Working Group II to the Fourth Assessment Report of the Intergovernmental panel on Climate Change. Cambridge University Press, Cambridge.
Jardine, T.D., B.J. Pusey, S.K. Hamilton, N.E. Pettit, P.M. Davies, M.M. Douglas, V. Sinnamon, I.A. Halliday and S.E. Bunn. 2011. Fish mediate high food web connectivity in the lower reaches of a tropical floodplain river. Oecologia 168: 829–838.
Jenkins, G.P., S.D. Conron and A.K. Morison. 2010. Highly variable recruitment in an estuarine fish is determined by salinity stratification and freshwater flow: implications of a changing climate. Mar. Ecol.: Prog. Ser. 417: 249–261.
Jonsson, N. 1991. Influence of water flow water temperature and light on fish migration in rivers. Nordic J. Freshwater Res. 66: 20–35.
Kapetsky, J.M. 2000. Present applications and future needs of meteorological and climatological data in inland fisheries and aquaculture. Agric. For. Meteorol. 103: 109–117.
Katersky, R.S. and C.G. Carter. 2007. A preliminary study on growth and protein synthesis of juvenile barramundi, *Lates calcarifer* at different temperatures. Aquaculture 267: 157–164.
Keenan, C. and J. Salini. 1989. The genetic implications of mixing barramundi stocks in Australia. Introduced and translocated fishes and their ecological effects. Proceedings of the Australian Society for Fish Biology Workshop No. 8. Bureau of Rural Resources, Australian Government Publishing Service, Magnetic Island, Townsville, Australia.
Liston, P., M.J. Furnas, A.W. Mitchell and E.A. Drew. 1992. Local and mesoscale variability of surface water temperature and chlorophyll in the northern Great Barrier Reef, Australia. Cont. Shelf Res. 12: 907–921.
Lloret, J., I. Palomera, J. Salt and I. Sole. 2004. Impact of freshwater input and wind on landings of anchovy (*Engraulisencrasicolus*) and sardine (*Sardinapilchardus*) in shelf waters surrounding the Ebre (Ebro) River delta (north-western Mediterranean). Fish. Oceanogr. 1: 102–110.
Loneragan, N.R. and S.E. Bunn. 1999. River flows and estuarine ecosystems: implications for coastal fisheries from a review and a case study of the Logan River, southeast Queensland. Aust. J. Ecol. 24: 431–440.
MacDonald, G.M. and R.A. Case. 2005. Variations in the Pacific Decadal Oscillation over the past millennium. Geophys. Res. Lett. 32: L08703.
Maceina, M.J. 1997. Simple application of using residuals from catch-curve regressions to assess year-class strength in fish. Fish. Res. 32: 115–121.
Marshall, C.R.E. 2005. Evolutionary genetics of barramundi (*Lates calcarifer*) in the Australian region. School of Biological Sciences, Murdoch University, Perth. 120pp.
McDougall, A. 2004. Assessing the use of sectioned otoliths and other methods to determine the age of the centropomid fish, barramundi (*Lates calcarifer*) (Bloch), using known-age fish. Fish. Res. 67: 129–141.
McGlennon, D., G.K. Jones, J. Baker, W.B. Jackson and M.A. Kinloch. 2000. Ageing, catch-age and relative year-class strength for snapper (*Pagrusauratus*) in northern Spencer gulf, South Australia. Mar. Freshwater Res. 51: 669–677.
Manson, F.J., N.R. Loneragan, G.A. Skilleter and S.R. Phinn. 2005. An evaluation of the evidence for linkages between mangroves and fisheries: A synthesis of the literature and identification of research directions. Oceanogr. Mar. Biol. 43: 485–515.
Meynecke, J.-O. and S.Y. Lee. 2011. Climate-coastal fisheries relationships and their spatial variation in Queensland, Australia. Fish. Res. 110: 365–376.

Meynecke, J.-O., S.Y. Lee, N.C. Duke and J. Warnken. 2006. Effect of rainfall as a component of climate change on estuarine fish production in Queensland, Australia. Estuar. Coast. Shelf Sci. 69: 491–504.

Milton, D., I. Halliday, M. Sellin, R. Marsh, J. Staunton-Smith, D. Smith, M. Norman and J. Woodhead. 2008. The effect of habitat and environmental history on otolith chemistry of barramundi Lates calcarifer in estuarine populations of a regulated tropical river. Estuar. Coast. Shelf Sci. 78: 301–315.

Moore, R. 1979. Natural sex inversion in the Giant Perch (Lates calcarifer). Aust. J. Mar. Freshwater Res. 30: 803–813.

Munday, P.L., D.L. Dixson, J.M. Donelson, G.P. Jones, M.S. Pratchett, G.V. Devitsina and K.B. Døving. 2009. Ocean acidification impairs olfactory discrimination and homing ability of a marine fish. Proc. Natl. Acad. Sci. USA, 106: 1848–1852.

Pender, P.J. and R.K. Griffin. 1996. Habitat history of barramundi Lates calcariferin a North Australian river system based on barium and strontium levels in scales. Trans. Am. Fish. Soc. 125: 679–689.

Pillay, T.V.R. 1993. Aquaculture principles and practices.Blackwell Sciences Ltd., Oxford.

Poloczanska, E.S., R.C. Babcock, A. Butler, A.J. Hobday, O. Hoegh-Guldberg, T.J. Kunz, R. Matear, D. Milton, T.A. Okey and A.J. Richardson. 2007. Climate change and Australian marine life. Oceanogr. Mar. Biol. Ann. Rev. 45: 409–480.

Pörtner, H.O. 2008. Ecosystem effects of ocean acidification in times of ocean warming: a physiologist's view. Mar. Ecol. Prog. Ser. 373: 203–207.

Pusey, B.J., M.J. Kennard and A.H. Arthington. 2004. Freshwater Fishes of North-eastern Australia.CSIRO Publishing, Melbourne.

Pyper, B.J. and R.M. Peterman. 1998. Comparison of methods to account for autocorrelation in correlation analyses of fish data. Can. J. Fish.Aquat. Sci. 55: 2127–2140.

Quiñones, R.A. and R.M. Montes. 2001. Relationship between freshwater input to the coastal zone and the historical landing of the benthic/demersal fish Eleginopsmaclovinusin central-south Chile. Fish. Oceanogr. 10:311–328.

Rajaguru, S. 2002. Critical thermal maximum of seven estuarine fishes. J. Therm. Biol. 27: 125–128.

Roberts, D.T., L.J. Duivenvorden and I.G. Stuart. 2008. Factors influencing recruitment patterns of Golden Perch (Macquariaambiguaoriens) within a hydrologically variable and regulated Australian tropical river system. Ecol. Freshwater Fish 17: 577–589.

Robins, J.B., I.A. Halliday, J. Staunton-Smith, D.G. Mayer and M.J. Sellin. 2005. Freshwater-flow requirements of estuarine fisheries in tropical Australia: a review of the state of knowledge and application of a suggested approach. Mar. Freshwater Res. 56: 343–360.

Robins, J., D. Mayer, J. Staunton-Smith, I.B.S. Halliday and M. Sellin. 2006. Variable growth rates of the tropical estuarine fish barramundi Lates calcarifer (Bloch) under different freshwater flow conditions. J. Fish Biol. 69: 379–391.

Russell, D.J. 1990. Reproduction, migration and growth in Lates calcarifer. Department of Biology and Environmental Science, Queensland University of Technology, University of Queensland, Brisbane. 194pp.

Russell, D.J. and R.N.Garrett. 1985. Early life history of barramundi, Lates calcarifer(Bloch), in north-eastern Queensland. Aust. J. Mar. Freshwater Res. 36:191–201.

Sawynok, B. 1998. Fitzroy river effects of freshwater flows on fish. Info Fish Services, Rockhampton. 56 pp.

Sawynok, W. and J.R. Platten. 2011. Effects of local climate on recreational fisheries in central Queensland Australia: a guide to the impacts of climate change. Am. Fish. Soc. Symp. 75: 201–206.

Schipp, G. 1996. Barramundi farming in the Northern Territory. Department of Primary Industry and Fisheries, Aquaculture Branch, Fisheries Division, Brisbane. pp?

Shaklee, J.B., J. Salini and R.N. Garrett. 1993. Electrophoretic characterization of multiple genetic stocks of barramundi perch in Queensland, Australia. Trans. Am. Fish. Soc. 122: 685–701.

Shapiro, S.S., M.B. Wilk and H.J. Chen. 1968. A comparative study of various tests of normality. J. Am. Stat. Assoc. 63: 1343–1372.
Solow, A.R. 2002. Fisheries recruitment and the North Atlantic Oscillation. Fish. Res. 54: 295–297.
Staunton-Smith, J., J.B. Robins, D.G. Mayer, M.J. Sellin and I.A. Halliday. 2004. Does the quantity and timing of fresh water flowing into a dry tropical estuary affect year-class strength of barramundi (*Lates calcarifer*)? Mar. Freshwater Res. 55: 787–797.
Stuart, I.G. and S. McKillup. 2002. The use of sectioned otoliths to age barramundi (*Lates calcarifer*) (Bloch, 1790)[Centropomidae]. Hydrobiologia 479: 231–236.
Tanimoto, M., J.B. Robins, M.F. O'Neil, I.A. Halliday and A.B. Campbell. 2012. Quantifying the effects of climate change and water abstraction on a population of barramundi (*Lates calcarifer*): a diadromous estuarine finfish. Mar. Freshwater Res. 63: 715–726.
Vance, D.J., M.D.E. Haywood, D.S. Heales, R.A. Kenyon and N.R. Loneragan. 1998. Seasonal and annual variation in abundance of postlarval and juvenile banana prawns *Penaeusmerguiensis* and environmental variation in two estuaries in tropical northeastern Australia: a six year study. Mar. Ecol. Prog. Ser. 163: 21–36.
Voogt, S., S. Giles, P. Harding, J.P. Vítkovský and A. Loy. 2009. *In:* A.C.T. Barton (ed.). Hydrologic Assessment of Regional Water Supply Strategies in Queensland Considering Climate Variability and Climate Change. Engineers Australia, Newcastle, New South Wales, pp. 1260–1271.
Welch, D., N. Gribble and R. Garrett. 2002. Assessment of the Barramundi Fishery in Queensland –2002. Department of Primary Industries and Fisheries, Brisbane. pp?
Williams, L.E. 2002. Queensland's fisheries resources: Current condition and recent trends 1988–2000. Queensland Department of Primary Industries and Fisheries, Brisbane. pp?
Woodroffe, C.D. 1995. Response of tide-dominated mangrove shorelines in Northern Australia to anticipated sea-level rise. Earth Surf. Proc. Land. 20: 65–85.
Xiao, Y. 2000. Use of the original von Bertalanffy growth model to describe the growth of barramundi, *Lates calcarifer*. Fish. Bull. 98: 835–841.

Reproductive Biology of the Asian Seabass, *Lates calcarifer*

*Evelyn Grace De Jesus-Ayson** and *Felix G. Ayson*

4.1 Introduction

The Asian seabass, *Lates calcarifer*, or popularly known as barramundi in Australia, belongs to the family Latidae. Seabass are considered as one of the most important species for fisheries and aquaculture and are widely distributed in the Indo-West Pacific. They are euryhaline and are found in freshwater, estuarine, and coastal areas and then migrate downstream for spawning. Seabass are protandrous hermaphrodites, first maturing as males and then undergoing sex inversion to become females later in life. Spawning of seabass is seasonal and varies depending on geographical location, with latitudinal variation in spawning season thought to be related to temperature and maybe photoperiod. Spawning occurs in river mouths, estuaries or coastal areas, and is usually associated with incoming tides and the lunar cycle. Seabass are tolerant of a wide range of physiological conditions and the females are highly fecund. These attributes have helped trigger interest in the aquaculture of the species which started in Thailand in the early 1970s and rapidly spread throughout Southeast Asia.

Aquaculture Department, Southeast Asian Fisheries Development Center, Tigbauan, Iloilo 5021.
*Corresponding author

4.2 Reproductive Biology

4.2.1 Maturation stages

Lates calcarifer is hermaphroditic and this species' protandry has been demonstrated by experimental studies involving biopsy and histology (Moore 1979, Davis 1982). In the Northern Territory and the southeastern Gulf of Carpentaria, Australia, fish with sizes ranging from 550 to 600 mm body length (BL) were observed to undergo first maturation as males, with most males around 850–900 mm BL observed to be changing sex. Females are therefore thought to be normally derived from post-spawning males. Generally, seabass males in the Gulf of Carpentaria undergo first maturation as well as sex inversion at smaller sizes compared to fish in the Northern Territory. These size differences were thought to be due to the slower growth rate of barramundi in the Gulf of Carpentaria, both processes being related to age rather than to size.

Davis (1982) described the maturation stages of male and female seabass by gross morphological observations and gonadal histology. Davis (1982) categorized male maturation stages as follows: immature or newly formed (Stage 1), developing-recovering spent (Stage 2), maturing (Stage 3), mature (Stage 4), ripe (Stage 5), and spent (Stage 6) (Table 4.1).

The different maturation stages of the testis can be distinguished macroscopically based on size and shape, and to some extent, by color. Immature testes are semi-transparent and strap-like. Developing or first maturing testes largely look like immature testis, but are bigger in size. Recovering-spent testes are strap-like, but opaque. Maturing testes become thicker and are surrounded by smaller quantities of fat. Mature testes are thick and wedge-shaped, the lateral margins becoming rounded and the ventral lobes pronounced. Spermatozoa flow freely from cut testis. Ripe testes are large, their lateral margins rounded and ventral lobes swollen. Direct pressure on the testes causes spermatozoa to be extruded through the main duct. Spent testes revert to the strap-like appearance of immature testes. Although there are limited reports of the age that seabass males become sexually mature, in a recirculation system at least, Szentes et al. (2012) report that spermatozoa was detectable in testes by 9 months of age.

In females, the paired ovaries are elongated bodies, pyriform in cross-section and crenated along the narrow, most dorsal edge. Like the male gonad, the different maturation stages of the ovaries can be distinguished macroscopically based on size and color. Developing or recovering-spent ovaries are pink to red in color. As the ovaries mature, they turn paler red to cream in color. Mature and ripe ovaries are creamy yellow to yellow in

Table 4.1. Description of the different maturation stages of male sea bass (Davis 1982).

Maturation Stage	Gonadal Description	Morphological Description	Histological Description
Stage 1	Immature or newly-formed	Testes are semi-transparent, thin and strap-like. The testes first become distinguishable by the appearance of furrows on the ventral side	Consist largely of undifferentiated stroma cells with large vacuoles and densely staining nuclei
Stage 2	Developing or recovering spent	Testes are opaque and strap-like, with deeper longitudinal furrows	The lobes of first maturing consist of undifferentiated vacuolated cells with no lumina or lobule walls; recovering-spent testes are composed largely of vacuolated cells in the walls and interstitial areas
Stage 3	Maturing	Maturing testes become thicker and are surrounded by smaller quantities of fat. The lobules increase in size; those towards the dorsal surface becoming proportionally larger	There is little interlobular tissue remaining, most interstitial areas having formed new lobules. Some spermatogonia are still present, and large nests of spermatocytes and spermatids extend into the lobule lumina. Spermatozoa are present in most lumina and sperm ducts
Stage 4	Mature	Mature testes are thick and wedge-shaped, the lateral margins becoming rounded and the ventral lobes pronounced. Spermatozoa flow freely from cut testis	The lobule walls have become very thin and their lumina and sperm ducts are packed with spermatozoa. Spermatogonia are generally absent. Nests of spermatocytes and spermatids are still present
Stage 5	Ripe	Ripe testes are large, their lateral margins rounded and ventral lobes swollen. Direct pressure on the testes causes spermatozoa to be extruded through the main duct	The lobule walls are thin and distended with spermatozoa. Occasional nests of spermatocytes remain attached to the lobule walls, particularly in the smaller ventral lobules
Stage 6	Spent	Spent testes become strap-like, their weight ranging from 50% of peak stage 5 weight to marginally heavier than stage 2 testes. Some spermatozoa can be squeezed from a cut testis	The lobule walls contract, becoming thicker and wrinkled. The vacuolated cells characteristic of early stage 2 testes begin to appear. Some residual nests of spermatocytes persist, and spermatozoa are present in lobules and ducts

color with the yellow mature oocytes becoming visible through the thin ovary wall. Spent ovaries look like developing or recovering-spent ovaries (Table 4.2).

4.2.2 Reproductive season

Spawning of seabass is seasonal and varies according to geographical location, with latitudinal variation in spawning season thought to be related to temperature and maybe photoperiod. Two main periods could be distinguished in the reproductive cycle of the Asian seabass in French Polynesia. The resting period (March to September) is characterized by very low gonadal-somatic index (GSI), whereas the reproductive season (October to February) is characterized by a high GSI and nearly all the broodstock had gonads undergoing spermiation or vitellogenesis (Guiguen et al. 1993, 1994, 1995). Spawning takes place in brackish waters, near the mouths of rivers, during the wet season. Eggs, embryos, larvae and juveniles first develop in coastal swamps, and the young of the year migrate upstream at the end of the wet season. They usually remain in freshwater until they reach sexual maturity as males (3–4 years). These maturing males migrate downstream at the start of the wet season and, after spawning, males and females can remain in tidal waters or move upstream again. In Papua New Guinea, seabass also has a single annual spawning period extending from October to February (Moore 1982). Papuan seabass are highly fecund. Fecundity can vary from 2 to 32 million eggs within the weight range (7.7 to 20.8 kg) and is related to the total weight of the fish. The average GSI increases markedly from September, reaching a peak from November to January, after which it declines sharply without further increases until August to September the following year. Spawning commences in October; the peak spawning is from November to January, with a few individuals spawning as late as February. Similar observations were noted for Australian stocks (Davis 1985). In equatorial countries like Singapore, reproduction seems to be continuous, with a peak of sexual activity from April/May to September/October (Lim et al. 1986). A similar reproductive pattern was also reported in Thailand (Ruangpanit 1987). The breeding season of captive broodstock in the Philippines is from June to October (Toledo et al. 1991). In the tropical northern hemisphere, the reproductive season is observed during the summer monsoon season (Patnaik and Jena 1976).

Implants of luteinizing hormone releasing hormone-analog A (LHRH-A), methyltestosterone (MT), and their combination, have been found to advance gonadal maturation and spawning in captive seabass (Garcia 1990a). A low dose (100 ug/kg BW) and/or high dose (200 ug/kg BW) of each hormone, or their combination, implanted into sexually quiescent seabass at 45 day intervals have been shown to induce mature

Table 4.2. Description of the different maturation stages of female seabass (Davis 1982).

Maturation Stage	Gonadal Description	Morphological Description	Histological Description
Stage 1	Immature or newly formed		Distinguishable from stage 2 ovaries for only a short period; oocytes diameter less than 80 µm
Stage 2	Developing or recovering spent	Ovaries are compact, thick-walled, pink-red and well vascularized	Contain perinucleolus-stage and a high proportion of chromatin-nucleolus-stage oocytes; average maximum oocyte diameter is less than 110 µm
Stage 3	Maturing	Maturing ovaries increase in size, become pyriform in cross-section, with the narrow edge crenate, paler-red to cream in color and the oocytes are not visible to the naked eye	Perinucleolus-stage and vesicle-stage oocytes are present in large numbers, with occasional clusters of chromatin-nucleolus-stage oocytes; average maximum oocyte diameter of perinucleolus-stage and vesicle-stage oocytes ranges from 110–230 µm
Stage 4	Mature	Mature ovaries are creamy yellow, with thinner walls and oocytes are visible to the naked eye	Maturing oocytes initially consisting of primary yolk oocytes and then secondary yolk oocytes with average maximum diameter of 230–500 µm dominate, though some perinucleolus-stage and vesicle-stage oocytes are still present
Stage 5	Ripe	Ovaries are distended and occupy most of the body cavity; yellow mature oocytes clearly visible through the thin ovary wall	Mature oocytes >500 µm in diameter and clearly visible even through the naked eye
Stage 6	Spent	Ovaries are flaccid, elongated and narrow; will undergo atresia and revert to Stage 2	Large numbers of collapsed follicles of ovulated oocytes, some residual mature oocytes and other oocytes stages are present

females. For example, high doses of LHRH-A either alone, or in combination with MT, induced a significant number of females (43–71%) to undergo final maturation 45 days after a single implantation. The lower dose of LHRH-A, alone or with MT, induced maturation after three implantations (Garcia 1990a). Year-round spawning of seabass broodstock in Australia was achieved by temperature manipulation (Garrett and O'Brien 1994). At day length fixed at 13 hr, water temperature at 28–29°C and salinity at 30–36 ppt, broodfish started to mature after 4 months and were successfully induced to spawn every month for a period of 15 months by injections of 19–27 microgram/kg BW of LHRH-A (Garrett and O'Brien 1994). Similarly, in the Philippines mature male and female seabass were observed throughout the year when water temperature in the tank was maintained between 29–31°C, although attempts to induce spawning using these stocks in January and February outside of the normal spawning season were unsuccessful (SEAFDEC/AQD, unpublished observations).

4.2.3 Sex inversion

Seabass are protandrous hermaphrodites, first maturing as males and then undergoing sex inversion to become females later in life. Fish in the transitional stage cannot be detected by any external morphological features (or for that sake can males or females), with the only reliable method to identify transitional fish being to collect gonadal samples by cannulation and check for the presence of milt or oocytes. Transition of the gonad from testis to ovary can be documented by histological examination (Davis 1982). Oocytes first appear in the ventral lobe of the gonads, either singly or in clumps, although no distributional or sequential pattern of oocyte formation could be detected during the gonadal transition. Once the change from testis to ovary is completed, no testicular tissue or structure remains that would indicate a recent transition, although structures that may have functioned as sperm ducts are present in the dorsal wall of the ovary, but these also produce oocytes and are thus functional parts of the ovary (Davis 1982).

In Australia, males attain sexual maturity in their 3rd to 5th year. They spawn at least once and sometimes for a number of years before changing sex, hence sex inversion usually takes place between 6–8 years of age (Davis 1984). Sex reversal is initiated as the male gonads ripen for the last time and the transition to ovary is completed within about a month of spawning (Guiguen et al. 1994). However, the actual environmental or physiological driver of sex change is unknown, although Athauda et al. (2012) found a positive correlation with possible sex change and warmer water temperatures in seabass reared in culture facilities. In Tahiti, the period of sex inversion, characterized by the presence of transitional animals occurs from January to April, that is, between the reproductive season (October to

February) and the resting period (March to September). However, fish that were undergoing sex inversion were more numerous at the beginning of the resting period (15% in March and 30% in April) than at the end of the reproductive period, and the percentage of males declined significantly only in April (Guiguen et al. 1994). Furthermore, the sex inversion process was not synchronous in the male population in as much as transitional stages were found in January and April, when day length decreased, although temperatures increased until April, when sex inversion was at its height (Guiguen et al. 1994).

A sexually precocious population of barramundi was identified in the Gulf of Carpentaria by Davis (1984). Unlike normal males that mature at 3–5 years, precocious males matured at 1–2 years of age. Sex change, which is linked to the maturity of males, also occurred earlier in this population. A 1:1 sex ratio, which is normally observed before seven years, occurred in 3-year old precocious barramundi. Stunting was also apparent in the sexually precocious population, most likely because of channeling of energy into gonad growth at the expense of somatic growth at a relatively early age. Moore (1979) also reported that a small proportion of primary females that have never matured as males may exist in natural populations and also the possibility that some males do not undergo sex change into females. Similar observations were made by Davis (1982). However, the proportion of primary females in a population is not known since identification of primary females depends upon detection of a female that is considered to be too young or too small to have been a functional and mature male during the previous breeding season.

4.3 Breeding

4.3.1 Spontaneous and induced spawning

Spontaneous spawning of seabass broodstock held in cages or land-based facilities has been observed. In the Philippines, seabass juveniles collected from the wild and reared in floating marine cages matured spontaneously after 2–3 years and spawned naturally the following year (Toledo et al. 1991). Spermiating males and maturing females were first sampled from the stocks after 2 and 3 years in captivity, respectively. Captive male and female broodstock in Singapore also mature between 2.5 and 3–4 years, respectively (Cheong and Yeng 1987).

While seabass broodstock spawn spontaneously in captivity (Toledo et al. 1991), seabass broodstock can and have been induced to spawn by manipulation of environmental parameters (e.g., salinity, temperature or water depth simulating migration and coastal conditions experienced by wild seabass during natural spawning), or hormonal manipulations using human chorionic gonadotropin (HCG), carp pituitary extracts or LHRH-A (see Grey 1987). LHRH-A administered via saline injection, as pelleted implants, or via osmotic pumps, induced mature female seabass to spawn singularly (Harvey et al. 1985, Lim et al. 1986, Nacario 1987), or consecutively over several days (Almendras et al. 1988, Garcia 1989a, b, Garcia 1990b, Garcia 1992).

4.3.2 Broodstock management

Seabass broodstock are held in either floating cages, or land-based tanks. They may be maintained in either fresh or seawater, but must be placed in seawater prior to the breeding season to allow final gonadal maturation and spawning. Seabass do not exhibit obvious external signs of gonadal development and must be examined by gonadal biopsy or cannulation to determine their gender and reproductive status, although milt can be easily expressed from mature male fish during the spawning season.

The breeders are usually fed trash fish, although some facilities, particularly in Australia have breeders that have been trained to accept pelleted feeds. In order to improve the nutrition of broodstock and consequently the quality of the spawned eggs and larvae, vitamin supplements may be included in the feeds, or the breeders are fed fresh squid a few days before the expected spawning.

4.3.3 Broodstock health management

Seabass broodstock are vulnerable to viral, bacterial and parasitic diseases (Alapide-Tendencia and de la Pena 2010, Cruz-Lacierda 2010, Lio-Po 2010). Currently, one of the major disease problems in seabass aquaculture is that of viral nervous necrosis (VNN). The first incidence of mass mortalities due to VNN infection in hatchery-reared seabass in the Philippines was thought to be due to transmission of the virus from trash fish-fed broodstock to eggs (Maeno et al. 2004). Pakingking et al. (2009) has shown that treatment with preparations of formalin inactivated betanodavirus vaccine conferred protective immunity in seabass. Based on these results, an immunization regimen for seabass broodstock for the production of VNN-free eggs/larvae is being developed (R. Pakingking, unpublished data).

4.5 Summary

Seabass aquaculture is a significant industry in countries like Australia, Thailand, Indonesia and Malaysia, and is rapidly gaining popularity in other countries partly because of relative ease in seed production and culture technologies for this species, as well as the high demand and hence the good price that seabass commands in the market. However, production is constrained by the seasonality of seabass reproduction in most countries. Strategies in environmental manipulation that will prolong, if not ensure year-round reproduction in seabass will help address the perennial problem of shortage of seabass fingerlings for grow-out culture.

References

Alapide-Tendencia, E. and L.D. de la Pena. 2010. Bacterial diseases. In: G.D. Lio-Po and Y. Inui (eds.). Health Management in Aquaculture (2nd edition). Southeast Asian Fisheries Development Center Aquaculture Department, Tigbauan, Iloilo, Philippines, pp. 52–76.
Almendras, J.M., C. Duenas, J. Nacario, N.M. Sherwood and L.W. Crim. 1988. Sustained hormone release III. Use of gonadotropin-releasing hormone analogues to induce multiple spawning in seabass, Later calcarifer. Aquaculture 74: 97–111.
Athauda, S., A. Anderson and R. De Nys. 2012. Effect of rearing water temperature on protandrous sex inversion in cultured Asian seabass (Lates calcarifer). Gen. Comp. Endocrinol. 175: 416–423.
Cheong, L. and L. Yeng. 1987. Status of seabass (Lates calcarifer) culture in Singapore. In: J.W. Copland and D.L. Grey (eds.). Management of wild and cultured seabass/barramundi (Lates calcarifer). Australian Centre for International Agricultural Research, Canberra, pp. 65–68
Cruz-Lacierda, E.R. 2010. Parasitic diseases and pests. In: G.D. Lio-Po and Y. Inui (eds.). Health Management in Aquaculture (2nd edition). Southeast Asian Fisheries Development Center Aquaculture Department, Tigbauan, Iloilo, Philippines, pp. 10–38.
Davis, T.L.O. 1982. Maturity and sexuality in barramundi Lates calcarifer (Bloch), in the Northern Territory and south-eastern Gulf of Carpentaria. Aust. J. Mar. Freshwater Res. 33: 529–545.
Davis, T.L.O. 1984. A population of sexually precocious barramundi Lates calcarifer in the Gulf of Carpentaria, Australia. Copeia 1984: 144–149.
Davis, T.L.O. 1985. Seasonal changes in gonad maturity and abundance of larvae and early juveniles of barramundi Lates calcarifer (Bloch), in Van Diemen Gulf and the Gulf of Carpentaria. Aust. J. Mar. Freshwater Res. 36: 177–190.
Garcia, L.M.B. 1989a. Dose-dependent spawning response of mature female seabass, Lates calcarifer (Bloch), to pelleted luteinizing hormone-releasing hormone analogue (LHRH-A). Aquaculture 77: 85–96.
Garcia, L.M.B. 1989b. Spawning response of mature female seabass, Lates calcarifer (Bloch) to a single injection of luteinizing hormone-releasing hormone analogue: effect of dose and initial oocyte size. J. Appl. Ichthyol. 5: 177–184.
Garcia, L.M.B. 1990a. Advancement of sexual maturation and spawning of seabass Lates calcarifer (Bloch), using pelleted luteinizing hormone-releasing hormone analogue and 17a-methyltestosterone. Aquaculture 86: 333–345.
Garcia, L.M.B. 1990b. Spawning response latency and egg production capacity of LHRHa-injected mature female seabass, Lates calcarifer Bloch. J. Appl. Ichthyol. 6: 167–172.
Garcia, L.M.B. 1992. Lunar synchronization of spawning in seabass, Lates calcarifer (Bloch): effect of luteinizing hormone-releasing hormone analogue (LHRHa) treatment. J. Fish Biol. 40: 359–370.
Garrett, R.N. and J.J. O'Brien. 1994. All-year round spawning of hatchery barramundi in Australia. Aust. Fish. 8: 40–42.
Grey, D.L. 1987. An overview of Lates calcarifer in Australia and Asia. In: J.W. Copland and D.L. Grey (eds.). Management of wild and cultured seabass/barramundi (Lates calcarifer). Australian Centre for International Agricultural Research, Canberra, pp. 15–29.

Guiguen, Y., B. Jalabert, E. Thouard and A. Fostier. 1993. Changes in plasma and gonadal steroid hormones in relation to the reproductive cycle and the sex inversion process in the protandrous seabass, *Lates calcarifer*. Gen. Comp. Endocrinol. 92: 327–338.

Guiguen, Y., C. Cauty, A. Fostier, J. Fuchs and B. Jalabert. 1994. Reproductive cycle and sex inversion of the seabass, *Lates calcarifer*, reared in sea cages in French Polynesia: histological and morphometric description. Environ. Biol. Fishes 39: 231–274.

Guiguen, Y., B. Jalabert, A. Benett and A. Fostier. 1995. Gonadal *in vitro* androstenedione metabolism and changes in some plasma and gonadal steroid hormones during sex inversion of the protandrous seabass, *Lates calcarifer*. Gen. Comp. Endocrinol. 100: 106–118.

Harvey, B., J. Nacario, L. Crim, J. Juario and C. Marte. 1985. Induced spawning of seabass, *Lates calcarifer*, and rabbitfish, *Siganus guttatus*, after implantation of pelleted LHRH analogue. Aquaculture 47: 53–59.

Lim, L.C., H.H. Heng and H.B. Lee. 1986. The induced breeding of seabass, *Lates calcarifer* (Bloch) in Singapore. Singapore J. Prim. Ind. 14: 81–95.

Lio-Po, GD. 2010. Viral diseases. *In*: G.D. Lio-Po and Y. Inui (eds.). Health Management in Aquaculture (2nd edition). Southeast Asian Fisheries Development Center Aquaculture Department, Tigbauan, Iloilo, Philippines, pp. 77–146.

Maeno, Y., L.D. de la Pena and E.R. Cruz-Lacierda. 2004. Mass mortalities associated with viral nervous necrosis in hatchery-reared seabass *Lates calcarifer* in the Philippines. JARQ 38: 69–73.

Moore, R. 1979. Natural sex inversion in the giant perch (*Lates calcarifer*). Aust. J. Mar. Freshwater Res. 30: 803–813.

Moore, R. 1982. Spawning and early life history of barramundi *Lates calcarifer* (Bloch), in Papua New Guinea. Aust. J. Mar. Freshwater Res. 33: 647–661.

Nacario, J.F. 1987. Releasing hormones as an effective agent in the induction of spawning in captivity of seabass (*Lates calcarifer*). *In:* J.W. and D.L. Grey (eds.). Management of wild and cultured seabass/barramundi (*Lates calcarifer*). Australian Centre for International Agricultural Research, Canberra, pp. 126–128.

Pakingking, R. Jr., R. Seron, L. de la Pena, K. Mori, H. Yamashita and T. Nakai. 2009. Immune responses of Asian seabass (*Lates calcarifer*) against the inactivated betanodavirus vaccine. J. Fish Dis. 32: 457–463.

Patnaik, S. and S. Jena. 1976. Some aspects of biology of *Lates calcarifer* (Bloch) from Chilka Lake. Indian J. Fish. 23: 65–71.

Ruangpanit, N. 1987. Biological characteristics of wild seabass/barramundi (*Lates calcarifer*). *In*: J.W. Copland and D.L. Grey (eds.). Management of wild and cultured seabass/barramundi (*Lates calcarifer*). Australian Centre for International Agricultural Research, Canberra, pp. 132–137.

Szentes, K., E. Mészáros, T. Szabó, B. Csorbai, G. Borbély, G. Bernáth, B. Urbányi and Á. Horváth. 2012. Gonad development and gametogenesis in the Asian seabass (*Lates calcarifer*) grown in an intensive aquaculture system J. Appl. Ichthyol. 28: 883–885.

Toledo, J.D., C.L. Marte and A.R. Castillo. 1991. Spontaneous maturation and spawning of seabass *Lates calcarifer* in floating net cages. J. Appl. Ichthyol. 7: 217–222.

5

Lates calcarifer Wildstocks: Their Biology, Ecology and Fishery

D.J. Russell

5.1 Introduction

Barramundi, or Asian seabass (*Lates calcarifer* (Bloch)), is an iconic food and sports fish species throughout much of the tropical Indo-west Pacific region. As well as being farmed throughout much of its range, *L. calcarifer* also forms the basis of well-known and valuable recreational and commercial wild fisheries, particularly in northern Australia and Papua New Guinea (Grey 1987, Grant 1997, Rimmer and Russell 1998a, Tucker et al. 2006). The economic and cultural significance of *L. calcarifer* and the potential for overexploitation of this fishery has stimulated research into aspects of this species' biology, ecology, stock enhancement and fishery. It is these aspects that are documented in this chapter.

5.2 Fishery

5.2.1 Wild fishery

This species is commercially fished across the majority of *L. calcarifer*'s Indo-Pacific range. Globally, the capture fishery has been increasing steadily since 1950 and in 2010 stood at 102 thousand tonnes (UN Food and Agriculture

---

Northern Fisheries Centre, PO Box 5396, Cairns, Queensland 4868, Australia.
Email: johnru2001@hotmail.com

Organisation 2012). In Australia, the *L. calcarifer* commercial fishery is relatively small by world standards, with a total wild catch in 2009–2010 of 1643 tonnes worth AUS$26.3 million (Skirtun et al. 2012). The Australian fishery is governed by a range of operational restrictions that vary depending on individual State and Territory regulations (Russell and Rimmer 2004). Common themes in these regulations include seasonal and area closures, gear restrictions, recreational angler bag limits and size restrictions on fish that can be taken. The commercial fisheries predominantly use gill nets as a capture technique and are limited to tidal waters. Recreational and customary indigenous fishers and commercial tour guides cannot use gill nets, but are also able to fish in most freshwater areas that are prohibited to commercial fishers. In the framing of regulations governing the fishery, consideration has been given to the complex life history of the species; for example, through the introduction of a closed season designed to protect fish during the period when spawning activity is greatest, or by limiting gill net mesh sizes to protect juvenile fish (Williams 2002, Russell and Rimmer 2004). In Queensland, there are upper (1200 mm TL) and lower fish size (580 mm TL) limits designed to allow *L. calcarifer* to spawn at least once as males before entering the fishery and also to protect larger breeding females (Queensland Fisheries Management Authority 1992, 1998).

The Papua New Guinean *L. calcarifer* fishery is also composed of commercial, recreational and customary indigenous elements. From about the 1970's and 1980's, the commercial component was made up mainly of a boat-based gill-net fishery in the Fly River and adjacent areas. In response to declining coastal catches, this boat-based fishery was progressively replaced by local land-based fish-processing plants, mainly in the middle Fly River (Milton et al. 2005). These reduced catch rates in the coastal fishery may have been related to both earlier exploitation in the Fly River commercial fishery and/or the El Niño climate pattern that occurred during that time (Milton et al. 2005).

5.2.2 Stocked fishery

Despite strict management measures designed to ensure the sustainability of the wild *L. calcarifer* fishery that have been progressively introduced in Australia, fish stocking has been undertaken in Queensland and, to a lesser degree, in the Northern Territory. Since the mid 1980's, hatchery-reared *L. calcarifer* have been released into impoundments to create new 'put and take' impoundment fisheries and into Queensland rivers to enhance existing wild fisheries (Cadwallader and Kerby 1995, Holloway and Hamlyn 1998). The first major stocking of hatchery-reared *L. calcarifer* in Australia was into Tinaroo Dam, a Queensland freshwater impoundment, in 1985 (McKinnon and Cooper 1987) and corresponded to local successes in developing and applying mass breeding technology for the species (McKinnon and Cooper 1987, Russell et al. 1987). Since then, stocking activities have expanded to include numerous reservoirs and some coastal streams throughout mostly the northern part of the State (Rutledge et al. 1990, Rimmer and Russell 1998b, Russell and Rimmer 1999, 2002, 2004, Russell et al. 2004). While most *L. calcarifer* stocking activity in Australia has been restricted to Queensland, at least one impoundment in the Northern Territory (Manton Dam) has also been stocked (Tucker et al. 2006).

To maximise benefits for the fishery in Australia, studies have been undertaken to improve stocking strategies for *L. calcarifer* in rivers and impoundments (Russell and Rimmer 1997, Rimmer and Russell 1998b, McDougall et al. 2008) and also on the efficacy and cost-benefits of stock enhancement (Russell and Rimmer 1997, 1999, 2000, Russell et al. 2002). The data obtained from these latter studies into stock enhancement suggest that, even after low to moderate augmentation, hatchery-reared fish can contribute between about 10 and 15% to the commercial and recreational catch, respectively (Rimmer and Russell 1998b).

The potential deleterious impacts of uncontrolled fish stocking have been recognised in Australian jurisdictions and safeguards progressively developed. For instance, in Queensland stocking protocols have been amended to prevent the occurrence of potential problems associated with the translocation of hatchery-reared *L. calcarifer* derived from one genetic stock into another discrete genetic stock (Cadwallader and Kerby 1995). To further assist in the development of stocking protocols, research into the genetic and some ecological impacts of stocking *L. calcarifer* into Queensland rivers and impoundments has recently been completed (Russell et al. 2013). The findings of this study suggest that, at current stocking rates, genetic and ecological impacts of riverine stocking of *L. calcarifer* are minimal, but caution that this result may differ in other locations or change if densities and other environmental conditions vary.

5.2.3 Range

Lates calcarifer is an euryhaline member of the family Latidae which is widely distributed in the Indo-west Pacific region from the Arabian Gulf to China, Taiwan, the Philippines, the Indonesian archipelago, Papua New Guinea and northern Australia (Grey 1987). Dunstan (1959) suggested that the Arabian Gulf, with its high water temperatures, high salinities and lack of freshwater inflow was an effective barrier preventing *L. calcarifer* colonising the African continent where there are at least a further 11 recognised freshwater species of the genus *Lates* (Otero 2004, Pethiyagoda and Gill 2013 see Chapter 1). In Australia, *L. calcarifer* is distributed from south-eastern Queensland on the east coast (Grant 1997) across the north of the continent to the Ashburton

River on the west coast (Grey 1987, Morrissy 1987). In Southeast Asia, the once widespread distribution of *L. calcarifer* has now been complicated by the recent description of two new species of *Lates*, *L. lakdiva* and *L. uwisara*, which have been described from western Sri Lanka and eastern Myanmar, respectively (Pethiyagoda and Gill 2012). There is also another closely related species, *L. japonicus*, whose distribution is restricted to the Pacific coast of southern Japan (Iwatsuki et al. 1993).

5.2.4 Reproduction and spawning

5.2.4.1 Seasonality

The timing of the spawning season of wild stocks of *L. calcarifer* varies depending on geographic location. In northern Australia and Papua New Guinea spawning occurs in the warmer months and there are distinct latitudinal variations in reproductive activity that are, presumably, in response to varying water temperatures (Dunstan 1959, Russell and Garrett 1983, Davis 1985b, Russell and Garrett 1985, Garrett 1987). In north-eastern Australia, Russell and Garrett (1985) found gonadal activity in *L. calcarifer* in the Austral spring and summer from October to February, with evidence of spawning from November. In Papua New Guinea, Moore (1982) observed that *L. calcarifer* spawning activity peaked from November to January. In the northern hemisphere, *L. calcarifer* in the Philippines spawn from late June to late October (Parazo et al. 1990), while in Thailand, Ruangpanit (1987) noted that *L. calcarifer* spawn all year round, with peak activity from April to September. However, Barlow (1981) suggests that Thai *L. calcarifer* spawn in association with the monsoon season, with two peaks during the north-east monsoon (August-October) and the southwestern monsoon (February-June).

Spawning of *L. calcarifer* in Australia and Papua New Guinea has also been associated with the occurrence of the northern monsoon (Dunstan 1959, Moore 1982). However, in some areas of Australia, spawning has been documented well before the onset of freshwater runoff resulting from monsoonal activity (Russell and Garrett 1985). Moore (1980, 1982) suggested that, under circumstances when the wet season is delayed, organic material carried from nursery swamps on high ebbing tides could provide the necessary stimulus for *L. calcarifer* spawning. In northern Australia, Davis (1985b) found that while the timing and duration of the spawning season varied between regions, rivers, and from year to year, it was essentially synchronized so that juveniles could take advantage of newly formed aquatic nursery habitats created during the wet season. Flooding appears to enhance recruitment for *L. calcarifer* through increased access to inundated off-stream nursery habitats (Staunton-Smith et al. 2004).

During the spawning season, actual spawning activity in *L. calcarifer*, as it is in many fish species (e.g., Johannes 1978, Takemura et al. 2004), is closely synchronised with the lunar cycle. *Lates calcarifer* spawn after the full and new moons, with most spawning activity associated with incoming tides that apparently assist the transport of eggs and larvae into estuarine habitats (Kungvankij et al. 1986, Garrett 1987, Ruangpanit 1987).

5.2.4.2 Spawning grounds

Whilst factors determining the location of *L. calcarifer* spawning grounds are complex, one of the key criteria appears to be high salinities (Moore 1982, Davis 1985b). While this condition may preclude many estuarine and hinterland habitats, spawning grounds may be located in a variety of other inshore habitats including estuaries, coastal mud flats headlands and other nearshore waters (Moore 1982, Davis 1985b, Kungvankij et al. 1986, Garrett 1987, Ruangpanit 1987). In north eastern Queensland, specific descriptions of spawning sites were given by Garrett (1987) who noted spawning fish in shallow (up to 2 m deep) side gutters near river mouths, but offset from the main channel. These gutters were characteristically protected from the strongest tidal run by sand and/or mud bars.

5.2.4.3 Sex change

In Papua New Guinea, Moore (1979) first identified *L. calcarifer* as protandrous hermaphrodites, with females being derived from sexually mature males. This trait was later confirmed in Australia in stocks in the Northern Territory and the Gulf of Carpentaria (Davis 1982). However, not all females are necessarily derived from males and in Australia Davis (1984b) reported finding primary females in a sexually precocious population of *L. calcarifer* in the north-eastern Gulf of Carpentaria. In Papua New Guinea, Moore (1979) also found some evidence of the presence of primary females by capturing a single two year old female that was smaller than at the size which males mature and much smaller than the next smallest female. While sex inversion has since been reported in sea cages in French Polynesia (Guiguen et al. 1994), in parts of Asia the occurrence and documentation of sex change appears to be less well known in wild stocks (Grey 1987) and indeed primary females are relatively common, at least, in aquaculture operations in Asia (Parazo et al. 1990).

5.2.4.4 Fecundity

Lates calcarifer is a very fecund fish species, with estimates of between 2.3 × 10^6 and 32 × 10^6 ova per individual in Papua New Guinea (Moore 1980) and up to 45.7×10^6 ova per female in Australia (Davis 1984a). Fecundity in larger fish appears much more variable than in smaller fish and Davis (1984a) found no apparent differences in estimates from widely spaced Australian geographic areas including in a sexually precocious population from the Gulf of Carpentaria. While not ruling out the possibility that *L. calcarifer* is a batch spawner, studies in the Northern Territory found no histological evidence of partial spawning (Davis 1982, 1985b, 1987). Similarly, in Chilka Lake in India, Patnaik and Jena (1976) concluded that each mature *L. calarifer* they examined had only a single batch of maturing ova. However, Moore (1982) sampled partially spent ovaries in Papua New Guinea, while in Thailand, Barlow (1981) reported that smaller fish shed all their eggs at once while larger fish exhibited multiple spawning. In aquaculture facilities, captive *L. calcarifer* broodstock regularly spawn for up to five consecutive nights (Parazo et al. 1990, Garrett and Connell 1991), suggesting that in this environment, at least, that *L. calcarifer* has the capacity to be batch spawners.

5.2.4.5 Size- and age-at-maturity

Davis (1982) observed that mature male *L. calcarifer* first began to appear from about 600 mm and 550 mm total length in populations in the Northern Territory and Gulf of Carpentaria respectively and by their third to fifth year most had matured. The lengths at which 50% of the male *L. calcarifer* were mature (Lm_{50}) was 700–750 mm and 600–650 mm total length for the Northern Territory and Gulf of Carpentaria populations, respectively (Davis 1982). In a sexually precocious population in the Northern Territory, Davis (1984b) reported that both maturation and sex change occurred at an early age (1–2 years). They suggested that 'stunting' in this population was most likely due to the channelling of energy into gonadal growth, at the expense of somatic growth, at a relatively early age. Variations in the age-at-maturity have been documented in some captive populations where fish have been reported as maturing as early as 2–2.5 years old (Aquacop and Nedelec 1989).

5.3 Recruitment and Nursery Habitats

Recruitment of *L. calcarifer* from spawning grounds into coastal nursery habitats appears to occur first as either post-larvae or as early juveniles, and these movements, either passive or active, may be assisted by tidal action (Russell and Garrett 1983, Davis 1985b, Russell and Garrett 1985, Kungvankij et al. 1986, Garrett 1987, Ruangpanit 1987). McCulloch et al. (2005), citing the results of isotopic analyses of otoliths, also suggest that immediately after spawning, juveniles have a relatively limited mobility and thus have a quite restricted geographic range. Juvenile *L. calcarifer* appear to be able to make use of a range of different habitats as nurseries. Studies in Asia, Papua New Guinea and Australia have referred to juvenile *L. calcarifer* utilising various estuarine habitats, particularly discrete wetland areas, as nursery habitats (De 1971, Ghosh 1973, Kowtal 1976, Patnaik and Jena 1976, Mukhopadhyay and Verghese 1978, Russell and Garrett 1983, Davis 1985b, Russell and Garrett 1985). In Asia, while a number of early studies have made passing references to *L. calcarifer* larvae and juveniles occurring in estuaries and associated tidal wetlands, most lack detail on the actual recruitment process and the specific habitats that were utilised. For example, in India, Kowtal (1976) documented collecting juvenile *L. calcarifer* from estuaries on the east coast, while Mukhopadhyay and Verghese (1978) noted that *L. calcarifer* juveniles 'ascended' into estuaries and brackishwater lagoons in the Hooghly system in search of food and shelter. In the same estuary, Ghosh (1973) recorded finding juvenile and larval *L. calcarifer* in estuarine creeks, canals and inundated low-lying areas from May to July. In the Bay of Bengal, Patnaik and Jena (1976) observed juvenile *L. calcarifer* entering the estuarine system of Lake Chilka from July to September. De (1971) gave some more detail of the preferred habitat of *L. calcarifer* in West Bengal. This author observed juvenile *L. calcarifer* as small as 10 mm in inundated tidal pools and swamps and suggested that they entered these habitats with tidal water, and apparently used the sheltered marginal areas as nurseries for only short periods before re-entering tidal waters. Similarly, in Thailand, Barlow (1981) reported studies that found *L. calcarifer* larvae and early juveniles entered a mangrove swamp in the Songkhla lakes region. Juveniles remained in this nursery area for 4–8 months before they re-entered the lake.

In Australia and Papua New Guinea, the early life history of *L. calcarifer* has been described in considerable detail by a number of authors (Dunstan 1959, Russell and Garrett 1983, Davis 1985b, Russell and Garrett 1985, Griffin 1987, Pender and Griffin 1996, Veitch and Sawynok 2004). These studies all indicate that larval *L. calcarifer* recruit into nursery swamps and lagoons adjacent to estuaries and the coast where they remain for several months before they return to nearby estuaries or coastal waters. These nursery grounds may be freshwater or saline (McCulloch et al. 2005) and include a wide range of ecotypes from temporary pools adjacent to salt pans (Russell and Garrett 1983, Davis 1985b), to coastal swamp systems (Moore 1982, Russell and Garrett 1985), to floodplain and billabong systems many kilometres from the coast (Davis 1985b). Figure 5.1 shows an example of one type of nursery habitat adjacent to an estuary in north-eastern Australia. In an early study of *L. calcarifer* in Australia, Dunstan (1959) observed fish of less than 25 mm swimming up shallow gutters into waterholes on flood plains, usually in close proximity to coastal waters. *Lates calcarifer* larvae

84 *Biology and Culture of Asian Seabass*

Figure 5.1. Typical nursery habitat of early stages of *Lates calcarifer* development in northeastern Australia. Photo: John Russell.

entered coastal swamp systems adjacent to coastal spawning grounds in Papua New Guinea when they were about 5 mm long (Moore 1982). Russell and Garrett (1983) found that juveniles as small as 9.5 mm moved into temporary pools created by high seasonal tides on salt flats adjacent to a major northern Australian estuary from about the end of December. In the Northern Territory, Davis (1985b) monitored a coastal swamp and feeder stream which juvenile *L. calcarifer* as small as 8.3 mm accessed on high spring tides. These fish apparently used this swamp as a nursery area during the northern Australian summer monsoon season (Davis 1985b). Given the diverse nature of *L. calcarifer* nursery habitats and the logistical difficulties posed in sampling some of them for fish fauna, it is not unexpected that *L. calcarifer* nursery habitats in some areas remain unidentified.

Many of the *L. calcarifer* nursery habitats documented in the various studies mentioned above are characterised by an abundance of prey and reduced predator numbers; conditions that would appear ideal for fostering rapid growth and enabling improved survival (Moore 1982, Russell and Garrett 1983, Davis 1985b, Russell and Garrett 1985). However, environmental conditions in some of the temporary pools can be, at times, seemingly harsh. Russell and Garrett (1983) recorded extreme fluctuations in temperatures (25–36°C) and salinity (< 1–94) in supra-littoral tidal pools in the Gulf of Carpentaria, but the juvenile *L. calcarifer* using these pools appeared to be tolerant of such conditions. Despite this, the limited mobility of larval and juvenile *L. calcarifer* that McCulloch et al. (2005) deduced using isotopic analyses of otoliths led these authors to speculate that juvenile *L. calcarifer* may be more sensitive to changes in their local ecosystems than had previously been suspected. In the Northern Territory, Davis (1985b) suggested that the juveniles of late-spawning *L. calcarifer* would find nursery habitats to be hostile environments due to the risk of cannibalism by larger fish. He also contended that in situations where the monsoon is interrupted or delayed, early-spawned fish could perish as nursery areas failed to form, leaving late-spawned *L. calcarifer* with the same predator-free, food-rich habitat that would normally benefit early-spawned fish. Some climatic conditions, for example high coastal rainfall and freshwater flows may act to enhance survival of the early life history stages of *L. calcarifer* by generating improved access to supra-littoral nursery areas, increasing their productivity and carrying capacity or increasing their suitability in other ways (Staunton-Smith et al. 2004).

After having benefited from the protection and sustenance provided by these nursery areas, most surviving juveniles move into coastal waters and estuaries with some subsequently ascending up into the freshwater reaches of coastal rivers and creeks. In Papua New Guinea, juvenile *L. calcarifer* remain in the nursery swamps until they attain a length of approximately 200–300 mm before then returning to coastal waters as the swamps begin to dry out as a result of seasonal conditions (Moore and Reynolds 1982). In the Gulf of Carpentaria, Russell and Garrett (1983) found that resident *L. calcarifer* must abandon these pools just before seasonal falls in tidal heights complete their isolation from the rest of the estuary, or perish as they dry out (see Fig. 5.2). Dunstan (1959) observed in eastern Queensland, that with the recession of flood waters, fish in river lagoons may become landlocked until the following wet season. However, under different conditions in north eastern Australia, Russell and Garrett (1985) found juvenile *L. calcarifer* in samples taken from nearby tidal creeks in April at the end of the wet season, suggesting that the egress from the nursery swamps had begun. These authors concluded that seasonal lowering of water levels in the coastal swamps and a depletion of food supplies were significant factors contributing to the exit of juvenile *L. calcarifer* to more permanent habitats. Moore (1982) discussed similar escape strategies for *L. calcarifer* resident in coastal swamps in Papua New Guinea. Moore (1982) noted that by the end of their first year, juvenile *L. calcarifer* had become widely distributed throughout adjacent coastal and estuarine regions and had dispersed into inland waters by their second and third year of life (Moore 1982, Moore and Reynolds 1982). Figure 5.3 provides a stylised summary of the life-cycle of *L. calcarifer* in Australia.

86 *Biology and Culture of Asian Seabass*

Figure 5.2. A nursery area in the south-east Gulf of Carpentaria that has dried out after the end of the wet season. Juvenile *Lates calcarifer* that had remained in this nursery after connection with the adjacent permanent water course was broken would have perished.

5.3.1 Movements and migrations

Early inferences on the movements of *L. calcarifer* wild stocks were derived primarily from observations of its occurrence rather than from structured tagging programs or other quantitative techniques. The conclusions of most of these studies were general, lacked detail and, in some cases, were even erroneous. For example, in Thailand Yingthavorn (1951) concluded that the species was anadromous, which conflicted with later studies in India (Jones and Sujansingani 1954), Thailand (Wongsumnuk and Maneewongsa 1974), Papua New Guinea (Dunstan 1962, Moore 1980) and Australia (Dunstan 1959, Garrett and Russell 1982, Davis 1986, Griffin 1987), all of which concluded that the species was in fact catadromous. More recently, it has been suggested that, particularly in areas that lack extensive freshwater hinterlands, *L. calcarifer* may be facultatively catadromous (Russell and Garrett 1988, Russell 1990). This contention is supported for fish in the Northern Territory where Pender and Griffin (1996) noted that most *L. calcarifer* found in marine habitats remote from freshwater areas probably did not have a freshwater phase to their life cycle. Even in areas where there are extensive, accessible freshwaters areas in the hinterlands, there may be a residential population of all age classes of *L. calcarifer* in coastal

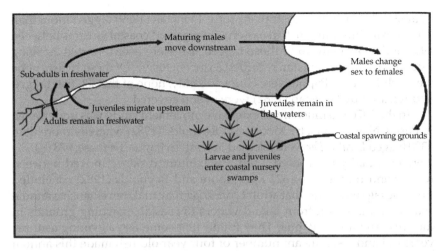

Figure 5.3. *Lates calcarifer* life history. This graphic shows a stylised life cycle of *L. calcarifer* in Australia and Papua New Guinea. After spawning, larvae and early juveniles move into coastal swamps, lagoons and other typically lotic habitats where they can remain for anything from several months to up to a year. Some of these habitats are temporary, being created by heavy seasonal rains or large tides. After leaving these nursery habitats, juveniles move into a range of available habitats including up rivers into freshwater, into estuaries or into other inshore coastal areas. Those that enter freshwaters may remain there until they are at least three years old after which some move downstream into coastal areas to participate in spawning. There is evidence that some fish may never return to coastal waters to spawn. Those fish that do make a seaward movement from freshwater hinterland areas don't return to freshwater areas although they may move along the coast and between river systems. At around 6–8 years old, most *L. calcarifer* change sex from males to female and remain female for the rest of their lives. Several sexually precocious populations of *L. calcarifer* have been identified in northern Australia; these fish change sex at a younger age (4–5 years) and at a smaller size (300–500 mm TL).

areas (Moore and Reynolds 1982). Using *in situ* laser ablation (UV) multi-collector inductively coupled plasma mass spectrometry, McCulloch et al. (2005) confirmed the flexibility in *L. calcarifer*'s life history in north-eastern Australia, finding that some individuals spent almost their entire life cycle in marine environments, while others moved between freshwater, estuarine and marine habitats. These authors also suggested that *L. calcarifer* may spend its entire life cycle in freshwater although, given the necessity of high salinity environments for reproduction (Moore 1982), this would exclude involvement in spawning activities. In Northern Australia, Davis (1986) proposed that *L. calcarifer* should be considered catadromous, as fish in freshwater must move seaward to spawn. However, he qualified this by noting that the majority of fish resident in tidal waters and which spawn early in the wet season cannot be regarded as truly catadromous.

The first documented tagging studies on the movements of *L. calcarifer* were undertaken in Australia (Dunstan 1959) and Papua New Guinea

(Dunstan 1962). In the Australian study, Dunstan (1959) tagged more than 2000 mostly immature fish in eastern Queensland coastal streams between 1953 and 1959. Only 16 of those tagged fish were ever reported as being recovered, a result Dunstan regarded as inconclusive. Of these 16 fish, none were adult. In the Papua New Guinean study, none of the 300 fish tagged and released by Dunstan (1962) were ever recovered.

In the 1970's, a much more extensive movement study was undertaken in Papua New Guinea by Moore and Reynolds (1982) whereby more than 6300 tagged *L. calcarifer* were released in western Papua. Of these, 978 (~15%) were eventually recaptured, with a maximum distance moved between release and recapture of 622 km. Moore and Reynolds (1982) concluded from the tag recaptures that adult *L. calcarifer* may make an extensive annual spawning migration from inland waters to coastal spawning grounds in the western Gulf of Papua. They found that some fish as young as three years old and a significant number of four year old fish made this annual spawning migration. The subsequent westerly movement along the coast was apparently because the huge freshwater discharge from the Fly River (water discharge averages ~ 6500 m^3 s^{-1} (Ogston et al. 2008)) resulted in unsuitably low salinities for spawning in most other local coastal areas (Moore 1982).

The return of adult fish from the spawning grounds identified by Moore and Reynolds (1982) in Papua New Guinea to freshwater habitats is a matter of some contention. Some of their tag recoveries suggest that *L. calcarifer*, having migrated to coastal spawning grounds, subsequently returned to inland waters to the same general area from where they originally migrated (Moore and Reynolds 1982). However, these authors found that over 70% of fish tagged in coastal areas made only local movements, with only a small proportion of *L. calcarifer* moving more than 15 km from their release location and with multiple local coastal recaptures during the non-spawning season. They interpreted this result as indicating the presence of a resident coastal population of *L. calcarifer*. Milton and Chenery (2005), using isotope ratios in otoliths to infer habitat usage, also contended that most of the fish that they examined from the coastal spawning grounds were predominantly marine residents, although some had spent short periods in the freshwater areas of rivers adjacent to their spawning grounds.

Moore and Reynolds (1982) also suggested that not all adult fish resident in freshwaters made an annual migration to coastal areas to spawn. Similarly, Milton and Chenery (2005) found that many adult *L. calcarifer* resident in Papua New Guinean rivers have never migrated to the coast and, by implication, had not participated in spawning. They estimated that only half of the adult fish in the Fly and nearby Kikori rivers had ever migrated back to the sea to spawn. In a short term study of movement and dispersal patterns of *L. calcarifer* in a Northern Territory billabong, Heupel et al. (2011) found that fish mostly remained in permanently flooded sections of the lagoon with only limited movements into seasonally flooded areas.

In Australia, most recent studies have found that mature fish remained in tidal waters, either after having made a spawning migration from freshwaters, or for their entire life cycle (Davis 1986, Pender and Griffin 1996, McCulloch et al. 2005). Both tagging and scale chemistry studies in northern Australia have found little evidence of coastal-resident adult *L. calcarifer* moving into freshwater (Davis 1986, Pender and Griffin 1996). Davis (1986) noted a general seaward movement of tagged *L. calcarifer* from freshwater which he associated with spawning activity, but no evidence of movement of spent fish back into freshwater. This author also found some evidence of coastal movements and exchange of *L. calcarifer* between river systems. In eastern Queensland, while most tagged fish were recovered after moving only short distances there have also been some significant movements recorded. For example, Sawynok (1998) recorded a *L. calcarifer* that moved 650 km to the north along the coast from the Fitzroy River and another that moved from the same river 350 km to the south-east and into the Gregory River.

A recent genetics study in Asia suggests that the dispersal of adult *L. calcarifer* may be more extensive than more traditional tagging studies have indicated. Yue et al. (2012) genotyped 549 adult *L. calcarifer* from four geographic locations in the South China Sea and detected significant genetic differentiation, both among and between fish at these sampling locations. They also found strong evidence of female-biased dispersal in *L. calcarifer* and suggested that this may be the result of various factors including inbreeding avoidance behaviour in females, selective female mate choice under the condition of low mate competition among males and male resource competition. Further studies will need to be conducted to determine if sex-biased dispersal is a common occurrence in *L. calcarifer* across this species' wider distribution.

5.3.2 Juvenile movements

As part of an extensive tagging program, Moore and Reynolds (1982) investigated the movement of juvenile *L. calcarifer* from coastal nursery habitats to inshore coastal and estuarine areas and then to freshwater habitats in the west-Papuan hinterland. Of the 893 juvenile *L. calcarifer* that were tagged and released in coastal areas as part of their study, 115 were later recaptured. Moore and Reynolds (1982) suggested that after leaving nursery swamps and entering coastal waters early in the year (i.e., March–June), most fish dispersed in an easterly direction along the coast, moving as far as the Fly River by the end of their first year. During their second and third year, juvenile *L. calcarifer* either moved into inland waters or remained resident in coastal areas. Similarly, Milton and Chenery (2005) found that in the middle Fly River, most fish do not enter freshwater until they are > 3 years old and that about 40% of the older fish examined (> 8 years old) had not left freshwater after they had entered it as 3–5 year old fish.

In Australia, a number of studies have investigated juvenile *L. calcarifer* movements in coastal waters after their departure from nursery habitats (Dunstan 1959, Moore and Reynolds 1982, Davis 1986). In a tagging study in north-eastern Queensland, Russell and Garrett (1988) marked and released 1268 juvenile *L. calcarifer* into two tidal creeks. Their results indicated that most juvenile *L. calcarifer* remained resident in the same tidal creeks until the end of their first year, when they moved out into the main estuary and then dispersed into nearby estuaries and coastal habitats. In an earlier study in eastern Queensland, Dunstan (1959) observed that early juveniles moved into fresh water during the dry Austral winter months and occasionally even penetrated into the headwaters of some short, coastal rivers.

In Asia, while there are no records of large scale and extensive movements of juveniles of the type documented in Papua New Guinea (Moore and Reynolds 1982), observational records of where *L. calcarifer* occur in river systems show movement of up to 130 km from the coast, wherefrom presumably juveniles have originated (James and Marichamy 1987).

5.3.3 Triggers for movements

In Papua New Guinea, Moore and Reynolds (1982) suggest that the trigger for movements of *L. calcarifer* resident in hinterland swamps is a response to changing water levels. They noted that during the dry season, when inland water levels began to fall, *L. calcarifer* moved from the extensive freshwater swamp systems into the deeper rivers and lakes. They further observed that, if there was also a following period of rising water levels, fish may move back into the same swamps thus leading them to contend that falling water levels may not necessarily lead to *L. calcarifer* moving into the coastal zone where spawning activity occurs. Dunstan (1959, 1962) suggests that in unusually dry wet seasons when floodwaters are insufficient to release land-locked adult *L. calcarifer*, the number of spawning fish at sea is greatly reduced. Further, under these circumstances he found no adult fish with well-developed gonads in non-flowing freshwater or in land-locked coastal lagoons.

Studies on the otolith chemistry of *L. calcarifer* in Papua New Guinea led Milton and Chenery (2005) to propose that the proximate cues for movements of *L. calcarifer* from freshwater areas are complex and/or are linked to rare climatic events and occur irregularly. They further suggest that fish resident in deeper lakes and swamps away from the coast, presumably similar to the ones mentioned by Moore and Reynolds (1982), may never migrate despite them being sexually active.

5.3.4 Impediments to Lates calcarifer movements

The construction of weirs and dams across watercourses that are naturally inhabited by *L. calcarifer* often provide effective barriers to their movements, both upstream and downstream. This is particularly the case in Australia where, to facilitate riverine fish movements, some planning authorities have made provision for fish passage by either incorporating structures into their barrier design or having them retrospectively fitted. The early fishways were, because of a lack of data on the specific requirements of Australian fish, often copied from European or American designs that subsequently proved ineffective to traverse by many native species, including *L. calcarifer* (Kowarsky and Ross 1981). These included variations of the pool-and-weir design that was later demonstrated as being ineffective in Australian river systems for the passage of many native species, including *L. calcarifer* (Kowarsky and Ross 1981, Russell 1991).

A number of studies (e.g., Stuart and Berghuis 1999, Stuart and Mallen-Cooper 1999, Stuart and Berghuis 2002) have shown that vertical-slot type fishways with wide pools, lower velocities and less turbulence are able to pass a diverse range of the migratory fish fauna present in Australian coastal rivers. The greatest diversity of fish species pass through vertical-slot fishways during periods of low river flows (Stuart and Berghuis 2002). Other designs, such as fish locks, have been trialled on Queensland impoundments, although at least one fish lock was shown to have had both operational problems and apparently minimal, or no use by *L. calcarifer* (Stuart et al. 2007). However, Stuart et al. (2007) regarded the lack of passage of less numerous fish species such as *L. calcarifer* at this particular lock as inconclusive and noted that it (and other species) were also recorded at two other fish locks in Queensland (Stuart and Berghuis 2002). Low ramp style fishways have proved very successful in facilitating the passage of *L. calcarifer* in the Northern Territory and other areas (de Lestang et al. 2001).

5.3.5 Movements of stocked fish

Stocking of juvenile *L. calcarifer* at various sizes up to ~300 mm total length into impoundments and rivers in northern Australia has been underway since the mid 1980's. While tagging fish prior to release is not generally the accepted practice when stocking *L. calcarifer* in Australia, it has been undertaken at a number of locations (e.g., Russell et al. 1991, Russell 1995, Russell and Rimmer 1997), thus providing a mechanism for determining post-stocking movements (Russell and Rimmer 1999).

In general, most records of the movements of tagged fish, either of those released into impoundments or into rivers and coastal areas, only involved short distances from their original release locations. Sawynok and Platten

92 *Biology and Culture of Asian Seabass*

(2009) noted that the majority of recaptured tagged *L. calcarifer* in the Fitzroy River in central Queensland moved only 20 km or less from their release locations. However, there are records of stocked fish moving both upstream and downstream in rivers and also along the coast and into adjacent river systems. Direction of movement in rivers also appears age-related. For example, in the Fitzroy River in central Queensland, juvenile fish were found to mostly move upstream in response to river flow events while adult fish primarily moved downstream (Sawynok and Platten 2009).

The largest distances moved by stocked fish appears to be either in a downstream direction, or downstream and along the coast, sometimes into adjacent river systems (Russell et al. 2011). For example, in the Fitzroy River system in Queensland (Fig. 5.4) there have been some significant downstream movements of stocked *L. calcarifer* recorded in the Sunfish

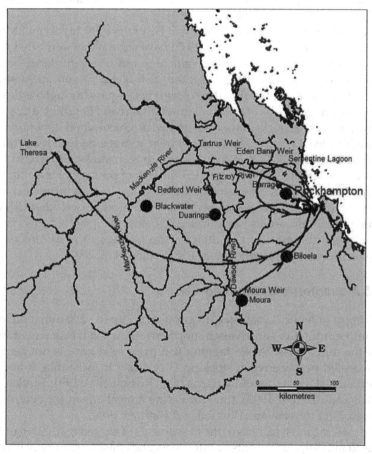

Figure 5.4. Large downstream movements of stocked *Lates calcarifer* in the Fitzroy River catchment in central Queensland.

recreational tagging database (Sawynok and Platten 2009, Russell et al. 2011). Between 2002 and 2009, there were 83 recaptures of stocked *L. calcarifer* that had moved 5 km or more from their original release location. Of these, 28 involved movements over multiple dam walls and/or river barrages. One fish that moved downstream to the Fitzroy River estuary from the Moura Weir (Fig. 5.4) successfully negotiated five stream barriers (Sawynok and Platten 2009). In the Fitzroy River system alone there were records of 23 stocked *L. calcarifer* travelling net distances of at least 700 km before being recaptured.

There are also records in the Suntag recreational fishing database (W. Sawynok, Infofish, pers comm., http://www.info-fish.net) of smaller numbers of stocked adult or sub-adult *L. calcarifer* making substantial upstream riverine movements. For example, in the upper Burdekin River in Queensland stocked *L. calcarifer* have made upstream movements of up to 120 km from their original release location. As well as riverine movements, there are also records from north Queensland of stocked *L. calcarifer* leaving rivers and moving distances in excess of 100 km along the coast and into adjacent river systems (Russell et al. 2011).

5.4 Growth

Various studies have been conducted on the age and growth of *L. calcarifer*, including in Australia (Dunstan 1959, Davis 1984b, Davis and Kirkwood 1984, McDougall 2004, Robins et al. 2006), Papua New Guinea (Dunstan 1962, Reynolds and Moore 1982) and India (Jhingran and Natarajan 1969, Marichamy et al. 2000). In Papua New Guinea, Reynolds and Moore (1982) estimated *L. calcarifer* growth rates using both modal progression analysis of length-frequency data and tag-recapture data and concluded that growth was rapid. They also surmised that, because of the complex life history of *L. calcarifer*, growth was likely to vary considerably during its different life history stages (Reynolds and Moore 1982). The growth rates for *L. calcarifer* in northern Australia, as determined by Dunstan (1959) using length-frequency analyses, were greater than those estimated by Reynolds and Moore (1982) for fish in Papua New Guinea. Using modal progression of length frequencies, they estimated that by the end of their first, second and third years *L. calcarifer* averaged 334, 462 and 575 mm total length (TL) respectively, whereas Dunstan (1959) estimated the lengths for same ages to be 450, 730 and 870 mm TL respectively. Reynolds and Moore (1982) speculated that this may have been because Dunstan erroneously assigned bimodal peaks in the data to first year fish instead of to first and second year fish. In northern Australia, Davis and Kirkwood (1984) noted that the growth of *L. calcarifer* varied markedly, both between and within rivers. These authors suggest that this variation was most likely a reflection of the different environmental and seasonal factors experienced by the fish under the different river conditions. They also suggested that growth achieved in different years within the same river varied for the same reasons. The growth of sexually precocious *L. calcarifer* present in streams in the north-eastern Gulf of Carpentaria was considerably slower than that of fish in other parts of northern Australia (Davis 1984b, 1987). When mean length-at-age of the sexually precocious group from northern Gulf of Carpentaria rivers was compared with more 'typical' populations in southern Gulf of Carpentaria streams, an analysis of variance showed the former was significantly smaller at all ages. Davis (1984b) regarded these fish as being 'stunted' and suggested that sexual precocity may be linked to local environmental conditions. Sawynok (1998) investigated the relationship between freshwater flows in a large Australian river system and growth of *L. calcarifer* and concluded that, after taking seasonal variations into consideration, fish grew faster under higher annual flow conditions (i.e., > 79.3 $m^3 s^{-1}$). In the same river Robins et al. (2006) established relationships between freshwater flows and *L. calcarifer* growth rates using data from a long-term tag-recapture program, thereby supporting the results of Sawynok's (1998) earlier study. They found that *L. calcarifer* growth rates were significantly and positively correlated to freshwater flowing into the estuary. While freshwater flows had only a minor influence on growth in winter, the greatest effect was in summer with median or greater flows resulting in about twice the growth rates of minimal flows. Robins et al. (2006) regarded this relationship as evidence to support the hypothesis that freshwater flows are important in driving the productivity of estuaries and can improve growth of species high in the trophic chain.

5.5 Diet

Various studies of the diet of *L. calcarifer* throughout its range have consistently shown that diet progressively changes with increasing size. In West Bengal, microcrustacea were the dominant component of the diet of 10–15 mm long *L. calcarifer* (De 1971). As the fish grew, their diet changed to become predominantly insectivorous in fish 16–45 mm TL and then to mainly insects, fish and crustaceans (at ~50–200 mm TL). In the Chilka Lake region of India, diet of juvenile *L. calcarifer* also changed with their size (Patnaik and Jena 1976). Fish 24–50 mm TL fed mainly on microcrustaceans and larger *L. calcarifer* (51–150 mm) consumed mostly mysids, prawns and smaller fish. The diet of *L. calcarifer* over 150 mm TL was predominantly fish and crustaceans. In Papua New Guinea, Moore (1982) described the advantage gained by juvenile *L. calcarifer* entering nursery swamps rich in prey species such as zooplankton and insect larvae. An earlier Australian study by Dunstan (1959) sampled the gut contents of 96 *L. calcarifer* of unspecified size, but probably subadult and adult fish, from eastern Queensland. He concluded that *L. calcarifer* is predatory throughout its life cycle, consuming approximately 60% teleosts and 40% crustaceans. In northeastern Queensland, gut contents of juvenile fish from both nursery swamp and tidal creek habitats varied with habitat usage (Russell and Garrett 1985). *L. calcarifer* from the swamp habitats had a varied diet, with fish and insects the main components. In tidal creeks, juvenile *L. calcarifer* were found to have fed primarily on fish and crustaceans, with insects entirely absent from their diet. Davis (1985a) sampled the stomach contents of juvenile and larval fish down to 4 mm TL in the Van Diemen Gulf area of the Northern Territory. Here it was shown that *L. calcarifer* is an opportunistic predator, with an ontogenetic progression in its diet from microcrustaceans to fish. Davis (1985a) showed that very early stage juveniles <40 mm TL fed exclusively on microcrustaceans. In larger fish, microcrustaceans became progressively less common in the stomach contents and absent in the diet of fish longer than 80 mm which fed on macrocrustaceans, other invertebrates and some fish. Fish and macrocrustaceans were equally important in the species' diet at about 300 mm TL and in larger *L. calcarifer* fish were the dominant component (Davis 1985a). In fresh, non-tidal waters of rivers flowing into the Van Diemen Gulf of the Northern Territory, crustaceans and fish made up 43% and 49% respectively of the diet of *L. calcarifer* between 200 and 400 mm TL (Davis 1985a). In freshwater reaches of the lower Ord and Fitzroy rivers of Western Australia, *L. calcarifer* prey primarily on teleosts (72%) and decapods (26%) (Morgan et al. 2004).

Cannibalism has been reported in a number of dietary studies of *L. calcarifer* in Australia (Davis 1985a), Papua New Guinea (Moore 1980, 1982) and also in Asia (De 1971). In the Van Diemen Gulf of the Northern Territory, conspecifics comprised 5% of the fish species consumed by juvenile *L. calcarifer* in the 50–200 mm TL size range (Davis 1985a), with some fish cannibalised up to half their own body size. Davis (1985a) also found that, while cannibalism was generally rare in fish in Gulf of Carpentaria streams, conspecifics did comprise 11.4% of the diet of larger *L. calcarifer* in the size range 1001–1200 mm (Davis 1985a). The reason that cannibalism was so rare in the Gulf of Carpentaria may be a reflection of prey availability, rather than because of food preference. In the stocked *L. calcarifer* fishery in Tinaroo Falls Dam in north-eastern Queensland, McDougall et al. (2008) found no evidence of cannibalism in the stomach contents of larger fish, but even so, suggested that it may have been a cause of low survivorship of stocked juveniles observed into the recreational fishery in the late 1990s and early 2000s.

5.6 Conclusions

The significance of the capture fisheries for *L. calcarifer*, particularly in Australia and Papua New Guinea, and the resultant need to develop appropriate strategies for its sustainable management has led to a plethora of relatively recent and increasingly more sophisticated studies on the biology and ecology of the species. It has become clear from these studies that, throughout its range, there are both many similarities and also some differences in the biological traits of the species. In Australia and elsewhere these studies have been vital in the framing of a range of management regulations designed to take into account the complex life history of the species, thereby contributing to the future viability of the fishery. However, of note is the Australia-Papua New Guinean focus of previous research on the wild *L. calcarifer* fishery, with wider, easy accessible scientifically documented studies relatively rare from throughout the majority of this species' Asian distribution. Given the increased pressure on all wild fisheries there is a need to expand out the scientific knowledge base on the ecology and recruitment of *L. calcarifer* to encompass its entire range and to ensure its sustainable exploitation.

References

Aquacop, F.J. and G. Nedelec. 1989. Larval rearing and weaning of Seabass, *Lates calcarifer* (Block), on experimental compounded diets. Advances in Tropical Aquaculture-Tahiti, February 20–March 4, 1989. Aquacop. IFREMER. Actes de Colloque 9: 677–697.

Barlow, C.G. 1981. Breeding and larval rearing of *Lates calcarifer* in Thailand. New South Wales Fisheries, Sydney, Australia 14 pp.

Cadwallader, P.L. and B. Kerby. 1995. Fish Stocking in Queensland—Getting it Right! Proceedings of a symposium held in Townsville, Queensland. Queensland Fisheries Management Authority, Townsville, Australia.

Davis, T.L.O. 1982. Maturity and sexuality in barramundi (*Lates calcarifer*), in the Northern Territory and south-eastern Gulf of Carpentaria. Aust. J. Mar. Freshwater Res. 33: 529–545.

Davis, T.L.O. 1984a. Estimation of fecundity in barramundi, *Lates calcarifer* (Bloch), using an automatic particle counter. Aust. J. Mar. Freshwater Res. 35: 111–118.

Davis, T.L.O. 1984b. A population of sexually precocious barramundi, *Lates calcarifer*, in the Gulf of Carpentaria, Australia. Copeia 1: 144–149.

Davis, T.L.O. 1985a. The food of barramundi, *Lates calcarifer*, in coastal and inland waters of van Diemen Gulf and the Gulf of Carpentaria. J. Fish Biol. 26: 669–682.

Davis, T.L.O. 1985b. Seasonal changes in gonad maturity, and abundance of larvae and early juveniles of barramundi, *Lates calcarifer* (Bloch), in Van Diemen Gulf and the Gulf of Carpentaria. Aust. J. Mar. Freshwater Res. 36: 177–190.

Davis, T.L.O. 1986. Migration patterns in barramundi, *Lates calcarifer* (Bloch), in Van Diemen Gulf, Australia, with estimates of fishing mortality in specific areas. Fish. Res. 4: 243–258.

Davis, T.L.O. 1987. Biology of wildstock *Lates calcarifer* in northern Australia. *In:* J.W. Copland and D.L. Grey. (eds.). Management of Wild and Cultured Seabass/Barramundi (*Lates calcarifer*). ACIAR, Darwin, Australia, pp. 22–29.

Davis, T.L.O. and G.P. Kirkwood. 1984. Age and growth studies on barramundi, *Lates calcarifer* (Bloch), in Northern Australia. Aust. J. Mar. Freshwater Res. 35: 673–689.

De, G.K. 1971. On the biology of post-larval and juvenile stages of *Lates calcarifer* (Bloch). J. Indian Fish. Assoc. 1: 51–64.

de Lestang, P., Q.A. Allsop and R.K. Griffin. 2001. Assessment of fish passage ways on fish migration. Department of Business, Industry and Resource Development, Darwin, Australia. 43 pp.

Dunstan, D.J. 1959. The barramundi *Lates calcarifer* (Bloch) in Queensland waters. CSIRO Technical Paper No. 5. Commonwealth Scientific and Industrial Research Organisation. Melbourne, Australia 22 pp.

Dunstan, D.J. 1962. The barramundi in Papua New Guinea waters. Papua New Guinea Agric. J. 15: 23–30.

Garrett, R.N. 1987. Reproduction in Queensland barramundi (*Lates calcarifer*). *In:* J.W. Copland and D.L. Grey. (eds.). Management of Wild and Cultured Seabass/Barramundi (*Lates calcarifer*). ACIAR, Darwin, Australia, pp. 38–43.

Garrett, R.N. and D.J. Russell. 1982. Premanagement investigations into the barramundi, *Lates calcarifer* (Bloch) in north-east Queensland waters. Queensland Department of Primary Industries, Brisbane, Australia 22 pp.

Garrett, R.N. and M.R.J. Connell. 1991. Induced breeding in barramundi. Austasia Aquacult. 8: 40–42.

Ghosh, A. 1973. Observations on the larvae and juveniles of 'bhekti' from the Hooghly-Matlah estuarine system. Indian J. Fish. 20: 372–379.

Grant, E.M. 1997. Guide to Fishes. EM Grant Pty. Ltd., Brisbane, Australia.

Grey, D.L. 1987. An overview of *Lates calcarifer* in Australia and Asia. *In:* J.W. Copland and D.L. Grey. (eds.). Management of Wild and Cultured Seabass/Barramundi (*Lates calcarifer*). ACIAR, Darwin, Australia, pp. 15–21.

Griffin, R.K. 1987. Life history, distribution, and seasonal migration of barramundi in the Daly River, Northern Territory, Australia. *In:* M.J. Dadswell, R.J. Klauda, C.M. Moffitt, R.L. Saunders, R.A. Rulifson and J.E. Cooper. (eds.). Common Strategies of Anadromous and Catadromous Fishes. American Fisheries Society, Boston, USA, pp. 358–363.

Guiguen, Y., C. Cauty, A. Fostier, J. Fuchs and B. Jalabert. 1994. Reproductive cycle and sex inversion of the seabass, *Lates calcarifer*, reared in sea cages in French Polynesia: histological and morphometric description. Environ. Biol. Fishes 39: 231–247.

Heupel, M.R., D.M. Knip, P. de Lestang and Q.A. Allsop. 2011. Short-term movement of barramundi in a seasonally closed freshwater habitat. T. Am. Soc. Agric. Biol. Eng. 12: 147–155.

Holloway, M. and A. Hamlyn. 1998. Freshwater fishing in Queensland: A guide to stocked waters. Queensland Department of Primary Industries, Brisbane, Australia 76 pp.

Iwatsuki, Y., K. Tashiro and T. Hamasaki. 1993. Distribution and fluctuations in occurrence of the Japanese centropomid fish, *Lates japonicus*. Japn. J. Ichthyol. 40: 327–332.

James, P.S.B.R. and R. Marichamy. 1987. Status of seabass (*Lates calcarifer*) culture in India. *In:* J.W. Copland and D.L. Grey. (eds.). Management of Wild and Cultured Seabass/Barramundi (*Lates calcarifer*). ACIAR, Darwin, Australia, pp. 74–79.

Jhingran, V.G. and A.V. Natarajan. 1969. A study of the fisheries and fish populations of the Chilka Lake during the period 1957–65. J. Inland Fish. Soc. India 1: 49–125.

Johannes, R.E. 1978. Reproductive strategies of coastal marine fishes in the tropics. Environ. Biol. Fishes 3: 65–84.

Jones, S. and K.H. Sujansingani. 1954. Fish and fisheries of Chilka Lakes with statistics of fish catches for the years 1948–50. Indian J. Fish 1: 256–344.

Kowarsky, J. and A.H. Ross. 1981. Fish movement upstream through a central Queensland (Fitzroy River) coastal fishway. Aust. J. Mar. Freshwater Res. 32: 93–109.

Kowtal, G.V. 1976. Studies on the juvenile fish stock of Chilka Lake. Indian J. Fish 23: 31–40.

Kungvankij, P., L.B. Tiro, B.J. Pudadera and I.O. Potestas. 1986. Biology and Culture of Seabass (*Lates calcarifer*). Training Manual Series No. 3. Network of Aquaculture Centres in Asia, Bangkok, Thailand.

Marichamy, R., H.M. Kasim, V.S. Rengaswamy and K.M.S.A. Hamsa. 2000. Culture of seabass *Lates calcarifer*. In: V.N. Pillai and N.G. Menon (eds.). Marine Fisheries Research and Management. Central Marine Fisheries Research Institute, Kerala, India, pp. 818–825.

McCulloch, M., M. Cappo, J. Aumend and W. Muller. 2005. Tracing the life history of individual barramundi using laser ablation MC-ICP-MS Sr-isotopic and Sr/Ba ratios in otoliths. Mar. Freshwater Res. 56: 637–644.

McDougall, A.J. 2004. Assessing the use of sectioned otoliths and other methods to determine the age of the centropomid fish, barramundi (*Lates calcarifer*) (Bloch) using known age fish. Fish. Res. 67: 129–141.

McDougall, A.J., M.G. Pearce and M. MacKinnon. 2008. Use of a fishery model (FAST) to explain declines in the stocked barramundi (*Lates calcarifer*) (Bloch) fishery in Lake Tinaroo, Australia. Lakes Reservoirs: Res. Manage. 13: 125–134.

McKinnon, M.R. and P.R. Cooper. 1987. Reservoir stocking of barramundi for enhancement of the recreational fishery. Aust. Fish. 46: 34–37.

Milton, D.A. and S.R. Chenery. 2005. Movement patterns of barramundi *Lates calcarifer*, inferred from Sr-87/Sr-86 and Sr/Ca ratios in otoliths, indicate non-participation in spawning. Mar. Ecol.: Prog. Ser. 301: 279–291.

Milton, D., M. Yarrao, G. Fry and C. Tenakanai. 2005. Response of barramundi, *Lates calcarifer*, populations in the Fly River, Papua New Guinea to mining, fishing and climate-related perturbation. Mar. Freshwater Res. 56: 969–981.

Moore, R. 1979. Natural sex inversion in the giant perch (*Lates calcarifer*). Aust. J. Mar. Freshwater Res. 30: 803–813.

Moore, R. 1980. Migration and Reproduction in the Percoid Fish *Lates calcarifer* (Bloch). Ph.D. Thesis, University of London, London, UK.

Moore, R. 1982. Spawning and early life history of barramundi, *Lates calcarifer* (Bloch), in Papua New Guinea. Aust. J. Marine Freshwater Res. 33: 647–661.

Moore, R. and L.F. Reynolds. 1982. Migration patterns of barramundi, *Lates calcarifer* (Bloch), in Papua New Guinea. Aust. J. Mar. Freshwater Res. 33: 671–682.

Morgan, D.L., A.J. Rowland, S.G. Gill and R.G. Doupe. 2004. The implications of introducing a large piscivore (*Lates calcarifer*) into a regulated northern Australian river (Lake Kununurra, Western Australia). Lakes Reservoirs: Res. Manage. 9: 181–193.

Morrissy, N.M. 1987. Status of the barramundi (*Lates calcarifer*) fishery in Western Australia. In: J.W. Copland and D.L. Grey (eds.). Management of Wild and Cultured Seabass/Barramundi (*Lates calcarifer*). ACIAR, Darwin, Australia, pp. 55–56.

Mukhopadhyay, M.K. and P.U. Verghese. 1978. Observations on the larvae of *Lates calcarifer* (Bloch) from Hooghly Estuary with a note on their collection. J. Inland Fish. Soc. India 10: 138–141.

Ogston, A.S., R.W. Sternberg, C.A. Nittrouer, D.P. Martin, C.D. Goñi and J.S. Crockett. 2008. Sediment delivery from the Fly River tidally dominated delta to the nearshore marine environment and the impact of El Niño. J. Geophys. Res. 113(F01S11): doi:10.1029/2006JF000669.

Otero, O. 2004. Anatomy, systematics and phylogeny of both Recent and fossil latid fishes (Teleostei, Perciformes, Latidae). J. Linn. Soc. London, Zool. 141: 81–133.

Parazo, M.M., L.M.B. Garcia, F.G. Ayson, A.C. Fermin, J.M.E. Almendras and D.M.J. Reyes. 1990. Seabass hatchery operations. Aquaculture extension manual No. 18. Aquaculture Department, Southeast Asian Fisheries Development Center, Iloilo, Phillipines.

Patnaik, S. and S. Jena. 1976. Some aspects of biology of *Lates calcarifer* (Bloch) from Chilka Lake. Indian J. Fish. 23: 65–71.

Pender, P.J. and R.K. Griffin. 1996. Habitat history of barramundi *Lates calcarifer* in a North Australian river system based on barium and strontium levels in scales. T. Am. Fish. Soc. 125: 679–689.

Pethiyagoda, R. and A.C. Gill. 2012. Description of two new species of seabass (Teleostei: Latidae: Lates) from Myanmar and Sri Lanka. Zootaxa 3314: 1–16.

Pethiyagoda, R. and A.C. Gill. 2013. Taxonomy and distribution of Indo-Pacific *Lates*. *In*: D.R. Jerry (ed.). Ecology and Culture of Asian Seabass *Lates calcarifer*. CRC Press.

Queensland Fisheries Management Authority. 1998. Draft Management Plan and Regulatory Impact Statement. Queensland Fisheries Management Authority, Brisbane, Australia. 100 pp.

Queensland Fisheries Management Authority. 1992. Management of Barramundi in Queensland. Queensland Fish Management Authority, Brisbane, Australia. 2 p.

Reynolds, L.F. and R. Moore. 1982. Growth rates of barramundi, *Lates calcarifer* (Bloch), in Papua New Guinea. Aust. J. Mar. Freshwater Res. 33: 663–670.

Rimmer, M.A. and D.J. Russell. 1998a. Aspects of the biology and culture of *Lates calcarifer*. *In*: S.S. De Silva (ed.). Tropical Mariculture. Academic Press, San Diego, USA, pp. 449–476.

Rimmer, M.A. and D.J. Russell. 1998b. Survival of stocked barramundi, *Lates calcarifer* (Bloch), in a coastal river system in far northern Queensland, Australia. Bull. Mar. Sci. 62: 325–336.

Robins, J., D. Mayer, J. Staunton-Smith, I. Halliday, B. Sawynok and M. Sellin. 2006. Variable growth rates of the tropical estuarine fish barramundi *Lates calcarifer* (Bloch) under different freshwater flow conditions. J. Fish Biol. 69: 379–391.

Ruangpanit, N. 1987. Biological characteristics of wild seabass (*Lates calcarifer*) in Thailand. *In*: J.W. Copland and D.L. Grey (eds.). Management of Wild and Cultured Seabass/ Barramundi (*Lates calcarifer*). ACIAR, Darwin, Australia, pp. 55–56.

Russell, D.J. 1990. Reproduction, migration and growth in *Lates calcarifer*. M. App. Sc. Thesis. Queensland University of Technology, Brisbane, Australia.

Russell, D.J. 1991. Fish movements through a fishway on a tidal barrage in sub-tropical Queensland. Proc. R. Soc. Queensl. 101: 109–118.

Russell, D.J. 1995. Measuring the success of stock enhancement programs. *In*: P.L. Cadwallader and B.M. Kerby (eds.). Fish Stocking in Queensland—Getting it Right!. Queensland Fisheries Management Authority, Townsville, Australia, pp. 96.

Russell, D.J. and R.N. Garrett. 1983. Use by juvenile barramundi, *Lates calcarifer* (Bloch), and other fishes of temporary supralittoral habitats in a tropical estuary in northern Australia. Aust. J. Mar. Freshwater Res. 34: 805–811.

Russell, D.J. and R.N. Garrett. 1985. Early life history of barramundi, *Lates calcarifer* (Bloch), in north-eastern Queensland. Aust. J. Mar. Freshwater Res. 36: 191–201.

Russell, D.J. and R.N Garrett. 1988. Movements of juvenile barramundi, *Lates calcarifer* (Bloch), in north-eastern Queensland. Aust. J. Mar. Freshwater Res. 39: 117–123.

Russell, D.J. and M.A. Rimmer. 1997. Assessment of stock enhancement of barramundi *Lates calcarifer* (Bloch) in a coastal river system in far northern Queensland, Australia. *In*: D.A. Hancock, D.C. Smith, A. Grant and J.P. Beumer (eds.). Developing and Sustaining World Fisheries Resources. CSIRO, Collingwood, Australia, pp. 498–503.

Russell, D.J. and M.A. Rimmer. 1999. Stock enhancement of barramundi (*Lates calcarifer*) in a coastal river in northern Queensland, Australia. Proceedings of the Annual International Conference of the World Aquaculture Society. World Aquaculture Society, Sydney, Australia.

Russell, D.J. and M.A. Rimmer. 2000. Measuring the success of stocking barramundi *Lates calcarifer* (Bloch) into a coastal river system in far northern Queensland, Australia. *In*: A. Moore and R. Hughes (eds.). Australian Society for Fish Biology, Albury, Australia, pp. 70–76.

Russell, D.J. and M.A. Rimmer. 2002. Importance of release habitat for survival of stocked barramundi in northern Australia. *In*: J.A. Lucy and A.L. Studholme (eds.). Catch and Release in Marine Recreational Fisheries Symposium 30. American Fisheries Society, Bethesda, Maryland, USA, pp. 237–240.

Russell, D.J. and M.A. Rimmer. 2004. Stock enhancement of barramundi in Australia. *In*: D.M. Bartley and K.M. Leber (eds.). FAO Fishery Technical Paper 429: Marine Ranching. FAO, Rome, Italy, pp. 73–108.

Russell, D.J., R.N. Garrett and M.R. McKinnon. 1987. Hatchery techniques for rearing barramundi (*Lates calcarifer*). Aust. Mar. Sci. Assoc. Proc. 55 pp.

Russell, D.J., P.W. Hales and B.A. Ingram. 1991. Coded-wire tags—a tool for use in enhancing coastal barramundi stocks. Austasia Aquacult. 5: 26–27.

Russell, D.J., M. Rimmer, A.J. McDougall, S.E. Kistle and W.L. Johnston. 2002. Stock enhancement of barramundi, *Lates calcarifer* (Bloch), in a coastal river system in northern Australia: stocking strategies, survival, biology and cost-benefits. *In*: K.M. Leber, S. Kitada, H.L. Blankenship and T. Svasand. Stock Enhancement and Sea Ranching: Developments, Pitfalls and Opportunities. Blackwell Publishing, Oxford, UK, pp. 490–500.

Russell, D.J., M.A. Rimmer, A.J. McDougall, S.E. Kistle and W.L. Johnston. 2004. Stock enhancement of barramundi, *Lates calcarifer* (Bloch), in a coastal river system in northern Australia: stocking strategies, survival and benefit-cost. *In*: K.M. Leber, S. Kitada, H.L. Blankenship and T. Svasand (eds.). Stock Enhancement and Sea Ranching: Developments, Pitfalls and Opportunities. Blackwell Publishing, Oxford, UK, pp. 490–500.

Russell, D.J., P.A. Thuesen and F.E. Thomson. 2011. Movements of stocked barramundi (*Lates calcarifer*) in Australia: a desktop study. Queensland Department of Agriculture, Fisheries and Forestry, Cairns, Australia. 16 pp.

Russell, D.J., D.R. Jerry, P.A. Thuesen, F.E. Thomson, T.N. Power and C.S.K. Smith-Keune. 2013. Fish stocking programs—assessing the benefits against potential long term genetic and ecological impacts. Report to the Fisheries Research and Development Corporation, Canberra, Australia. 104 pp.

Rutledge, W., M. Rimmer, D.J. Russell, R. Garrett and C. Barlow. 1990. Cost benefit of hatchery-reared barramundi, *Lates calcarifer* (Bloch), in Queensland. Aquacult. Fish. Manage. 21: 443–448.

Sawynok, B. 1998. Fitzroy River Effects of Freshwater Flows on Fish. National Fishcare Project 97/003753. Fishcare Australia, Rockhampton, Australia. 59 pp.

Sawynok, B. and J. Platten. 2009. Growth, movement and survival of stocked fish in impoundments and waterways of Queensland 1987–2008. Infofish Services and Australian National Sportsfishing Association, Rockhampton, Australia. 78 pp.

Skirtun, M., P. Sahlqvist, R. Curtotti and P. Hobsbawn. ABARES 2012. Australian Fisheries Statistics 2011, Canberra, December.

Staunton-Smith, J., J.B. Robins, D.G. Mayer, M.J. Sellin and I.A. Halliday. 2004. Does the quantity and timing of fresh water flowing into a dry tropical estuary affect year-class strength of barramundi (*Lates calcarifer*)? Mar. Freshwater Res. 55: 787–797.

Stuart, I.G. and A.P. Berghuis. 1999. Passage of native fish in a modified vertical-slot fishway on the Burnett River barrage, south-eastern Queensland. Queensland Department of Primary Industries, Bundaberg, Australia 53 pp.

Stuart, I.G. and A.P. Berghuis. 2002. Upstream passage of fish through a vertical-slot fishway in an Australian subtropical river. Fish. Manage. Ecol. 9: 111–122.

Stuart, I.G. and M. Mallen-Cooper. 1999. An assessment of the effectiveness of a vertical-slot fishway for non-salmonid fish at a tidal barrier on a large tropical/subtropical river. Regulated Rivers: Res. Manage. 15: 575–590.

Stuart, I.G., A.P. Berghuis, P.E. Long and M. Mallen-Cooper. 2007. Do fish locks have potential in tropical rivers? River Res. Appl. 23: 269–286.

Takemura, A., S. Rahman, S. Nakamura, Y.J. Park and K. Takano. 2004. Lunar cycles and reproductive activity in reef fishes with particular attention to rabbitfishes. Fish and Fish. 5: 317–328.

Tucker, J.W., D.J. Russell and M.A. Rimmer. 2006. Barramundi culture. *In*: A.M. Kelley and J.T. Silverstein. (eds.). Aquaculture in the 21st century (vol. 2). American Fisheries Society, Bethedsa, USA, pp. 643.

UN Food and Agriculture Organisation. 2012. 'Species fact sheets—*Lates calcarifer*'. Available at http://http://www.fao.org/fishery/species/3068 [Accessed 11 October 2012].
Veitch, V. and B. Sawynok. 2004. Freshwater Wetlands and Fish Importance of Freshwater Wetlands to Marine Fisheries Resources in the Great Barrier Reef. Report No. SQ200401. Great Barrier Reef Marine Park Authority, Townsville, Australia 116 pp.
Williams, L.E. 2002. Queensland's fisheries resources—current condition and trends 1988–2000. Queensland Department of Primary Industries, Brisbane, Australia 180 pp.
Wongsumnuk, S. and S. Maneewongsa. 1974. Biology and artificial propagation of seabass *Lates calcarifer* Bloch. First Mangrove Ecology Workshop 2: 645–664.
Yingthavorn, P. 1951. Notes on pla-kapong (*Lates calcarifer* Bloch) culturing in Thailand. Technical Paper No. 20. FAO Fisheries Biology, City?
Yue, G.H., J.H. Xia, F. Liu and G.C. Lin. 2012. Evidence for female-biased dispersal in the protandrous hermaphroditic Asian seabass, *Lates calcarifer*. PLoS ONE 7: e37976.

6

Infectious Diseases of Asian Seabass and Health Management

Kate S. Hutson

6.1 Introduction

Disease problems constitute the largest single cause of economic loss in aquaculture. Despite reliable hatchery production and commercial expansion of the Asian seabass industry, a number of viral, bacterial, fungal and parasitic diseases threaten sustained production. Consequently, the focus and content of this chapter is on major viral, bacterial, fungal and parasitic diseases which have occurred in, and which may impact adversely on, production of Asian seabass. Where available, information is also provided on diseases inflicting wild seabass and a list of known infectious diseases (pathogens and parasites) of both wild and farmed Asian seabass has been compiled in Table 6.1 as a reference.

Asian seabass can be cultured in a variety of aquaculture systems (open, semi-closed and closed) in freshwater, brackish and marine environments, each which presents its own unique disease challenges. However, most of the losses caused by pathogens and parasites in Asian seabass aquaculture

---

Marine Parasitology Laboratory, Centre for Sustainable Tropical Fisheries and Aquaculture and the School of Marine and Tropical Biology, James Cook University, Townsville, QLD 4811, Australia.
Email: Kate.Hutson@jcu.edu.au

Infectious Diseases of Asian Seabass and Health Management 103

Table 6.1. Pathogens and parasites of wild and farmed *Lates calcarifer*. Abbreviations: N/g = Not given; N/d = Not determined; NT, Northern Territory; Qld, Queensland; ^{vr} = validation required; ^{f} = farmed fish; ^{w} = wild fish; ^{w*} = recent captive fish; ^Incorrect spelling in Rückert et al. 2008 (as *celebensis*); **Likely to be the same taxon, requires further study

Pathogen (Group, Family, *Species*, authority)	Microhabitat	Reference	Location
Viruses			
Betanodaviridae	Brain, retina	Glazebrook et al. 1990	Qld, Australia[f]
	N/g	See Munday et al. 2002	China[N/g]
	N/d	Zafran et al. 1998	Indonesia[f]
	N/g	See Munday et al. 2002	Israel[N/g]
	N/d	Awang 1987	Malaysia[f]
	Brain, retina	Maeno et al. 2004	Philippines[f]
	N/d	Chang et al. 1997	Singapore[f]
	Brain, retina	Renault et al. 1991	Tahiti[f]
	N/d	Chi et al. 2001	Taiwan[N/d]
	Brain, retina	Glazebrook et al. 1990	Thailand[f]
	Brain, retina, spinal cord	Azad et al. 2005	India[f]
Lymphocystis	Skin, fins, gills, internal organs	Pearce et al. 1990	Australia[w*]
	N/d	Tonguthai and Chinabut 1987	Thailand[f]
	N/d	Cheong et al. 1983	Singapore[f]
Bacteria			
Streptococcus spp.	Brain, spleen, kidney, liver, eyes	Bromage et al. 1999	Qld, Australia[f]
	Brain, spleen, kidney, liver	Huang et al. 1990	China[f]
	N/g	Humphrey 2006	NT, Australia[f]

Table 6.1. contd....

Table 6.1. contd.

Pathogen (Group, Family, *Species*, authority)	Microhabitat	Reference	Location
Vibrio spp.	Spleen, kidney, eyes, brain and external lesions	Tendencia 2002	Philippines[f]
Pseudomonas spp.	Liver, kidney	Kumaran 2010	India[f]
Flexibacter	Gill, skin	Anderson and Prior 1992	Qld, Australia[f]
Cytophaga johnsonae (Stanier 1947)	Skin	Carson et al. 1993	Qld, Australia[f]
Epitheliocystis	Gill, skin	Anderson and Prior 1992	Qld, Australia[f]
	Gills	Humphrey 2006	NT, Australia[f]
Oomycetes			
Aphanomyces invadans (David and Kirk 1997)	Skin	Humphrey and Pearce 2004	NT, Australia[f]
Protozoa			
Eimeria	Intestine	Gibson-Kueh et al. 2011a	Vietnam[f]
Cryptosporidium spp.	Stomach, small intestine	Gabor et al. 2011	Australia[f]
	Tissues, not specified	Gibson-Kueh et al. 2011b	Vietnam[f]
Trypanosoma	Blood	Humphrey 2006	NT, Australia[f]
Amyloodinium ocellatum (Brown and Hovasse 1946)	Gills	K. Hutson, pers. obs.	Qld, Australia[f]
Trichodina spp.	N/g	Leong and Wong 1990	Thailand[f]
	N/g	Leong and Wong 1990	Malaysia[f]
	Gills, operculum	Rückert et al. 2008	Sumatra[f]
Chilodonella sp.	N/g	Leigh Owens, pers. comm.	Qld, Australia[f]
Cryptocarion irritans (Brown 1951)	N/g	Leong and Wong 1990	Thailand[f]
	Skin, fins	Kua 2008	Malaysia[f]

Myxozoa			
Myxozoa gen. et sp. indet.	Gall bladder	Rückert et al. 2008	Sumatra[f]
Cestoda			
Cestoda gen. sp.	N/g	Leong and Wong 1990	Thailand[f]
Order Tetraphyllidea			
Tetraphyllidia	N/g	Leong and Wong 1992a, b	Sumatra[f]
Emothrium sp.	N/g	Leong and Wong 1992a, b	Sumatra[f]
Scolex pleuronectis (Müller 1787)	Intestine, stomach, pyloric caeca	Rückert et al. 2008	Sumatra[f]
Tentaculariidae			
Nybelinia indica (Chandra 1986)	Stomach, stomach wall	Rückert et al. 2008	Sumatra[f]
Dasyrhynchidae			
Callitetrarhynchus gracilis (Rudolphi 1819)	Body cavity, viscera, muscle	Arthur and Ahmed 2002	Bangladesh[w]
	Around intestine and anus	Palm 1995	N/g[w]
Dasyrhynchus indicus (Chandra and Rao 1986)	Body cavity, viscera, muscle	Arthur and Ahmed 2002	Bangladesh[w]
Gymnorhynchidae			
Gymnorhynchus gigas (Cuvier 1817)	Body cavity, viscera, muscle	Arthur and Ahmed 2002	Bangladesh[w]
Mustelicolidae			
Bombycirhynchus sphyraenaicum (Pintner 1930)	Around intestine and gonads	Palm et al. 1998	New Guinea[w]
Patellobothrium quinquecatenatum (Beveridge and Campbell, 1989)	Viscera	Beveridge and Campbell 1989	NT, Australia[w]
	Around intestine and gonads	Palm 1995	New Guinea[w]

Table 6.1. contd....

Table 6.1. contd.

Pathogen (Group, Family, Species, authority)	Microhabitat	Reference	Location
Otobothriidae			
Otobothrium balli (Southwell 1915)	N/g	Bilqees 1995	Pakistan[w]
Poecilancistrum caryophyllum (Diesing 1850)	Muscle	Palm 1995	N/g[w]
Pterobothriidae			
Pterobothrium acanthotuncatum (Escalente and Carvajal 1984)	Body cavity, mesentery	Campbell and Beveridge 1996	NT, Australia[w]
Pterobothrium lintoni (MacCallum 1916) (Syn. *Gymnorhynchus malleus* (Linton 1924)	Body cavity, viscera, muscle	Arthur and Ahmed 2002	Bangladesh[w]
Digenea			
Acanthocolpidae			
Stephanostomum cloacum (Srivastava 1938)	N/d	Srivastava 1938	Pakistan[w]
Aporocotylidae	N/d	Saoud et al. 2002	Indian Ocean[w]
Cruoricola lates Herbert et al. 1994 (Syn. *Cardicola* sp. of Leong and Wong 1986, 1990)	Blood vessels	Leong and Wong 1986, Leong and Wong 1990	Thailand[f]
	Blood vessels	Herbert et al. 1994	Malaysia[f]
	Blood vessels	Herbert et al. 1994	Qld, Australia[f]
Parasanguinicola vastispina (Herbert and Shaharom-Harrison 1995)	Branchial arteries, dorsal aorta, mesenteric venules and renal artery	Herbert and Shaharom-Harrison 1995	Qld, Australia[f]
		Herbert and Shaharom-Harrison 1995	Malaysia[f]

Bucephalidae			
Bucephalus margaritae	N/g	Leong and Wong 1992a, b	Thailand[f]
	N/g	Leong and Wong 1992b	Malaysia[f]
Prosorhynchus luzonicus (Velasquez 1959)	Intestine	Arthur and Lumanlan-Mayo 1997	Philippines[w]
Prosorhynchus sp.	Stomach	Rückert et al. 2008	Sumatra[w]
Rhipidocotyle sp.	N/g	Leong and Wong 1992a, b	Thailand[f]
Callodistomidae			
Callodistomum minutus (Zaidi and Khan 1977)	N/g	Bilqees 1995	Pakistan[w]
Cryptogonimidae			
Cryptogonimidae gen. sp.	N/g	Leong and Wong 1992a, b	Sumatra[f]
Pseudometadena celebesensis^ (Yamaguti 1952)	Intestine & pyloric caeca	Rückert et al. 2008	Malaysia[f]
		Leong 1997	Indonesia[f]
		Leong and Wong 1990	Thailand[f]
	N/g	Leong and Wong 1992a	Sumatra[f]
		Leong and Wong 1992b	Malaysia[f]
Pseudometadena sp.	N/d	Arthur and Lumanlan-Mayo 1997	Philippines[w]
Lecithochirium sp.	N/d	Arthur and Lumanlan-Mayo 1997	Philippines[w]
Hemiuridae			
Hemiuridae gen. sp.	Intestine	Arthur and Ahmed 2002	Bangladesh[w]
	N/g	Leong and Wong 1990	Thailand[f]
Ectenurus sp.	N/g	Leong and Wong 1992b	Malaysia[f]

Table 6.1. contd....

Table 6.1. contd.

Pathogen (Group, Family, Species, authority)	Microhabitat	Reference	Location
Erilepturus hamati (Yamaguti 1934) (Manter 1947)	N/g	Bray et al. 1993	Philippines[N/g]
	Stomach	Bray et al. 1993	NT, Australia[w]
Lecithocladium grandulosum	N/g	Leong and Wong 1990	Malaysia[f]
Lecithocladium neopacificum	N/g	Leong and Wong 1990	Thailand[f]
Lissorchiidae			
Complexobursa magna (Bilqees 1980)	N/g	Bilqees 1995	Pakistan[w]
Psilostomidae			
Psilostomum sp. (metacercaria)	Intestine	Arthur and Ahmed 2002	Bangladesh[w]
Transversotrematidae			
Prototransversotrema steeri (Angel 1969)	Beneath scales	Cribb et al. 1992	Qld, Australia[w]
	Beneath scales	Rodgers and Burke 1988	Qld, Australia[w]
Transversotrema patialense (Soparkar 1924) (Syn. *T. laruei* Velasquez)	Beneath scales	Cribb et al. 1992	Philippines[w]
Monogenea			
Capsalidae			
Benedenia sp.	Body surface	Leong 1997	Indonesia[f]
	Body surface	Leong 1997	Malaysia[f]
Benedenia epinepheli (Yamaguti 1937) (Meserve 1938)	Gills, body surface	Rückert et al. 2008*	Sumatra[f]
Capsalidae gen. et sp. indet.	Gills, body surface	Rückert et al. 2008*	Sumatra[f]
Neobenedenia sp.**	Body surface	Hutson et al. 2012	Qld, Australia[f]

Neobenedenia melleni (MacCallum 1927) (Yamaguti 1963)**	Gills, body surface	Deveney et al. 2001	Qld, Australia[f]
	Gills, body surface	Rückert et al. 2008	Sumatra[f]
Diplectanidae			
Diplectanidae gen. sp.	N/g	Leong and Wong 1992a	Sumatra[f]
Diplectanum sp.	N/g	Leong and Wong 1990	Thailand[f]
Diplectanum narimeen Unnithan 1964	Gills	Tingbao et al. 2006	N/g
Diplectanum penangi (Laing and Leong 1991)	Gills	Tingbao et al. 2006	China[f]
	N/g	Leong and Wong 1992b	Malaysia[f]
Diplectanum setosum (Nagibina 1976)	Gills	Tingbao et al. 2006	East China Sea[w]
Laticola lingaoensis (Tingbao et al. 2006)	Gills	Tingbao et al. 2006	China[f]
Laticola latesi Tripathi 1957 (see Tingboa et al. 2006 for list of synonyms)	Gills	Tingbao et al. 2006	China[f]
		Tripathi 1957	India
	N/g	Leong and Wong 1990	Thailand[f]
	N/g	Leong and Wong 1992a	Sumatra[f]
	N/g	Leong and Wong 1992b	Malaysia[f]
Laticola paralatesi (Nagibina 1976) (syn. *Diplectanum latesi*)	Gills	Tingbao et al. 2006	China[f]

Table 6.1. contd....

Table 6.1. contd.

Pathogen (Group, Family, Species, authority)	Microhabitat	Reference	Location
Pseudorhabdosynochus epinepheli (Yamaguti 1938)	Gills	Rückert et al. 2008	Sumatra[f]
	Gills	Leong and Wong 1990	Malaysia[f]
	Gills	Leong and Wong 1990	Thailand[f]
	Gills	Leong and Wong 1990	Philippines
Pseudorhabdosynochus lantauensis Beverley (Burton and Suriano 1981)	Gills	Rückert et al. 2008	Sumatra[f]
Nematoda			
Anisakidae			
Anisakis sp.	N/g	Leong and Wong 1990	Thailand[f]
Hysterothylacium sp.	Intestine, liver, mesenteries, pyloric caeca	Rückert et al. 2008	Sumatra[f]
Raphidascaris sp.	Stomach wall	Rückert et al. 2008	Sumatra[f]
	N/g	Leong and Wong 1992a, b	Sumatra[f]
		Leong and Wong 1990	Thailand[f]
		Leong and Wong 1992b	Malaysia[f]
Raphidascaris sp. II	Intestine	Rückert et al. 2008	Sumatra[f]
Terranova sp.	Intestine, liver, mesenteries, pyloric caeca	Rückert et al. 2008	Sumatra[f]
Camallanoidea			
Procamallanus sparus (Akram 1975)		Akram 1995	Pakistan[w]
Dracunculoidea			

Infectious Diseases of Asian Seabass and Health Management 111

Philometra lateolabracis (Yamaguti 1935)	Gonads	Moravec et al. 1988	Indian Ocean[w]
Gnathostomatoidea			
Echinocephalus sp.		Akram 1995	Pakistan[w]
Echinocephalus uncinatus (Molin 1858)		Gibson et al. 2005	Arabian Sea[w]
Seuratoidea			
Cucullanus hians (Dujardin 1845)	Intestine	Akram 1994	Pakistan[w]
Cucullanus sp.		Akram 1995	Pakistan[w]
		Akram 1995	Pakistan[w]
Indocucullanus calcariferii (Zaidi and Khan 1975)		Arya 1991	Indian Ocean[w]
	N/g	Bilqees 1995	Pakistan[w]
Indocucullanus longispiculum (Khan 1969)	N/g	Bilqees 1995	Pakistan[w]
Acanthocephala			
Rhadinorhynchidae			
Acanthocephalus sp.	N/g	Leong and Wong 1992a, b	Malaysia[f]
Serrasentis sagittifer (Linton 1889) (Van Cleave 1923)	Mesenteries	Rückert et al. 2008	Sumatra[f]
	Intestine	Arthur and Ahmed 2002	Bangladesh[N/g]
Tenuiproboscis sp.	Intestine	Sanil et al. 2011	India[w]
Hirudinea			
Piscicolidae			
Zeylanicobdella arugamensis	Body surface	Kua et al. 2010	Malaysia[f]

Table 6.1. contd....

Table 6.1. contd.

Pathogen (Group, Family, *Species*, authority)	Microhabitat	Reference	Location
Copepoda			
Caligidae			
Caligus sp.	N/g	Leong and Wong 1992a, b	Sumatra[f]
Caligus epidemicus (Hewitt 1971)	N/g	Johnson et al. 2004	Thailand[f]
	Body surface, gill cavities	Venmathi Maran et al. 2009	Malaysia[f]
	Body surface, gill and oral cavities	Ho and Lin 2004	Taiwan[f]
C. chiastos (Lin and Ho 1993)	N/g	Muhd-Faizul et al. 2012	Malaysia[f]
C. orientalis (Gusev 1951) (syn. *C. communis* and *C. laticorpus*)	Body surface	Ho and Lin 2004	Taiwan[f]
C. pagrosomi (Yamaguti 1939)	Gills filaments & oral cavity	Ho and Lin 2004	Taiwan[f]
C. punctatus (Shiino 1955)	Body surface	Ho and Lin 2004	Taiwan[f]
	Body surface & gill cavities	Venmathi Maran et al. 2009	Malaysia[f]
	N/g	Muhd-Faizul et al. 2012	Malaysia[f]
C. rotundigenitalis (Yü 1933)	Gills, gill cavities & body surface	Ho and Lin 2004	Taiwan
	N/g	Muhd-Faizul et al. 2012	Malaysia[f]
Caligus sp.	N/g	Muhd-Faizul et al. 2012	Malaysia[f]
Lernanthropidae			
Lernanthropus latis (Yamaguti 1954)	Gill filaments	Ho and Kim 2004	Gulf of Thailand[w]
	Gill filaments	Leong and Wong 1986	Thailand
	Gill filaments	Tripathi 1962	India
	Gill filaments	Pillai 1985	Sri Lanka

			Gill filaments	Kua et al. 2012	Malaysia[f]
			Gill filaments	Brazenor and Hutson 2013	WA, Australia[f]
			Gill filaments	Brazenor and Hutson 2013	Qld, Australia[f,w]
[vt] *Lernanthropus kroyeri*			Gills	Vinoth et al. 2010	India[w]
Isopoda					
Cymothoidae					
Aegathoa sp.			N/g	Leong and Wong 1990	Thailand[f]
Cymothoa sp.			N/g	Leong and Wong 1990	Thailand[f]
Cymothoa indica (Schioedte and Meinert 1884)			Body surface		Thailand[f]
				Rajkumar 2005	India[f]
Nerocila barramundae (Bruce 1987)			Dorsal fin	Bruce 1987	Qld, Australia[w]
Rocinela latis (Southwell 1915)			Skin	Southwell 1915	Calcutta, India[w]

can be prevented by adopting appropriate biosecurity protocols. Common biosecurity measures may include routine health inspections, quarantine and treatment of wild-caught brood stock, egg disinfection, strict equipment sanitation, human traffic control, intake water treatment, effluent treatment, clean feeds, restricting movement of stock, appropriate disposal of mortalities and limiting interactions between wild and farmed organisms. Difficulties can arise once a disease has established and spread within a facility because eradication may be impossible. Following a disease outbreak, vaccination and chemoprophylaxis may need to be implemented to reduce outbreaks and minimise losses. Appropriate biosecurity is the most effective method of prevention of the numerous disease causing organisms described herein.

In this chapter a brief description of important diseases is given, together with diagnosis and possible treatment and/or control mechanisms that may be applied in aquaculture. Information on management and treatment is provided as a guide only, and the author recommends consultation should be made with government authorities to ascertain regulations associated with the use of chemotherapeutants. The author recognises that non-infectious disease and abnormalities due to environmental contaminants, or nutritional deficiencies, are equally important problems, but these were beyond the scope of this chapter which focuses on infectious disease issues. Furthermore, continued intensive production of Asian seabass will likely result in further occurrences of previously unknown or undescribed diseases.

6.2 Viral Disease

Asian seabass culture is inflicted by several important classes of viruses, including those from the viral families Nodaviridae and Iridoviridae. A virus is a small infectious agent that replicates only inside the living cells of an organism. Virus particles consist of two or three parts: the genetic material made from either DNA or RNA, long molecules that carry genetic information; a protein coat that protects these genes; and in some cases an envelope of lipids that surrounds the protein coat when they are outside a cell. The shapes of viruses range from simple helical and icosahedral forms to more complex structures. The average virus is about one one-hundredth the size of the average bacterium. Most viruses are too small to be seen directly with an optical microscope. Viruses spread through vectors (e.g., blood-sucking parasites), via the fecal–oral route, physical contact and/or entering the body in food or water. Viral infections in animals provoke an immune response that usually eliminates the infecting virus. Immune responses can also be produced by vaccines, which confer an artificially acquired immunity to the specific viral infection.

6.2.1 Nodavirus

Nodaviruses (Genus *Betanodavirus*: Family Nodaviridae) are microscopic single stranded RNA viruses which cause a disease known as viral nervous necrosis (VNN), or viral encephalopathy and retinopathy (VER). Nodaviruses infect the central nervous system of fish, including the brain, eyes and spinal cord, resulting in cellular vacuolation and degeneration (Azad et al. 2006a). Infected fish are generally pale or show dark colouration with redness around the head (Humphrey 2006). Asian seabass are very susceptible to VNN and the disease has been reported to occur in most regions where it is cultured, including Australia, China, India, Indonesia, Israel, Malaysia, Philippines, Singapore, Tahiti, Taiwan and Thailand (Table 6.1; see Munday et al. 2002 for review). Most fish are affected as larvae or juveniles and outbreaks tend to be associated with high intensity culture. Indeed, VNN could be considered the most significant viral disease affecting Asian seabass hatchery and nursery production.

Both vertical and horizontal transmission have been implicated in VNN outbreaks. Vertical transmission implies that the virus is shed in the reproductive fluids of male and female brood stock and is present on, or in, fertilised eggs (Johansen et al. 2002, Azad et al. 2006b), while horizontal transmission can occur though the water supply, cohabitation with infected individuals and feeding contaminated fish to cultured stock (Hick et al. 2011, Manin and Ransangan 2011). The cannibalistic nature of juvenile Asian seabass may also facilitate horizontal transmission (Manin and Ransangan 2011). Behaviour associated with infection can include anorexia, pale-grey pigmentation of body, whirling swimming and loss of equilibrium (Parameswaran et al. 2008). The disease is most prevalent in hatchery and juvenile fish cultured in intensive aquaculture systems and infections can result in up to 100% mortality (Hick et al. 2011).

Identification of VNN infection is fundamental to hatchery management. Viral nervous necrosis can be diagnosed by: demonstrating characteristic lesions in the brain and/or retina by light microscopy; detection of virions, viral antigens or viral nucleotides by electron microscopy, serology or molecular techniques (including immunofluorescence antibody test (IFAT), enzyme-linked immuno sorbent assay (ELISA) and polymerase chain reaction (PCR)); or detection of specific antibodies in sera or body fluids and tissue culture of virus (see Munday et al. 2002). However, subclinical infection may be challenging to identify because the quantity of virus and the prevalence of infection can be low. Scientists are attempting to improve the sensitivity of diagnostic tests in order to reduce the number of false negative results and enable more accurate population-level tests (Hick et al. 2010). There is no known treatment for VNN, although recent work has shown that vaccination of seabass juveniles with formalin-deactivated betanodavirus conferred protective immunity to seabass lasting for several months (Pakingking et al. 2009). Other current approaches to VNN control are largely based on preventing vertical transmission which may be achieved by excluding *Betanodavirus* infected brood stock (Mushiake et al. 1994) and by disinfecting fertilised eggs by washing them with ozone in seawater. Horizontal transmission can be reduced by maintaining good water quality and low stocking densities.

6.2.2 Lymphocystis

Lymphocystis is caused by viruses in the family Iridoviridae and occurs in marine and freshwater fish worldwide. Infection has been reported in Asian seabass in Australia, Thailand and Singapore (Table 6.1). A morbidity of 70% and mortality of 1% were reported for sea caged Asian seabass with lymphocystis disease in Thailand (Tonguthai and Chinabut 1987), while 100% mortality resulted from the disease in Asian seabass fry cultured in Singapore (Cheong et al. 1983). The disease is characterised by the development of single or multiple nodular lesions on the fins, but can also occur on the skin and internal organs (Pearce et al. 1990). The lesions may resemble small clutches of cauliflower and, as a consequence, the disease is sometimes called 'cauliflower disease'. The virus can be transferred horizontally (Pearce et al. 1990). Lymphocystis reduces the marketability of affected fish and can cause serious economic loss in the aquaculture industry. Eventually the lesions inhibit physiological processes and secondary bacterial infections kill the fish. There is no vaccine or known treatment for this virus. Fish exhibiting symptoms of the virus should be contained and humanely euthanased to prevent virus spread. Restocking of disease-free certified fish should follow a period of fallowing.

6.3 Bacterial Disease

Several bacterial species have been identified as important disease pathogens in Asian seabass, including those from the genera *Streptococcus, Vibrio and Flexibacter*. Bacteria constitute a large domain of prokaryotic microorganisms. Typically a few micrometres in length, bacteria have a wide range of shapes, ranging from spheres to rods and spirals. The vast majority of the bacteria are rendered harmless by the protective effects of the immune system, and a few are beneficial. However, a few species of bacteria are pathogenic and cause infectious diseases. Antibiotics are used to treat bacterial infections in aquaculture, and antibiotic resistance is becoming common. Although infection of humans is rare, bacteria are the most common agents transmitted from fish or culture water to humans (Austin 2010).

6.3.1 Streptococcus iniae

Streptococcossis is a severe infection caused by the gram positive coccoid bacterium *Streptococcus iniae*. The disease may occur in freshwater and marine environments and result in high mortalities. There are two clinical forms of the disease; sub-acute and acute. The sub-acute form displays signs typical of streptococcal infections including protrusion of the eyeball (exophthalmia), darkened colouration and erratic swimming (Perera et al. 1998, Evans et al. 2000). This form of the disease is responsible for ~1% of losses observed (Bromage and Owens 2002). There are limited clinical signs observed in fish dying of the acute form of the disease, with the only indication being mild corneal opacity in some cases. This form is more devastating with heavy losses occurring primarily overnight (Bromage and Owens 2002).

The culture of Asian seabass in marine cages in Australia has suffered losses due to *S. iniae* of between 8 and 15% of production per year, but in severe outbreaks it resulted in losses of up to 70% of production (Bromage et al. 1999). Farmers may be alerted to a pending outbreak from mortalities in wild fish found co-habiting around and in Asian seabass cages in the days before an outbreak (Bromage et al. 1999). Bromage et al. (1999) propose that wild fish co-habiting sea-caged Asian seabass may serve as an important reservoir of *S. iniae*. Subsequently, management practices such as excluding wild fish from inside sea cages by using more effective barrier netting, diligent removal of moribund fish, and reduction in stocking densities may help control the disease. In closed recirculating culture systems, depopulation, disinfection and restocking with disease free fish are the best means of elimination. Vaccination has been met with limited success as it only provides up to six months immunity, however, infected fish may respond to the administration of oral or injectable antibiotic (Agnew and Barnes 2007). Vaccine failure appears to result in part from multiple genotypes being found between and within different farms (Nawawi et al. 2008). *S. iniae* is known to be a human zoonotic agent and individuals who have handled infected fish have developed cellulitis of the hands and endocarditis (Agnew and Barnes 2007).

6.3.2 Vibrio

Members of the genus *Vibrio* and other related genera are the causative agent of vibriosis, a deadly haemorrhagic septicemia disease. In the Philippines, net caged culture of Asian seabass has been affected with 2–3% daily mortality associated with *Vibrio* infections following heavy rainfall (Tendencia 2002). *Vibrio* affects a diverse range of marine, estuarine and freshwater fish species and is frequently secondary to poor water quality, stress, poor nutrition and parasite infection. Infection is characterised by extensive cutaneous (skin) and systemic haemorrhages and localized cutaneous ulceration may occur (Krupesha Sharma et al. 2013). Clinical signs include abnormal swimming behavior, opaque eyes, exophthalmia and reddening of the abdomen. Internally, necrosis and haemorrhage in the kidney, liver and spleen may be observed. Because of its high morbidity and mortality rates, substantial research has been undertaken to elucidate the virulence mechanisms of this pathogen and to develop rapid detection techniques and effective disease-prevention strategies (see Frans et al. 2011 for review). *Vibrio* may be treated by administration of antibiotics, including oxytetracyline, which can be prescribed by a veterinarian.

6.3.3 *Flexibacter* spp.

Filamentous bacteria are well-recognised marine and freshwater fish pathogens. *Flexibacter* is a common bacterial infection most likely to infect fish that have been stressed by conditions such as poor water quality, inadequate diet, or stress from handling and shipping. The bacteria can enter the fish through the gills, mouth, or via small wounds on the skin. The disease is highly contagious and may be spread through contaminated nets, specimen containers, and even food. In challenge trials, Soltani et al. (1996) found that Asian seabass were extremely susceptible to *F. columnaris* infection and can die before external signs are observed.

6.3.4 Epitheliocystis

Epitheliocystis, or 'gill Chlamydia', is a bacterial disease thought to be caused by *Chlamydia* or rickettsia-like microorganisms. They are obligate intracellular bacteria that are not known to function well outside of host cells. The reservoir and mode of transmission amongst fish remains unknown, but horizontal transmission occurs within some host species. Research has speculated that epitheliocystis may exist as a chronic, dormant, yet opportunistic condition within Asian seabass populations (Anderson and Prior 1992). Preliminary diagnosis of epitheliocystis is made by observation of white to yellow cysts on the gills or skin of affected fish. The thick capsule and granular contents are characteristic of cysts and can be seen in wet mounts, but histology and electron microscopy is recommended for definitive diagnosis (Nowak and LaPatra 2006). The occurrence, prevalence and associated mortalities for this disease have not been reported for Asian seabass. There is no treatment once fish are infected.

6.4 Oomycetes

Oomycetes form a distinct phylogenetic lineage of fungus-like eukaryotic microorganisms. They are filamentous, microscopic, absorptive organisms that reproduce both sexually and asexually. They are also often referred to as water molds.

6.4.1 Aphanomyces invadans

Epizootic Ulcerative Syndrome (EUS) is caused by *Aphanomyces invadans* which has motile spores that invade the skin of fish. The disease occurs in freshwater and estuarine conditions and is generally not observed in marine environments. Infection begins as a small area of reddening over a single scale, which subsequently spreads to involve a number of adjacent scales resulting in severe ulcers and is commonly known as 'red spot disease' (Pearce 1989, Humphrey and Pearce 2004). Asian seabass can develop cloudiness of the cornea which may or may not be accompanied by lesions in the skin. Some cases of EUS heal spontaneously, but infection can result in high losses of juvenile Asian seabass (Humphrey and Pearce 2004). EUS may cause unsightly injuries in fish and marketability problems. Captive fish may respond to treatment with an antiseptic iodophore solution or increasing the salinity of holding waters (Humphrey and Pearce 2004). Infected fish should not be translocated to prevent further spread of disease.

6.5 Protozoa

Protozoans are a diverse group of unicellular eukaryotic organisms, many of which are motile. Protozoans commonly range from 10 to 52 μm, but can grow as large as 1 mm, and can be easily observed using a microscope. Some parasitic protozoa have life stages alternating between proliferative stages and dormant cysts. As cysts, protozoa can survive harsh conditions, such as exposure to extreme temperatures or harmful chemicals, or long periods without access to nutrients, water, or oxygen for a period of time.

6.5.1 Flagellates

6.5.1.1 Trypanosoma

Parasitic protozoan trypanosomes of the genus *Trypanosoma* are flagellates that infect the vascular system (blood) of fishes and can cause anemia and lethargy. Leeches and other blood feeding parasites are intermediate hosts and transfer flagellates when biting fish. The parasite may also be transferred directly between fish (Buchmann and Bresciani 2009). Affected seabass show signs of anorexia, lethargy, anemia, scale loss, intra-ocular haemorrhage, splenomegaly and exophthalmia (Humphrey et al. 2010). Stained blood smears on slides show the presence of tapered trypanosomes by light microscopy. Young fish are vulnerable and mortality can result from infection. In 2004, *Trypanosoma* sp. was diagnosed as the primary cause of death in Australian sea-caged Asian seabass (Humphrey et al. 2010). There is no practical treatment for management of trypanosomes, although eradication of leeches (see 6.6.1.5 below) may have a protective effect.

6.5.1.2 Oodinium species

Oodinium species are dinoflagellates that cause 'velvet disease' in marine (*Amyloodinium ocellatum*) and freshwater environments (*Piscinoodinium pillulare*). Outbreaks are usually associated with stress including poor water quality and/or poor fish health (Kuo and Humphrey 2008). These organisms bear two flagella for locomotion and infect the skin, fins and gills. *Oodinium* spp. have life cycles similar to that of *Cryptocaryon irritans* and *Ichthyopthirius multifiliis* (see 6.5.2.3). Pyriform trophonts (<160 μm) attach to the epidermis (Fig. 6.1) with an attachment disc containing root-like rhizocysts which penetrate epithelial cells and destroy them causing

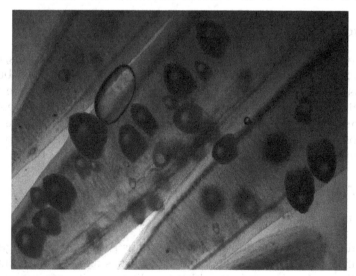

Figure 6.1. An infection of *Amyloodinium ocellatum* showing characteristic pyriform trophonts attached to the gill lamellae. Photo: Kate Hutson.

hyperplasia, fusion of lamellae and petechial haemorrahages (Humphrey 2006). Extreme infections result in a velvet-like layer of parasites on external surfaces. Infected fish are observed to rub their body against objects (i.e., 'flashing'), and may show signs of anorexia, hyperventilation and mass mortality. Infection may be diagnosed through microscopy. Bath treatments with copper sulphate (less than 2 ppm) or benzalkonium chloride (up to 0.5 ppm) for up to 3 days may be effective (Buchmann and Bresciani 2009).

6.5.2 Ciliates

6.5.2.1 Trichodina

Trichodina is a genus of ciliate protists whose species are ectocommensal or parasitic on aquatic animals, particularly fish. *Trichodina* spp. are characterised by the presence of a ring of interlocking cytoskeletal denticles, which provide support for the cell and allow for adhesion to surfaces, including the skin and gills of Asian seabass (Fig. 6.2). Trichodinids have a direct life cycle and reproduce by binary fission and occur in freshwater

Figure 6.2. A silver stain of *Trichodina* sp., showing characteristic ring of interlocking cytoskeletal denticles, isolated from fresh water pond cultured Asian sea bass. Photo: Terrence Miller.

and marine environments. *Trichodina* feed by filtration and remove bacteria and organic particles from water. *Trichodina* spp. are frequently found on the gills and inner operculum of Asian seabass (Rückert et al. 2008, Table 6.1) and high population densities can impair osmoregulation and respiration. Clinical signs of infection include lethargy, anorexia, hyperventilation and excessive mucous production. Diagnosis can be made through microscopic detection in scrapings from the skin or gills. *Trichodina* can be treated using formalin bathing, copper sulphate, sodium chloride and sodium percarbonate (Buchmann and Bresciani 2009). Given that *Trichodina* feeds on organic matter, reducing the organic load and improving water quality can reduce the population. Methods to reduce organic load include mechanical filtration combined with UV-irradiation and ozonation.

6.5.2.2 Chilodonella

Chilodonella are single-celled ciliates and the causative agent of disease that affects the gills and skin of freshwater Asian seabass. This parasite remains a relatively common cause of disease of larger Asian seabass in freshwater pond systems in northern Queensland (Anonymous 2007). Symptoms of infection include morbidity, mortality, lethargy and anorexia. Diagnosis can be made through microscopic detection in scrapings from the skin or gills. Dense populations on the host epithelium inhibit normal physiological function, including osmoregulation, gas exchange and excretion. *Chilodonella* can be treated with formalin, copper sulphate, malachite green and methylene blue (Buchmann and Bresciani 2009).

6.5.2.3 Cryptocaryon irritans

White spot disease is a common disease caused by infection of the skin, fins, and gills with a ciliated protozoan parasite that occurs in marine (*Cryptocaryon irritans*; saltwater 'Ich') or freshwater (*Ichthyophthirius multifiliis*; freshwater 'Ich') environments. White spots (0.4–0.8 mm) on external fish surfaces are encysted trophonts, which feed on host epidermis. The trophont escapes from the epidermis as the free-swimming tomont stage. Tomonts encyst by producing a gelatinous capsule and attach to substrate. The tomocyst undergoes numerous cell divisions. One tomocyst can produce up to 1000 free-swimming theronts, which escape from the tomocyst and search for a suitable host. A recent study showed that prevalence of infection increases with temperature (Karvonen et al. 2010).

Cryptocaryon irritans is one of the major parasitic diseases that affect Asian seabass kept in tanks for breeding (Humphrey 1996). Infection is easily introduced by translocation of other infected fishes. Parasites are persistent due to their low host-specificity and rapid life cycle. Once the organism establishes in a large fish culture facility, it is difficult to manage due to its rapid life cycle and there can be 100% mortality if left untreated. Infected Asian seabass may scratch or 'flash' against the tank bottom or walls and have visible white spots on the body (Fig. 6.3). Microscopic diagnosis can be made by making skin scrapings and examining for trophonts. Trophonts are characterized by horse-shoe-shaped macronucleus and numerous moving cilia. Heavy infections reduce the osmoregulatory ability eliciting hyperplasia and cellular infiltration, ultimately leading to mortality. Malachite green (0.2 ppm) is effective as a bath treatment, however, this treatment is not permitted for use in fish production facilities in all countries. Regular treatment of pond water with formalin, copper sulphate, sodium per-carbonate, sodium chloride, hydrogen peroxide, or other disinfectants will kill infective theronts and trophonts leaving the fish skin. Alternatively, salinity treatments (lowering or increasing salinity) can be used. Quarantine measures should be taken when translocating stock. Filtration of water (mesh size 10 um) traps trophonts and tomonts and can ease infection intensity (Buchmann and Bresciani 2009).

Figure 6.3. An intense infection of *Cryptocaryon irritans* on Asian sea bass showing characteristic white spots on the fins and body surface. Photo: Robert Lester.

6.6 Metazoa

Numerous species of metazoan parasites have been implicated in disease outbreaks in Asian seabass. Metazoan parasites include multi-celled organisms that live on and within the body of their host. Common metazoan parasites include monogeneans (haptor-worms), digeneans (flukes), cestodes (tape-worms), nematodes (round worms), acanthocepahalans (spiny-headed worms), leeches and crustaceans.

6.6.1 Helminthes

6.6.1.1 Monogeneans

Capsalid monogean infections are the most serious and pathogenic amongst all of the parasitic diseases known from Asian seabass and if left untreated fish quickly develop secondary infections (Leong 1997, Anonymous 2011). In Indonesia and Australia, monogenean infections have been associated with large losses and mortality rates of 30–40% are common (Deveney et al. 2001, Rückert et al. 2008, Anonymous 2011). The monogeneans *Neobenedenia melleni* and *Benedenia epinepheli* reported to infect Asian seabass are notorious, allegedly generalist, pathogens of tropical and subtropical fishes in aquaria and aquaculture worldwide (Table 6.1). Both species attach to the skin of their host and graze on mucous and epidermis, continuously laying eggs into the water which hatch into ciliated larvae that directly re-infect fish (Fig. 6.4). Capsalid monogeneans also irritate the eyes, causing opacity and exophthalmia and gradually cause the caudal and pectoral fins to become frayed. High infection intensities on fish lead to secondary infections by bacteria, ultimately resulting in emaciation and death.

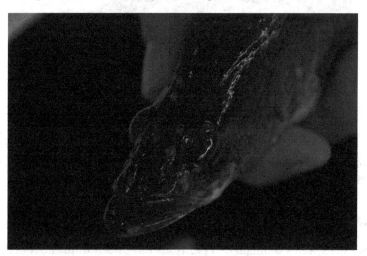

Figure 6.4. Monogeneans (*Neobenedenia* sp.) attached to the body surface of Asian sea bass. Parasites are transparent when they are alive, but turn opaque when they die; in this instance following immersion in fresh water. Photo: Alejandro Trujillo Gonzalez.

At least 10 diplectanid monogenean species (including *Diplectanum* spp., *Laticola* spp. and *Pseudorhabdosynochus* spp.) infect the gills of maricultured Asian seabass (Table 6.1). Infected fish exhibit a darkened body, pale gills, lethargy, loss of appetite and excess mucus production. Leong and Wong (1990) recorded that a large proportion of diseased Asian seabass in Malaysia were infected with *Laticola latesi* (as *Pseudorhabdosynochus latesi*) and *Diplectanum* spp.

The established method of monogenean treatment involves recurrent acute bathing of infected stock primarily in either formalin or freshwater solutions (Kaneko II et al. 1988, Thoney and Hargis 1991, Fajer-Ávila et al. 2008). While these treatments kill attached juveniles and adults, eggs are generally resistant and may exist outside the treatment area (Mueller et al. 1992, Ellis and Watanabe 1993). More recently, natural alternatives to chemical treatment have been investigated (Hutson et al. 2012, Militz et al. 2013).

6.6.1.2 Digeneans

A number of digenean species have been documented from wild and captive Asian seabass (Table 6.1). Of these, the blood flukes (Family Aporocotylidae) have the most potential to become serious pathogens of hosts in mariculture. Adult parasites release eggs into the fish's vascular system which may be sequestered in the gill, heart, kidney, liver, spleen, pancreas, or other organs, where they cause inflammation and decrease the physiological and mechanical efficiency of these organs. Blood flukes can be problematic in open or semi-closed aquaculture systems because their intermediate invertebrate host may inhabit areas close to farmed fish, such as on cage structures or in sediment (e.g., Cribb et al. 2011). Detection of adult flukes can be made in the blood vessels, while thin shelled eggs can be detected in microscopy or histology.

Cruoricola lates is known to infect farmed Asian seabass in Malaysia, Thailand and Australia (Herbert et al. 1994) and *Parasanguinicola vastispina* infects cultured fish in Malaysia (Herbert and Shaharom-Harrison 1995). Although these species have no known pathology, other fish blood fluke are considered a major threat to sustainable fish production and have been associated with mortalities of farmed amberjacks (*Seriola* spp.) in the Spanish Mediterranean (Crespo et al. 1992) and Japan (Ogawa and Fukudome 1994) and in farmed tuna (*Thunnus* spp.) in Japan (Shirakashi et al. 2012a) and Australia (Hayward et al. 2010). Recent research indicates that orally delivered praziquantel may be the most effective treatment against blood fluke infections (Hardy-Smith et al. 2012, Shirakashi et al. 2012b).

6.6.1.3 Cestodes

A number of cestode species, or tape worms, infect the intestine, stomach, stomach wall, mesenteries and pyloric caeca of Asian seabass as plerocercoids; the last larval form found in the second intermediate host (Table 6.1, Fig. 6.5). Of these, only two species, *Scolex pleuronectis* and *Nybelinia indica* are known to infect farmed Asian seabass (Rückert et al. 2008, Table 6.1), the remaining have been recorded from wild fish. Cestode larvae are common in wild Asian seabass and farmed fish may become infected if fed fresh or frozen seafood, or by feeding opportunistically on wild species harboring infective stages. Plerocercoid cysts in the muscle tissue may have implications for marketability, but this can be minimised by managing trophic transmission in captive populations by feeding using extruded pellets (Ogawa 1996).

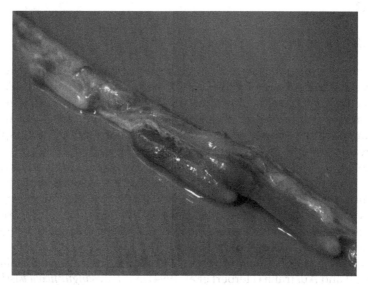

Figure 6.5. Cestode plerocercoids in the mesenteries alongside the intestine in wild Asian seabass. Photo: Kate Hutson.

6.6.1.4 Nematodes

Nematodes, or round worms, are long, slender and cylindrical un-segmented worms tapered at each extremity. A number of larval nematodes have been documented from wild and farmed Asian seabass (Table 6.1). Larval nematodes have to complete their life cycle in a specific bird, fish or mammal species, but their larval stages may be able to survive in a large variety of intermediate hosts. The impacts of nematodes on host fishes can generally be described as benign or in need of further study (McClelland 2005). However, members of Anisakidae, which have been reported from farmed Asian seabass, can present a potential human health issue if consumed in raw or undercooked seafood (Sabater 2000, Table 6.1). Nematodes generally exhibit complex life cycles and may be excluded from farmed fishes by maintaining clean feed sources, such as an extruded pellet diet.

6.6.1.5 Leeches

Leeches (Subclass Hirudinea) occur in freshwater, brackish and marine environments. Leeches that feed exclusively on the blood of fishes can also be encountered living freely since they may leave their host after feeding and not reattach to a new host until the last meal has been digested (Kearn 2004). Leeches attach to their host using anterior and posterior suckers. Fish leeches use their jaws to gain access to blood and may prevent clotting while feeding by injecting saliva that inhibits the host's clotting enzyme, thrombin. Most adult leeches detach from the host fish in order to lay cocoons on a chosen substrate, including aquaculture structures such as moorings and nets. Cocoons contain a ring shaped compartment that is effectively sealed from the environment which protects the developing embryo.

Heavy infestations of the leech, *Zeylanicobdella arugamensis*, have caused mortality in Asian seabass fingerlings reared in sea cages in Malaysia (Kua et al. 2010). Clinical symptoms include anemia, body discolouration, scale loss, frayed fins and restless swimming (Kua et al. 2010). Leeches can also serve as a vector for other parasites and pathogens, including bacteria (e.g., *Vibrio algniolyticus*, see Kua et al. 2010), viruses and protozoa (see Burreson 1995). Severe infestations render fish unmarketable due to unsightly leeches, frayed fins, haemorrhages and swelling at attachment and feeding sites (Cruz-Lacierda et al. 2000, Kua et al. 2010, Fig. 6.6). Cruz-Lacierda et al. (2000) found a 50 ppm formalin bath treatment effective for managing leeches and suggested drying culture facilities in order to desiccate leech cocoons.

6.6.2 Copepods

Several potentially harmful caligid copepod species (= sea-lice) are known from wild and farmed Asian seabass including *Caligus epidemicus*, *C. chiastos*, *C. orientalis*, *C. pagrosomi*, *C. rotundigenitalis* and *C. punctatus* (Table 6.1). Unlike most other parasitic copepods, *Caligus* adults are capable of swimming and can leave one host to become attached to another. Caligids present a predominant threat to aquaculture due to their broad distribution, direct life cycle and low host specificity. For example, *C. epidemicus* has been

Figure 6.6. Persistent leech epidemics can result in mortality. In this photo an infection is shown on the body surface and fins of a Goldspotted Rockcod (*Epinephelus coioides*). Photo: Richard Knuckey.

associated with mass mortality of mullet (Mugilidae) and porgies (Sparidae) in Australia (Hewitt 1971) and tilapia in Taiwan (Lin and Ho 1993). This species is known from sea caged Asian seabass in Malaysia (Venmathi Maran et al. 2009, Muhd-Faizul et al. 2012). *Caligus chiastos* is also known from sea-caged Asian seabass in Malaysia (Muhd-Faizul et al. 2012). This species has been implicated in epizootics in southern bluefin tuna sea cage culture and can cause gross corneal damage (Hayward et al. 2008).

The haematophagous copepod, *Lernanthropus latis* (Fig. 6.7), is of concern in brackish pond culture and sea cage culture of Asian seabass in Australia. Adult females attach to the primary gill filaments, while smaller males are found on the gills or attached to females (Brazenor and Hutson 2013). Although *L. latis* has not caused significant mortality in aquaculture, their presence is usually associated with poor fish health (Kuo and Humphrey 2008). Parasites cause irreparable damage to the gills by way of their mode of attachment and feeding activity (Small et al. 2002). Disease problems include haemorrhages, hyperplasia and necrosis along the secondary lamellae of gill filaments (Kua et al. 2012). There are no known treatments, although hydrogen peroxide bathing is currently being trialed in the Australian mariculture industry.

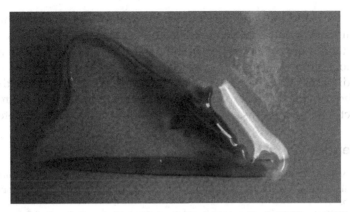

Figure 6.7. Adult female *Lernanthropus latis* attached to an excised gill filament. Photo: Kate Hutson.

6.6.3 Isopods

Cymothoids (Family Cymothoidae) are ectoparasitic isopods on marine, brackish and freshwater fish. In most cases, parasitic isopods are blood feeders and occur on the body, in the mouth or in the branchial cavity of the host fish. Five cymothoid isopod parasite species infect wild and/or farmed Asian seabass (Table 6.1). Infections of farmed hatchery seabass in the branchial and anterodorsal regions by *Cymothoa indica* resulted in skin lesions and were associated with lowered growth rates and mortality (Rajkumar et al. 2005). In this instance, the parasite was believed to have been introduced to hatchery-reared fish through wild zooplankton used as feed (Rajkumar et al. 2005). Consequently, infection could be reduced by filtering wild zooplankton to remove the infectious swimming larvae of *C. indica*, or by using alternative live feeds. *Nerocila barramundaei* and *Rocinela latis* are only known from wild fish and the associated pathology for these species is not known.

6.7 Summary

The biodiversity of pathogens and parasites that may infect Asian seabass and appropriate mechanisms to treat infections, is an area of continued research effort. Effective parasite management focuses on reducing stress, preventing introduction of pathogens and parasites, and use of effective drugs and vaccines (where available). Understanding of the source and transmission of infectious pathogens and parasites can enable proactive management strategies. Continued reliance on wild conspecifics for brood stock and/or stocking enables vertical and horizontal pathogen transmission. As our knowledge of Asian seabass disease increases and we understand more of the biology of the pathogens and the optimal rearing conditions, improved management strategies will be developed to help avoid diseases and increase productivity. Thus, it is crucial to further our knowledge of the diseases, pathogens and parasites that infect, or are likely to infect Asian seabass as a basis for developing management strategies.

References

Agnew, W. and A.C. Barnes. 2007. *Streptococcus iniae*: An aquatic pathogen of global veterinary significance and a challenging candidate for reliable vaccination. Vet. Microbiol. 122: 1–15.

Akram, M. 1994. Studies on host parasite relationships between marine food fishes of Sindh creeks and cucllanid nematode parasites (Cucullanidae: Nematoda) in Pakistan. Pakistan J. Zool. 26: 182–185.

Akram, M. 1995. Occurrence of nematode parasites in marine food fishes of Sindh Creeks in Pakistan. Biologia 41: 117–127.

Anderson, I.G. and H.C. Prior. 1992. Subclinical epitheliocystis in barramundi, *Lates calcarifer*, reared in sea cages. Aust. Vet. J. 69: 226–227.

Anonymous, 2007. NAAHTWG Slide of the Quarter–July–September 2007. Chilodenellosis or *Chilodenella* sp. infestation in barramundi (*Lates calcarifer*) 06-44312. Electronic publication http://www.fish.wa.gov.au/ (accessed January 2013).

Anonymous, 2011. Diseases of farmed barramundi in Asia. Electronic publication http://www.thefishsite.com/articles/1086/diseases-of-farmed-barramundi-in-asia (accessed April 2013).

Arthur, J.R. and A.T.A. Ahmed. 2002. Checklist of the parasites of fishes from Bangladesh. FAO Fisheries Technical Paper. No. 369/1. Rome 77 pp.

Arthur, J.R. and S. Lumanlan-Mayo. 1997. Checklist of the parasites of fishes of the Philippines. FAO Fisheries Technical Paper. Rome. 369: 1–102.

Arya, S.N. 1991. Synonymy of the genus *Oceanocucullanus* Schmidt and Kuntz, 1969 with the genus *Indocucullanus* Ali, 1956 with the proposal of two new subgenera. *Rivista di Parassitologia*. 7: 179–182.

Austin, B. 2010. Vibrios as causal agents of zoonoses. Vet. Microbiol. 140: 310–317.

Awang, A.B. 1987. Seabass (*Lates calcarifer*) larvae and fry production in Malaysia. In: J.W. Copland and D.L. Grey (eds.). Management of Wild and Cultured Seabass/Barramundi (*Lates calcarifer*). Australian Centre for International Agricultural Research, Canberra, pp. 144–147.

Azad, I.S., M.S. Shekhar, A.R. Thirunavukkarasu, M. Poornima, M. Kailasam, J.J.S. Rajan, S.A. Ali, M. Abraham and P. Ravichandran. 2005. Nodavirus infection causes mortalities in hatchery produced larvae of *Lates calcarifer*: first report from India. Dis. Aquat. Organ. 63: 113–118.

Azad, I.S., M.S. Shekhar, A.R. Thirunavukkarasi and K.P. Jithendran. 2006a. Viral nerve necrosis in hatchery-produced fry of Asian Seabass *Lates calcarifer*: sequential microscopic analysis of histopathology. Dis. Aquat. Organ. 73: 123–130.

Azad, I.S., K.P. Jithendran, M.S. Shekhar, A.R. Thirunavukkarasu and L.D. de la Pena. 2006b. Immunolocalisation of nerve necrosis virus indicates vertical transmission in hatchery produced Asian seabass (*Lates calcarifer* Bloch)—a case study. Aquaculture 255: 39–47.

Beveridge, I. and R.A. Campbell. 1989. *Chimaerarhynchus* n. g. and *Patellobothrium* n. g., two new genera of trypanorhynch cestodes with unique poeciloacanthous armatures, and a reorganisation of the poeciloacanthons trypanorhynch families. Syst. Parasitol. 14: 209–225.

Bilqees, F.M. 1995. Teaching and research in marine parasitology—problems and solutions. Proceedings of Parasitology. Karachi. *In:* D.I. Gibson, R.A. Bray and E.A. Harris (eds.). 2005. Host-Parasite Database of the Natural History Museum, London. 20: 63–89.

Bray, R.A., T.H. Cribb and S.C. Barker. 1993. Hemiuridae (Digenea) from marine fishes of the Great Barrier Reef, Queensland, Australia. Syst. Parasitol. 25: 37–62.

Brazenor, A.K. and K.S. Hutson. 2013. Effect of temperature and salinity on egg hatching and description of the life cycle of *Lernanthropus latis* (Copepoda: Lernanthropidae) infecting barramundi, *Lates calcarifer*. Parasitol. Int., in press.

Bromage, E.S. and L. Owens. 2002. Infection of barramundi *Lates calcarifer* with *Streptococcus iniae*: effects of different routes of exposure. Dis. Aquat. Organ. 52: 199–205.

Bromage, E.S., A. Thomas and L. Owens. 1999. *Streptococcus iniae*, a bacterial infection in barramundi *Lates calcarifer*. Dis. Aquat. Organ. 36: 177–181.

Bruce, N.L. 1987. Australian species of *Nerocila* Leach, 1818, and *Creniola* n. gen. (Isopoda: Cymothoidae), crustacean parasites of marine fishes. Rec. Aust. Mus. 39: 355–412.

Buchmann, K. and J. Bresciani. 2009. Parasites. *In*: K. Buchmann (ed.). Fish Diseases—An Introduction. Biofolia.

Burreson, E.M. 1995. Phylum Annelida: Hirudinea as vectors and disease agents. *In*: P.T.K. Woo (ed.). Fish Diseases and Disorders: Vol. 1. Protozoan and Metazoan Infections. CAB International, United Kingdom, pp. 599–629.

Campbell, R.A. and I. Beveridge. 1996. Revision of the Family Pterobothriidae Pintner, 1931 (Cestoda: Trypanorhyncha). Invertebr. Taxon. 10: 617–62.

Carson, J., L.M. Schmidtke and B.L. Munday. 1993. *Cytophaga johnsonae*: a putative skin pathogen of juvenile farmed barramundi, *Lates calcarifer* Bloch. J. Fish Dis. 16: 209–218.

Chang, S.F., G.H. Ngoh and S. Kueh. 1997. Detection of viral nervous necrosis nodavirus by reverse transcription polymerase chain reaction in locally farmed marine food fish. Singapore Vet. J. 21: 39–44.

Cheong, L., R. Chou, R. Sing and C.T. Mee. 1983. Fish Quarantine and Fish Diseases in Southeast Asia. IDRC, Ottawa 47 pp.

Chi, S.C., K.W. Lee and S.J. Hwang. 2001. Investigation of host range of fish nodavirus in Taiwan. Tenth International Conference of the European Association of Fish Pathologists, 9–14 September 2001, Dublin.

Crespo, S., A. Grau and F. Padros. 1992. Sanguinicoliasis in the cultured amberjack *Seriola dumerili* Risso, from the Spanish Mediterranean area. Bull. Eur. Assn. Fish. P. 12: 157–159.

Cribb, T.H., R.A. Bray and S.C. Barker. 1992. A review of the Family Transversotrematidae (Trematoda: Digenea) with the description of a new genus, *Crusziella*. Invertebr. Taxon. 6: 909–935.

Cribb, T.H., R.D. Adlard, C.J. Hayward, N.J. Bott, D. Ellis, D. Evans and B.F. Nowak. 2011. The life cycle of *Cardicola forsteri* (Trematoda: Aporocotylidae), a pathogen of ranched southern bluefin tuna, *Thunnus maccoyi*. Int. J. Parasitol. 41: 861–870.

Cruz-Lacierda, E.R., J.D. Toledo, J.D. Tan-Fermin and E.M. Burreson. 2000. Marine leech (*Zeylanicobdella arugamensis*) infestation in cultured orange-spotted grouper, *Epinephelus coioides*. Aquaculture 185: 191–196.

Deveney, M.R., L.A. Chisholm and I.D. Whittington. 2001. First published record of the pathogenic monogenean parasite *Neobenedenia melleni* (Capsalidae) from Australia. Dis. Aquat. Organ. 46: 79–82.

Ellis, E.P. and W.O. Watanabe. 1993. The effects of hyposalinity on eggs, juveniles and adults of the marine monogenean, *Neobenedenia melleni*. Treatment of ecto-parasitosis in seawater-cultured tilapia. Aquaculture 117: 15–27.

Evans, J., C. Shoemaker and P. Klesius. 2000. Experimental *Streptococcus iniae* infection of hybrid striped bass (*Morone chrysops* × *Morone saxatilis*) and tilapia (*Oreochromis niloticus*) by nares inoculation. Aquaculture 189: 197–210.

Fajer-Ávila, E.J., I. Martínez-Rodríguez, M.I. Abdo de la Parra, L. Álvarez-Lajonchere and M. Betancourt-Lozano. 2008. Effectiveness of freshwater treatment against *Lepeophtheirus simplex* (Copepoda: Caligidae) and *Neobenedenia* sp. (Monogenea: Capsalidae), skin parasites of bullseye puffer fish, *Sphoeroides annulatus* reared in tanks. Aquaculture 284: 277–280.

Frans, I., C.W. Michiels, P. Bossier, K.A. Willems, B. Lievens and H. Rediers. 2011. *Vibrio anguillarum* as a fish pathogen: virulence factors, diagnosis and prevention. J. Fish Dis. 34: 643–661.

Gabor, L.J., M. Srivastava, J. Titmarsh, M. Dennis, M. Gabor and M. Landos. 2011. Cryptosporidiosis in intensively reared Barramundi (*Lates Calcarifer*). J. Vet. Diagn. Invest. 23: 383–386.

Gibson, D.I., R.A. Bray and E.A. Harris (eds.). 2005. Host-Parasite Database of the Natural History Museum, London.

Gibson-Kueh, S., N.T.N. Thuyb, A. Elliot, J.B. Jones, P.K. Nicholls and R.C.A.Thompson. 2011a. An intestinal *Eimeria* infection in juvenile Asian seabass (*Lates calcarifer*) cultured in Vietnam—A first report. Vet. Parasitol. 181: 106–112.

Gibson-Kueh, S., R. Yanga, N.T.N. Thuy, J.B. Jones, P.K. Nicholls and U. Ryan. 2011b. The molecular characterization of an *Eimeria* and *Cryptosporidium* detected in Asian seabass (*Lates calcarifer*) cultured in Vietnam. Vet. Parasitol. 181: 91–96.

Glazebrook, J.S., M.P. Heasman and S.W. de Beer. 1990. Picorna-like viral particles associated with mass mortalities in larval barramundi, *Lates calcarifer* Bloch. J. Fish Dis. 13: 245–249.

Hardy-Smith, P., D. Ellis, J. Humphrey, M. Evans, D. Evans, K. Rough, V. Valdenegro and B.F. Nowak. 2012. In vitro and in vivo efficacy of anthelmintic compounds against blood fluke (*Cardicola forsteri*). Aquaculture 334: 39–44.

Hayward, C.J., H.M. Aiken and B.F. Nowak. 2008. An epizootic of *Caligus chiastos* on farmed southern bluefin tuna *Thunnus maccoyii* off South Australia. Dis. Aquat. Organ 79: 57–63.

Hayward, C.J., D. Ellis, D. Footec, R.J. Wilkinson, P.B.B. Crosbie, N.J. Bott and B.F. Nowak. 2010. Concurrent epizootic hyperinfections of sea lice (predominantly *Caligus chiastos*) and blood flukes (*Cardicola forsteri*) in ranched Southern Bluefin tuna. Vet. Parasitol. 173: 107–115.

Herbert, B.W. and F.M. Shaharom-Harrison. 1995. A new blood fluke, *Parasanguinicola vastispina* gen. nov., sp. nov. (Trematoda: Sanguinicolidae), from seabass *Lates calcarifer* (Centropomidae) cultured in Malaysia. Parasitol. Res. 81: 349–354.

Herbert, B.W., F.M. Shaharom-Harrison and R.M. Overstreet. 1994. Description of a new blood fluke, *Cruoricola lates* n. g., n. sp. (Digenea: Sanguinicolidae), from sea-bass *Lates calcarifer* (Bloch, 1790) (Centropomidae). Syst. Parasitol. 29: 51–60.

Hewitt, G.C. 1971. Two species of *Caligus* (Copepoda, Caligidae) from Australian waters, with a description of some developmental stages. Pac. Sci. 25: 145–164.

Hick, P., A. Tweedie and R. Whittington. 2010. Preparation of fish tissues for optimal detection of betanodavirus. Aquaculture 310: 20–26.

Hick, P., G. Schipp, J. Bosmans, J. Humphrey and R. Whittington. 2011. Recurrent outbreaks of viral nervous necrosis in intensively cultured barramundi (*Lates calcarifer*) due to horizontal transmission of betanodavirus and recommendations for disease control. Aquaculture 319: 41–52.

Ho, J. and I. Kim. 2004. Lernanthropid copepods (Siphonostomatoida) parasitic on fishes of the Gulf of Thailand. Syst. Parasitol. 58: 17–21.

Ho, J. and C. Lin. 2004. Sea lice of Taiwan. Sueichan Press, Taiwan. 388 pp.

Huang, H.T., J.P. Hsu, H.H. Hung and W.F. Chang. 1990. Streptococcal infection in maricultured seabass *Lates calcarifer*. J. Chinese Soc. Vet. Sci. 16: 171–180.

Humphrey, J.D. 1996. Disease. *In*: G. Schipp, J. Bosmans and J. Humphrey (eds.). Northern Territory Barramundi Farming Handbook. Department of Primary Industries, Fisheries and Mines, Darwin.

Humphrey, J.D. 2006. NAAHTWG Slide of the Quarter (October – December 2006)—Barramundi (*Lates calcarifer*) with lamellar epithelial hyperplasia associated with *Amyloodinium* sp. Department of Fisheries, Government of Western Australia. Electronic publication http://www.fish.wa.gov.au (accessed January 2013).

Humphrey, J.D. and M. Pearce. 2004. Epizootic Ulcerative Syndrome (Red-spot Disease). Fishnote No. 01. Northern Territory Government, Australia. 4 pp.

Humphrey, J.D., S. Benedict and L. Small. 2010. Streptococcosis, Trypanosomiasis, Vibriosis and Bacterial Gill Disease in sea-caged Barramundi at Port Hurd, Bathurst Island, July –August 2005. Fishery Report No. 98. Northern Territory Government, Australia.

Hutson, K.S., L. Mata, N. Paul and R. de Nys. 2012. Seaweed extracts as a natural control against the monogenean ectoparasite *Neobenedenia* sp. infecting farmed barramundi (*Lates calcarifer*). Int. J. Parasitol. 42: 1135–1141.

Johansen, R., T. Ranheim, M.K. Hansen, T. Taksdal and G.K. Totland. 2002. Pathological changes in the juvenile Atlantic halibut *Hippoglossus hippoglossus* persistently infected with nodavirus. Dis. Aquat. Organ. 50: 161–169.

Johnson, S.C., J.W. Treasurer, S. Bravo, K. Nagasawa and Z. Kabata. 2004. A review of the impact of parasitic copepods on marine aquaculture. Zool. Stud. 43: 229–243.

Kaneko II J.J., R. Yamada, J.A. Brock and R.M. Nakamura. 1988. Infection of tilapia, *Oreochromis mossambicus* (Trewavas), by a marine monogenean, *Neobenedenia melleni* (MacCallum, 1927) Yamaguti, 1963 in Kaneohe Bay, Hawaii, USA, and its treatment. J. Fish Dis. 11: 295–300.

Karvonen, A., P. Rintamaki, J. Jokela and E.T. Valtonen. 2010. Increasing water temperature and disease risks in aquatic systems: Climate change increases the risk of some, but not all, diseases. Int. J. Parasitol. 40: 1483–1488.

Kearn, G.C. 2004. Leeches, Lice and Lampreys. Springer, The Netherlands, 432 pp.

Krupesha Sharma, S.R., G. Rathore, D.K. Verma, N. Sadhu and K.K. Philipose. 2013. *Vibrio alginolyticus* infection in Asian Seabass (*Lates calcarifer*, Bloch) reared in open sea floating cages in India. Aquacult. Res. 44: 86–92.

Kua, B.C. 2008. The internal transcribed spacer (ribosomal DNA) of fish protozoan, *Cryptocaryon irritans*, isolated from cultured seabass, *Lates calcarifer*, in Penang waters. Asian Fish. Sci. 21: 285–292.

Kua, B.C., M.A. Azmi and N.K.A. Hamid. 2010. Life cycle of the marine leech (*Zeylanicobdella arugamensis*) isolated from seabass (*Lates calcarifer*) under laboratory conditions. Aquaculture 302: 153–157.

Kua, B.C., M.R. Noraziah and R.A. Nik. 2012. Infestation of gill copepod *Lernanthropus latis* (Copepoda: Lernanthropidae) and its effect on cage-cultured Asian seabass *Lates calcarifer*. Trop. Biomed. 29: 443–450.

Kumaran, S., B. Deivasigamani, K.M. Alagappan, M. Sakthivel and S. Guru Prasad. 2010. Isolation and characterization of *Pseudomonas* sp. KUMS3 from Asian seabass (Lates calcarifer) with fin rot. World J. of Microb. Biot. 26: 359–363.

Kuo, C. and J.D. Humphrey. 2008. Monitoring the Health of Prawns, Barramundi and Mud Crabs on Aquaculture Farms in the Northern Territory. Fishnote, Northern Territory Government, Darwin. 13 pp.

Leong, T.S. 1997. Control of parasites in cultured marine finfishes in Southeast Asia—an overview. Int. J. Parasitol. 27: 1177–1184.

Leong, T.S. and S.Y. Wong. 1986. Parasite fauna of seabass, *Lates calcarifer* Bloch, from Thailand and from floating cage culture in Penang, Malaysia. In: J.L. Maclean, L.B. Dizon and L.V. HosiUos (eds.). The First Asian Fisheries Forum, Manila, pp. 251–254.

Leong, T.S. and S.Y. Wong. 1990. Parasites of health and diseased juvenile grouper (*Epinephelus malabaricus* (Bloch and Schneider)) and Seabass (*Lates calcarifer* (Bloch)) in floating cages in Penang, Malaysia. Asian Fish. Sci. 3: 319–327.

Leong, T.S. and S.Y. Wong. 1992a. Parasites of marine finfishes cultured in ponds and cages in Indonesia. Symposium on Tropical Fish Health Management, Biotrop Special Publication No. 48: 119–124.

Leong, T.S. and S.Y. Wong. 1992b. The parasite fauna of cultured Seabass, *Lates calcarifer* Bloch from Kelantan and Penang, Malaysia. J. Biosci. 3: 27–29.

Lin, C.L. and J.S. Ho. 1993. Life history of *Caligus epidemicus* Hewitt parasitic on the tilapia (*Oreochromis mossambica*) cultured in brackish water. *In*: G.A. Boxshall and D. Defaye (eds.). Pathogens of Wild and Farmed Fish: Sea Lice. Ellis Horwood, London, pp. 5–15.

Maeno, Y., L.D. de la Pena and E.R. Cruz-Lacierda. 2004. Mass mortalities associated with viral nervous necrosis in hatchery-reared seabass *Lates calcarifer* in the Philippines. JARQ 38: 69.

Manin, B.O. and J. Ransangan. 2011. Experimental evidence of horizontal transmission of *Betanodavirus* in hatchery-produced Asian seabass, *Lates calcarifer* and brown-marbled grouper, *Epinephelus fuscoguttatus* fingerling. Aquaculture 321: 157–165.

McClelland, G. 2005. Nematoda (roundworms). *In*: Marine Parasitology K. Rohde (ed.). CABI Publishing, Australia, pp. 104–115.

Militz, T.A., P.C. Southgate, A.G. Carton and K.S. Hutson. 2013. Efficacy of garlic (*Allium sativum*) extract applied as a therapeutic immersion treatment for *Neobenedenia* sp. management in aquaculture. J. Fish. Dis., in press.

Moravec, F., P. Orecchia and L. Paggi. 1988. Three interesting nematodes from the fish *Parupeneus indicus* (Mullidae, Perciformes) of the Indian Ocean, including a new species, *Ascarophis parupenei* sp.n. (Habronematoidea). Folia Parasit. 35: 47–57.

Mueller, K.W., W.O. Watanabe and W.D. Head. 1992. Effect of salinity on hatching in *Neobenedenia melleni*, a monogenean ectoparasite of seawater-cultured tilapia. J. World. Aquacult. Soc. 23: 199–204.

Muhd-Faizul, H.A.H., B.C. Kuab and Y.Y. Leaw. 2012. Caligidae infestation in Asian seabass, *Lates calcarifer*, Bloch 1790 cultured at different salinity in Malaysia. Vet. Parasitol. 184: 68–72.

Munday, B.L., J. Kwang and N. Moody. 2002. Betanodavirus infections of teleost fish: a review. J. Fish Dis. 25: 127–142.

Mushiake, K., T. Nishizawa, T. Nakai, I. Furusawa and K. Muroga. 1994. Control of VNN in Striped Jack: Selection of spawners based on the detection of SJNNV gene by polymerase chain reaction (PCR). Fish Pathol. 29: 177–182.

Nawawi, R.A., J. Baiano and A.C. Barnes. 2008. Genetic variability amongst *Streptococcus iniae* isolates from Australia. J. Fish Dis. 31: 305–9.

Nowak, B.F. and S.E. LaPatra. 2006. Epitheliocystis in fish. J. Fish Dis. 29: 573–588.

Ogawa, K. 1996. Marine parasitology with special reference to Japanese fisheries and mariculture. Vet. Parasitol. 64: 95–105.

Ogawa, K. and M. Fukudome. 1994. Mass mortality caused by blood fluke (*Paradeontacylix*) among amberjack (*Seriola dumerili*) imported to Japan. Fish Pathol. 29: 265–269.

Pakingking, R. Jr., R. Seron, L.D. de la Pena, K. Mori, H. Yamashita and T. Nakai. 2009. Immune responses of Asian seabass (*Lates calcarifer*) against the inactivated betanodavirus vaccine. J. Fish. Dis. 32: 457–463.

Palm, H.W. 1995. Untersuchengen zur Systematik von Russelbandwurmen (Cestoda: Trypanorhyncha) aus atlantischen Fischen.

Palm, H.W., S.L Poynton and P. Rutledge. 1998. Surface ultrastructure of plerocercoids of *Bombycirhynchus sphyraenaicum* (Pintner 1930) (Cestoda: Trypanorhyncha). Parasitol. Res. 84: 195–204.

Parameswaran, V., S.R. Kumar, V.P.I. Ahmed and A.S.S. Hameed. 2008. A fish nodavirus associated with mass mortality in hatchery-reared Asian Seabass, *Lates calcarifer*. Aquaculture 275: 366–369.

Pearce, M. 1989. Epizootic ulcerative syndrome. Technical Report Dec. 1987–Sept 1989. Fisheries Report No. 22. Northern Territory Department of Primary Industry and Fisheries, Darwin? 82 pp.

Pearce, M., J.D. Humphrey, A.D. Hyatt and L.M. Williams. 1990. Lymphocystis disease in captive barramundi *Lates calcarifer*. Aust. Vet. J. 67: 144–145.

Perera, R., R. Fiske and S. Johnson. 1998. Histopathology of hybrid tilapias infected with a biotype of *Streptococcus iniae*. J. Aquat. Anim. Health 10: 294–299.

Pillai, N.K. 1985. The fauna of India. Copepod parasites of marine fishes. Zoological Society of India, Calcutta.

Rajkumar, M., P. Perumal and J.P. Trilles. 2005. *Cymothoa indica* (Crustacea, Isopoda, Cymothoidae) parasitizes the cultured larvae of the Asian seabass *Lates calcarifer* under laboratory conditions. Dis. Aquat. Organ. 66: 87–90.

Renault, T., P. Haffner, F. Baudin-Laurencin, G. Breuil and J.R. Bonami. 1991. Mass mortalities in hatchery-reared seabass (*Lates calcarifer*) larvae associated with the presence in the brain and retina of virus-like particles. Bull. Eur. Assn. Fish. P. 11: 68–73.

Rodgers, L.J. and J.B. Burke. 1988. Aetiology of red spot disease (vibriosis) with special reference to the ectoparasitic digenean *Prototronsversotrema steeri* (Angel) and the sea mullet, *Mugil cephalus* (Linnaeus). J. Fish Biol. 32: 655–63.

Rückert, S., H.W. Palm and S. Klimpel. 2008. Parasite fauna of seabass (*Lates calcarifer*) under mariculture conditions in Lampung Bay, Indonesia. J. Appl. Ichthyol. 24: 321–327.

Sabater, L. 2000. Health hazards related to occurrence of parasites of the genera *Anisakis* and *Pseudoterranova* in fish. Food Sci. Technol. Int. 6: 183–195.

Sanil, N.K., P.K. Asokan, J. Lijo and K.K. Vijayan. 2011. Pathological manifestations of the acanthocephalan parasite, *Tenuiproboscis* sp. in the mangrove red snapper (*Lutjanus argentimaculatus*) (Forsskål 1775), a candidate species for aquaculture from Southern India. Aquaculture 310: 259–266.

Saoud, M.F.A., F.M Nahhas, K.S.R. Al Kuwairi and M.M. Ramadan. 2002. Helminth parasites of fishes from the Arabian Gulf: 10. Trematodes of the genus *Stephanostomum* Looss, 1899 (Digenea: Acanthocolpidae Lühe 1901), with description of *Stephanostomum qatarense* n. sp., and redescription of *Stephanostomum triacanthi* Madhavi. Riv. Parassitol. 19: 87–103.

Shirakashi, S., Y. Kishimoto, R. Kinami, H. Katano, K. Ishimaru, O. Murata, N. Itoh and K. Ogawa. 2012a. Morphology and distribution of blood fluke eggs and associated pathology in the gills of cultured Pacific bluefin tuna, *Thunnus orientalis*. Parasitol. Int. 61: 242–249.

Shirakashi, S., M. Andrews, Y. Kishimoto, K. Ishimaru, T. Okada, Y. Sawada and K. Ogawa. 2012b. Oral treatment of praziquantel as an effective control measure against blood fluke infection in Pacific bluefin tuna (*Thunnus orientalis*). Aquaculture 326: 15–19.

Small, L., J. Humphrey and C. Shilton. 2002. Report on investigation into gill ectoparasites in sea-caged barramundi at Port Hurd. Internal Report. Northern Territory Government, Darwin? 15 pp.

Soltani, M., B.L. Munday and C.M. Burke. 1996. The relative susceptinility of fish to infections by Flexibacter columnaris and Flexibacter maritmus. Aquaculture 140: 259–264.

Southwell, T. 1915. Notes from the Bengal Fisheries Laboratory, Indian Museum. No. 2. On some Indian parasites of fish, with a note on carcinoma in trout. Records of the Indian Museum, Calcutta 11: 311–330, pls 26–28.

Srivastava, H.D. 1938. A new parasite of the family Acanthocolpidae Luhe, 1909, from an Indian host. Indian J. Vet. Sci. Anim. Husb. 8: 247–248.

Tendencia, E. 2002. *Vibrio harveyi* isolated from cage-cultured seabass *Lates calcarifer* Bloch in the Philippines. Aquacult. Res. 33: 455–458.

Thoney, D.A. and W.J. Hargis. 1991. Monogenea (Platyhelminthes) as hazards for fish in confinement. Annu. Rev. Fish Dis. 1991: 133–153.

Tingbao, Y., D.C. Kritsky, S. Yuan, Z. Jianying, S. Suhua and N. Agrawal. 2006. Diplectanids infesting the gills of the barramundi *Lates calcarifer* (Bloch) (Perciformes: Centropomidae), with the proposal of *Laticola* n. g. (Monogenoidea: Diplectanidae). Syst. Parasitol. 63: 127–141.

Tonguthai, K. and S. Chinabut. 1987. Fish Quarantine and Fish Diseases in South and South East Asia; 1986 Update. Asian Fisheries Society, Manila. 41 pp.

Tripathi, Y.R. 1957. Studies on the parasites of Indian fishes II. Monogenea, Family: Dactylogyridae. Indian J. Helminthol. 7: 5–24.

Tripathi, Y.R. 1962. Parasitic copepods from Indian fishes. III. Family Anthosomatidae and Dichelesthiidae. First All-India Congress on Zoology 1959: 191–217.

Venmathi Maran, B.A., T.S. Leong, S. Ohtsuka and K. Nagasawa. 2009. Records of *Caligus* (Crustacea: Copepoda: Caligidae) from marine fish cultured in floating cages in Malaysia with a redescription of the male of *Caligus longipedis* Bassett-Smith, 1898. Zool. Stud. 48: 797–807.

Vinoth, R., T.T. Ajith Kumar, S. Ravichandran, M. Gopi and G. Rameshkumar. 2010. Infestation of copepod parasites in food fishes of Vellar Estuary, Southeast Coast of India. Acta Parasitol. Globalis 1: 1–5.

Zafran, T. Harada, I. Koesharyani, K. Yuasa and K. Hatai. 1998. Indonesian hatchery reared seabass larvae (*Lates calcarifer*) associated with viral nervous necrosis (VNN). Indonesian Fish. Res. J. 4: 19–22.

7

The Genetics of Asian Seabass

Dean R. Jerry* and Carolyn Smith-Keune

7.1 Introduction

From a geneticist's perspective, *L. calcarifer* is an interesting fish species to study due to its euryhaline, protandrous life-strategy, broad tropical geographic distribution, mass-spawning reproductive behaviour and economic importance to communities denizen to the area it naturally inhabits. The species is commercially exploited, both via wild fishery captures and through industrialized farming, and is also becoming a model tropical fish species for understanding the impacts of future climate change. As a result, various aspects of the genetics of this species have received considerable attention, both towards using genetic approaches to understand population structure, diversity and genetic connectivity for fishery management purposes, through to deciphering the genome and mapping of genes influencing traits of commercial importance to assist future aquaculture breeding programs.

This chapter summarizes the current state of genetic knowledge of *L. calcarifer*.

Centre for Sustainable Tropical Fisheries and Aquaculture, School Marine and Tropical Biology, James Cook University, Townsville, Queensland, Australia.
Email: Dean.Jerry@jcu.edu.au
*Corresponding author

7.2 Karyotype

Lates calcarifer cells contain a compact genome with a haploid C-value of ~700 Mb (Animal Genome Size Database—www.genomesize.com) packaged into 48 chromosomes (Table 7.1, Fig. 7.1). To date there have only been two studies reporting the chromosome number in *L. calcarifer*, one based on Australian populations and the other from Indian seabass (Sudhesh et al. 1992, Carey and Mather 1999). Interestingly, whilst both populations have 48 chromosomes, chromosome formulae differs between Australian and Indian *L. calcarifer*, with Australian fish having more subtelocentric (3 vs. 1) and less submetacentric chromosomal subtypes (1 vs. 3) compared to Indian specimens. These differences in chromosome structure, when viewed with the recent description of two new species of *Lates* (Pethiyagoda and Gill 2012, Chapter 1), are suggestive of deep genetic differentiation between these two populations and may even be reflective of possible further cryptic

Table 7.1. Chromosome formulae of Australian and Indian *Lates calcarifer*.

Population	Telocentric	Subtelocentric	Submetacentric	Metacentric	Reference
Australia	19	3	1	1	(Carey and Mather 1999)
India	19	1	3	1	(Sudhesh et al. 1992)

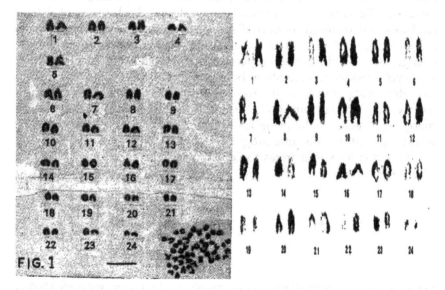

Figure 7.1. Metaphase and karyotype of *Lates calcarifer* from Cochin, India (left) (bar = 10 μm) (Sudheshet al. 1992) and Mary River, Queensland (Carey and Mather 1999).

speciation (Carey and Mather 1999). Distinctive sex chromosomes have not been identified to date in *L. calcarifer* and may not be present given this species' protandric hermaphroditism.

7.3 Population Genetics

The Asian seabass was until recently considered to comprise a single species with a distribution encompassing south-east Asia and the Indo-west Pacific, though, it is now recognized that multiple sister taxa are present within this broader distribution. As Chapter 1 has highlighted, a number of what were formally considered *L. calcarifer* populations, including those in Myanmar and Sri Lanka, have now been described as congeneric species based on mitochondrial DNA sequences (Ward et al. 2008) and morphological characters (Pethiyagoda and Gill 2012). Fortunately (or unfortunately), all population genetic studies conducted on *L. calcarifer* have so far been restricted to the part of the species' distribution east of Myanmar and thus these studies are likely to reflect the genetic structure of *L. calcarifer* as we currently know this species. However, wider inferences on the stock structure of the species cannot be drawn outside these few studies, as no genetic studies to date have examined wild populations from the north-eastern extreme of *L. calcarifer*'s range (i.e., eastern Indonesia, Philippines, Taiwan) and it is possible further extreme genetic divergence, or even cryptic speciation, may be present.

The below section summarizes current knowledge of broad-scale Asian seabass stock structure across south-east Asia and Australasia (Australia and Papua New Guinea, PNG), as well as population structure within each of these regions.

7.3.1 Regional population genetic structure of Lates calcarifer

Several population genetic analyses across varying spatial scales have been conducted for Asian seabass since the development of molecular methods as a tool to determine fishery stock structure. A complete summary of population genetic studies for *L. calcarifer* is provided in Table 7.2. The type of molecular markers and statistical analyses applied has varied widely and has evolved as new molecular markers have become available. Despite differences in the methods applied what has emerged from these studies is that *L. calcarifer* exhibits strong regional stock structure at various spatial scales. In particular, significant genetic divergence between south-east Asian and Australian/Papuan New Guinean populations is evident. For example, DNA sequencing and phylogenetic analysis of a hypervariable portion of mitochondrial DNA (the mtDNA control region) revealed 18 fixed base pair differences between Singaporean and Australian *L. calcarifer*

Table 7.2. Summary of population genetic studies of Asian sea bass (*Lates calcarifer*) in the Asia/Pacific region.

Study	Sample size (N)	No. of locations	India	SE Asia	Indonesia	PNG	Australia-West	Australia-North	Australia-East	Type of marker	Number of loci Examined[1]	Heterogeneity χ^2	Contingency G-tests	AMOVA	Pairwise F_{ST}	Bayesian (STRUCTURE)	Isolation by distance	Phylogenetic[2]	Other
Shaklee and Salini 1985	589	3							•	Allozymes	11	✓							
Salini and Shaklee 1988	595	8							•	allozymes	12	✓							
Shaklee et al. 1993	2912	26				•	•	•	•	allozymes	11	✓							
Keenan 1994	6000	44					•	•	•	allozymes	13		✓						✓
Chenoweth et al. 1998a	270	9						•	•	mtDNA CR	290bp			✓	✓		✓	NJ	
Chenoweth et al. 1998b	270	9						•	•	mtDNA CR	290bp			✓				ML	✓
Doupe et al. 1999	43	4					•			mtDNA CR	290bp			✓				FM	
Marshall 2005	284	12				•	•	•	•	mtDNA CR	290bp			✓			✓	NJ	✓
Marshall 2005	284	12				•	•	•	•	mtDNAcytb	340bp			✓	✓			NJ	✓
Marshall 2005	91	6						•	•	microsatellites	5			✓	✓				✓
Lin et al. 2006	25	2		•						mtDNA CR	767bp							NJ	
Norfatimah et al. 2009	132	5		•						mtDNAcytb	312bp			✓	✓			NJ	
Yue et al. 2009	772	9		•						microsatellites	14			✓	✓				
Yue et al. 2012	549	4		•						microsatellites	18			✓	✓	✓			
Smith-Keune et al. in prep	595	23						•	•	mtDNA	251bp			✓	✓	✓	✓	NJ	✓
Smith-Keune et al. in prep	588	12	•							microsatellites	7				✓	✓	✓		✓
Smith-Keune et al. in prep	1205	43						•	•	microsatellites	16			✓	✓	✓	✓		✓

1: bp base pairs of sequence examined is given for mtDNA markers. 2: NJ neighbor joining; ML maximum likelihood; FM Fitch-Margolisash mitochondrial haplotypes which clustered into distinct genetic clades (Lin et al. 2006). A broader study of nuclear DNA markers (14 microsatellite loci) utilizing a combination of wild caught and farmed fish from Singapore, Australia, Malaysia, Thailand, Indonesia and Taiwan, further supported the distinction of Asian and Australian stocks, with an overall F_{ST} among populations of 0.124. The greatest level of genetic divergence was between cultured Singaporean and Australian fish. This study also suggested that Australian stocks may exhibit lower levels of genetic diversity than Asian stocks (Yue et al. 2009). However, farmed stocks of fish usually exhibit lower/disparate levels of genetic diversity than wild populations so the use of only farmed stocks from Australia and their comparison to wild stocks in south-east Asia provides no indication of the true genetic diversity of wild Australian stocks. Therefore this study left some question on whether mean levels of divergence detected between regions reflected actual levels of divergence among wild seabass stocks, or whether the high level of genetic divergence between Asian and Australian stocks was influenced by the use of farmed fish with low genetic diversity.

Recently, the chapter authors conducted a broad-scale genetic analysis based purely on wild *L. calcarifer* samples to obtain a more accurate picture of stock structure in south-east Asian and Australian seabass. Seven microsatellite loci used by Yue et al. (2009) were screened against wild caught *L. calcarifer* samples from India, central Indonesia (Sulawesi and Kalimantan), Papua New Guinea and Australia (unpublished data). This data was then incorporated with genetic data from only those wild populations sampled by Yue et al. (2009). This meta-analysis showed that across *L. calcarifer*'s distribution it indeed exhibits strong levels of genetic divergence, with an overall F_{ST} of 0.162 evident among populations. Both regional and intra-regional stock structure was also detected, with modest pairwise F_{ST} values between populations within Asia ranging from 0.014–0.053 and between populations within Australia F_{ST} = 0.010–0.072. In confirmation of Yue et al.'s (2009) study, high levels of genetic divergence were detected between south-east Asian wild stocks and those in Australia/Papua New Guinea, with pairwise F_{ST} amongst these broader regional stocks ranging from 0.130–0.356. Bayesian clustering of the resultant multilocus genotypes implemented in the software STRUCTURE (v2.3.1) supports the existence of at least four ($K = 4$) different regional stocks of *L. calcarifer* in the Indo-West Pacific region, with clear differentiation between Australian/Papua New Guinean and south-east Asian stocks (Fig. 7.2). A further distinction is also evident between *L. calcarifer* in both of these regions and those sampled from the central Indonesian islands of Sulawesi and Kalimantan, suggesting at least one additional distinctive genetic stock occurs in this region (Fig. 7.2). A high level of admixture is evident between *L. calcarifer* stocks in India and those in Thailand, Malaysia, and to a lesser

142 Biology and Culture of Asian Seabass

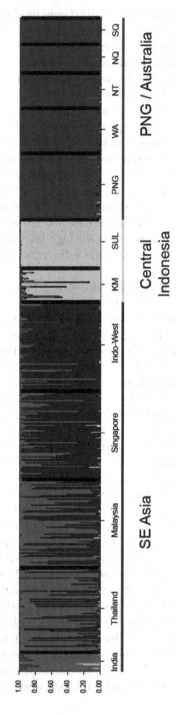

Figure 7.2. STRUCTURE ancestry membership plot for 558 individual Asian seabass sampled from India, SE Asia, the Indo-Pacific and Australasian regions and genotyped for 7 hypervariable microsatellite markers. The plot was generated using sampling locations (indicated on x-axis) as population priors and assuming a correlated allele frequency and admixture model. Some sampling location names are abbreviated (KM = Kalimantan, SUL = Sulawesi, PNG = Papua New Guinea, WA = Western Australia, NT = Northern Territory (Australia), NQ = northeast Queensland, SQ = southeast Queensland). Ten independent runs of each value of K from 1–10 were run, the most likely K (4) following the methods of Evanno et al. (2005) was selected and the proportion of ancestry to each of the four clusters is indicated by different colors.

Color image of this figure appears in the color plate section at the end of the book.

extent Singapore. This admixture is suggested by the occurrence in these later populations of numerous fish with high membership coefficients to the genetic cluster which is dominant in India. Given the possibility that populations in India may contain cryptic species (Pethiyagoda and Gill 2012, see Chapter 1) further molecular work is required to determine the precise taxonomy of the Indian individuals sampled and to explore the deeper phylogenetic relationships of populations in this region.

To further explore the genetic relationships between Indo-West Pacific *L. calcarifer* a phylogenetic analysis utilizing newly isolated and published mtDNA control region sequences has also recently been completed. Published sequences of a 258 base pair fragment from Australian (Chenoweth et al. 1998a, Marshall 2005) and Singaporean (Lin et al. 2006) *L. calcarifer*, along with newly isolated sequences from Sulawesi, Kalimantan and Vietnam, were analysed (Smith-Keune and Jerry unpublished data). A total of 213 haplotypes were identified from 595 individuals sequenced. Phylogenetic analyses again confirmed the deep genetic divergence between Asian and Australian/Papua New Guinean populations, with two well supported and geographically restricted clades resolved (Fig. 7.3). Further regional phylogenetic patterns of divergence with south-east Asia are also evident, with haplotypes originating from Singapore and Vietnam forming one phylogenetic branch and those from central Indonesia (Bali, Sulawesi and Kalimantan) forming a second distinct branch within the overall Asian clade (Fig. 7.3). Australian fish represented a third clade comprised of three sub-branches, however, haplotypic sorting based purely on geographical location of sampling was not strong although haplotypes from the Pilbara region in south-western Australia largely clustered together on one sub-branch.

7.3.2 Intra-regional genetic variability

Consistent with the above analysis, wild populations of Asian seabass have also been documented to exhibit significant stock structure at an intra-regional level, both within south-east Asia (section 7.3.2.1) and more extensively in Australia (section 7.3.2.2) (see also Table 7.2 and references therein).

7.3.2.1 Asian Lates calcarifer populations

Significant, but modest, genetic divergence was found in an examination of four wild populations within south-east Asia, including samples from Thailand, Malaysia, Singapore and western Indonesia (Yue et al. 2009). Based on 14 microsatellite loci, pairwise F_{ST} ranged from 0.012–0.055 among these stocks, with the highest divergence found between Indonesian and

144 *Biology and Culture of Asian Seabass*

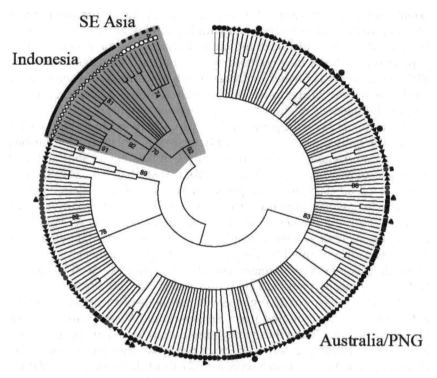

Figure 7.3. Neighbor-Joining bootstrap consensus tree inferred from 1000 replicates conducted in MEGA5. Branches corresponding to partitions reproduced in less than 50% bootstrap replicates are collapsed. The evolutionary distances were computed using the Tamura 3-parameter method. The analysis involved 213 haplotype sequences from 595 individuals sampled across major regions of SE Asia (including Singapore ○ and Vietnam □), Indonesia (including Kalimantan ◊, Sulawesi ∇ and Bali ∆), Australia (including the Pilbara ● and Kimberley Regions ■ of Western Australia, the Arafura Sea ▼, the Gulf of Carpentaria ▲ and the East coast of Queensland ♦) as well as Papua New Guinea (●). Haplotypes shared among locations are indicated by multiple markers. All positions containing gaps and missing data were eliminated. There were a total of 251 positions in the final dataset.

Malaysian seabass (Yue et al. 2009). A later comparison between these same populations using an additional four microsatellite loci found similarly modest levels of genetic differentiation (Yue et al. 2012). These significant, but low, pairwise F_{ST} estimates indicated moderate to high levels of gene flow, possibly via marine dispersal of larger female fish (Yue et al. 2012).

Interestingly, strong divergence (overall F_{ST} = 0.3557) was also observed among nine Malaysian populations of Asian seabass by DNA sequencing of a fragment of the mitochondrial cytochrome b locus (Norfatimah et al. 2009). In this case each population had its own unique haplotypes differing only by a few bases, but included in the study were wild, wild captive and pure cultured populations and levels of divergence among only the pure wild stocks was not reported in the paper. The original source location of the wild caught, but captive, populations also went unreported and there was uncertainty as to the precise origin of cultured fish. Examination of the neighbor joining tree shows evidence of geographic sorting of haplotypes in populations east and west of the Malaysian Peninsular, however, the true level of genetic structuring is difficult to interpret due to the inclusion of cultured stock in their analysis. Further detailed surveys are needed in this area to better elucidate the local and regional population structure, especially as Norfatimah et al. (2009) suggest that genetic pollution from cultured stocks may already be affecting wild population structure.

7.3.2.2 Australian Lates calcarifer populations

In contrary to Asian populations, stock structure in Australian seabass has been more extensively examined, with studies finding varying, but significant levels of population divergence over a wide variety of spatial scales. Allozyme electrophoresis was at first utilized to identify genetically differentiated stocks (putative management units) within the Australian *L. calcarifer* fishery (Table 7.2). These studies indicated the presence of distinct stocks at broad spatial scales and the important management implications of these findings led to several subsequent allozyme studies in different fisheries management areas (Shaklee and Salini 1985) (Table 7.2). Initially, seven distinct stocks of *L. calcarifer* were described from a sample of eight locations from northern Australia (Salini and Shaklee 1988), while an additional seven distinct stocks were later identified in Queensland (Shaklee et al. 1993). The most comprehensive of these early allozyme studies compared 50 separate collections of seabass from the Ord River on the north-west coast of Australia, northern Australia, the Gulf of Carpentaria and the species distributional range within Queensland (Keenan 1994). This study supported the presence of distinct genetic stocks maintained in the face of substantial gene flow (F_{ST} = 0.064, Keenan 1994). Sixteen distinct genetic stocks were recognized. The pattern of genetic differentiation among identified *L. calcarifer* stocks conformed to expectations under an isolation-by-distance model of gene flow, whereby increasing genetic differentiation was observed with increasing geographic separation between populations (Keenan 1994). The 16 stocks identified by Keenan (1994) comprised related stocks within three broader genetic stocks; a reduced genetic diversity eastern Queensland stock, a northern to northwestern genetic stock, and a central stock exhibiting relatively higher genetic diversity within the Gulf of Carpentaria.

As molecular methods of assessing DNA sequence variation became more readily available and interest in the phylogeographic relationships among different stocks grew, researchers turned to mitochondrial DNA (mtDNA) for a complementary view of Australian *L. calcarifer* population structure. Maternally inherited mtDNA markers are more sensitive to population bottlenecks and genetic drift than nuclear markers such as allozymes and, as a result, greater levels of population divergence (as indicated by F_{ST} or related distance measures) are often found with mtDNA. An analysis of mtDNA control region haplotypes at nine locations spread among the north-northwestern, central and eastern groups previously identified by allozymes indicated a mean Φ_{ST} of 0.328 (Table 7.2). A pattern of isolation-by-distance was again supported, as were regional differences among population groups (Chenoweth et al. 1998a). A vicariance event involving the repeated closure of the Torres Strait during Pleistocene sea level changes was invoked as the primary cause of an apparently non-random distribution of mtDNA clades either side of the Torres Strait (Chenoweth et al. 1998b). Subsequent mtDNA studies of *L. calcarifer* in Australia extended the sampling of stocks into the previously neglected western parts of the species' distribution (Doupe et al. 1999, Marshall 2005), although this was performed with only minimal sampling in the east. Restricted gene flow between the more western of the populations sampled in the Fitzroy and Ord Rivers of Western Australia and that immediately to the north in Darwin in the Northern Territory raised the possibility of an additional vicariance event in the western part of the Australian species distribution (Doupe et al. 1999). This was supported by a comprehensive examination of additional *L. calcarifer* stocks including samples from the Pilbara region in the southerly part of the species' Western Australian distribution (Marshall 2005). The presence of a strong biogeographic break coincident with the Great Sandy Desert was indicated and the Asian seabass populations in the Pilbara region were described as comprising an Evolutionarily Significant Unit (ESU) (Marshall 2005). Interestingly, the influence of the Torres Strait as a biogeographic barrier influencing population structure in this species in Australia is debated by Marshall (2005).

In an attempt to bring some consistency to the assessment of *L. calcarifer* stock structure in Australia we recently initiated a large-scale microsatellite analysis utilizing 16 microsatellite loci and covering the extent of the species' distribution from Western Australia across the northern tropics and down to the southern extremities in eastern Australia (Smith-Keune et al. unpublished data). Using an arbitrary cut off of 1% divergence between adjacent samples a total of 21 different stocks have thus far been identified among the 43 locations sampled with modest divergence overall indicated by a low $F_{ST} = 0.077$ (Smith-Keune et al. unpublished data). An initial analysis of the current dataset using Bayesian clustering of individual multilocus genotypes supports Chenoweth et al.'s (1998b) mtDNA data and Keenan's (1994) earlier allozyme study in highlighting the influence of the Torres Strait as an important historical biogeographic barrier. High membership coefficients to alternative genetic clusters are found in individuals sampled either side of this region and a clear pattern of reciprocal admixture within the Gulf of Carpentaria also supports contemporary gene flow across the Torres Strait (Level 1 Fig. 7.4). Further geographical substructure within regions either side of the Torres Strait, including between the most south-western and more northern or north-western samples, is also clearly evident in a hierarchical analysis supporting the existence of multiple genetic stocks of Asian seabass throughout Australia (Level 2 and 3 Fig. 7.4).

7.3.2.3 Application of population genetics to fisheries management

The plethora of genetic studies of stock structure among Asian seabass populations in Australian has largely been driven by intense discussions surrounding the often contentious issue of stock translocation for fisheries enhancement and aquaculture activities (Shaklee et al. 1993, Cross 2000, Keenan 2000, Marshall 2005). The rapid expansion of aquaculture activities through-out south-east Asia and the mounting evidence for strong regional stock structure make it imperative that more detailed studies of stock structure and genetic diversity of *L. calcarifer* occur through-out the species' natural range. This is especially true given signs of declining recruitment to wild fisheries in some areas and the possible ramifications of genetic pollution due to importation of foreign stocks for aquaculture (Norfatimah et al. 2009). Population genetic information will be critical for evaluating the risk to wild stocks of large scale release or on-going (prolonged) smaller scale releases from aquaculture ventures and will be especially useful if non-native stocks are being cultured or released. Detailed knowledge of the genetic distinctions between both wild and cultured stocks is also valuable for seafood product traceability (Lin et al. 2006, Jerry and Smith-Keune 2010) and will be important for detection of aquaculture escapees (Noble 2012). The development of cost-effective multiplexed assays for highly polymorphic microsatellite loci (see for example Zhu et al. 2010) means that consistent marker suites can now easily be applied across the species range for better informed stock enhancement of wild populations and for monitoring and risk assessment of aquaculture programs (see also Table 7.7).

7.4 Hatchery and Aquaculture Populations

The maintenance of genetic diversity to ensure the future vitality and capacity to make genetic gains through selection is essential if aquaculture is going to expand and grow. However, despite the best intentions of

148 Biology and Culture of Asian Seabass

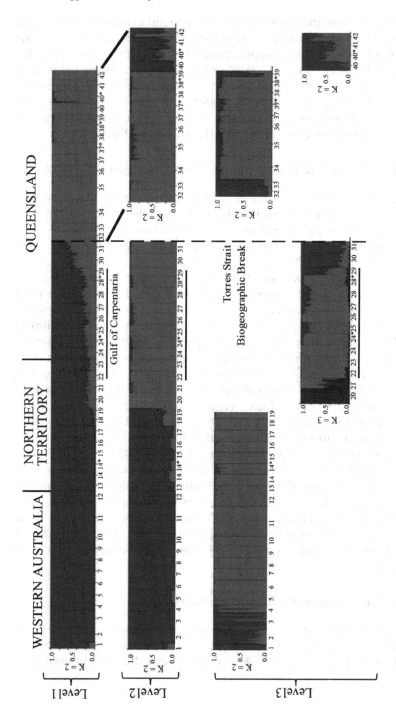

Figure 7.4. Hierarchical STRUCTURE ancestry membership bar plot for 1205 individual Asian seabass sampled from 43 locations around Australia and genotyped for 16 microsatellite loci. The plots were generated using sampling locations as population priors and assuming a correlated allele frequency and admixture model. Ten independent runs of each value of K from 1–10 were performed, the most likely K (number of clusters) for each level of the hierarchical analysis was determined by the methods of Evanno et al. (2005) and is indicated on the Y-axis. The proportion of ancestry to each of the identified clusters is indicated for each individual fish by different colors within each bar. The 43 sampling locations are indicated by differing numbers along the x-axis and are displayed from left to right in geographical order from Broome in Western Australia along the northern and eastern coasts of Australia to the Mary River in south-east Queensland. Five populations (14, 24, 28, 37, 38 and 40) have been sampled twice at different times (up to 20 years apart) and the earliest temporal sample is indicated by *. State and Territory boundaries (corresponding to different fisheries management jurisdictions) are indicated above the level 1 plot, while populations within the Gulf of Carpentaria are joined by a solid bar underneath each applicable plot. The most likely position of the Torres Strait biogeographic break is indicated by the vertical dashed line. Population codes are as follows. 1 Broome, 2 St George Basin, 3 Admiralty Gulf, 4 Swift Bay, 5 Drysdale River, 6 Salmon Bay, 7 King George, 8 Berkley River, 9 Helby River, 10 Nulla Nulla Creek, 11 Ord River, 12 Bonaparte Gulf, 13 Moyle River, 14 Daly River – 2008, 14* Daly River - 1990, 15 Bathurst Island, 16 Darwin Harbour, 17 Shoal Bay, 18 Mary River, 19 Alligator River, 20 Liverpool River, 21 Arnhem Bay, 22 Roper River, 23 McArthur river, 24 Albert River – 2011, 24* Albert/Leichhardt River – 1990/91, 25 Gilbert River, 26 Mitchell River, 27 Holroyd River, 28 Archer River – 2011, 28* Archer River – 1993, 29 Jardine River, 30 Jacky Jacky Creek - Kennedy Inlet, 31 Escape River, 32 Princess Charlotte Bay, 33 Bizant River, 34 Johnstone River, 35 Hinchinbrook Channel, 36 Cleveland Bay, 37 Bowling Green Bay – 2008, 37* Bowling Green Bay – 1988, 33 Burdekin River – 2008, 38* Burdekin River – 1989/90, 39 Broad Sound, 40 Fitzroy River – 2008, 40* Fitzroy River – 1988/90, 41 Port Alma, 42 Mary River.

Color image of this figure appears in the color plate section at the end of the book.

farmers, many aquaculture practices are detrimental to the long-term conservation of genetic diversity and levels of diversity initially present in populations can be rapidly eroded. Small broodstock effective population sizes, differential broodstock contribution, differential larval/juvenile survival during metamorphosis, size-based grading, and targeted selection all have the potential to drastically reduce the level of genetic variation remaining in hatchery populations (Frost et al. 2006). Monitoring levels of genetic variation and maintaining detailed pedigrees on progeny is the key to circumventing problems with the loss of diversity in aquaculture breeding programs. However, monitoring genetic diversity is very difficult to achieve for species like *L. calcarifer*, where progeny are usually produced through mass spawns involving several broodstock, with resultant larvae being very small and fragile, and where physical determination of pedigree is impractical. Fortunately for this species there are large numbers of microsatellites available that are suitable for DNA parentage assignment and the pedigree of individual progeny can be determined retrospectively (Yue et al. 2002, Frost et al. 2006, Zhu et al. 2006, 2009, 2010). It is these DNA based parentage markers that have been applied to understand what is happening in regard to genetic diversity within hatchery Asian seabass populations.

The effects of hatchery processes and early rearing of larvae on the retention of genetic diversity in *L. calcarifer* has been examined in several studies. Frost et al. (2006) applied DNA parentage techniques to track the loss of genetic diversity in two independent commercial hatcheries over three mass spawning events, where up to two females and seven males had the opportunity to participate in spawning (Fig. 7.5). Here, wide variance in the number of males and females genetically contributing to the resultant progeny cohort was first observed. Many broodstock in the tank were found to not have contributed to the spawn, resulting in estimated effective breeding sizes half that of the hormonally induced census size of fish in the tank. Of equal importance, Frost et al. (2006) also found large skews in the representation of progeny derived from those broodstock that had engaged in mating. Frost et al. (2006) further tracked genetic contribution of broodstock from that initially represented in spawns through to metamorphosis of juveniles (27 days post-hatch) and after the first size grading. Parentage analyses of these subsequent samples showed that an additional loss of genetic diversity had occurred, whereby some families that had been initially well represented in cohorts had decreased in their contribution, some even to the point where they were undetectable. Size grading also impacted on genetic diversity, with data suggesting that family representation in each of the size grades was non-uniform and that some families were on average growing faster, or slower, than others. If this differential growth pattern and loss of family diversity

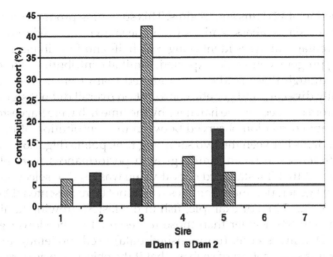

Figure 7.5. Contribution of sire broodstock to larvae in an Asian seabass mass spawning. Contributions to the cohort were significantly differentiated between dams ($\chi^2_{3.841\,d.f.\,1}$ = 10.348, p<0.05), between sires mating with Dam 1 ($\chi^2_{5.991\,d.f.\,2}$ =11.841, p<0.01), between sires mating with Dam 2 ($\chi^2_{7.815\,d.f.\,3}$ = 91.826, p<0.001) and between sires overall regardless of mating pair ($\chi^2_{9.488\,d.f.\,4}$ = 103.667, p<0.001). The effective population size (N_e) for this cohort was 4.5 (Frost et al. 2006).

continued through to harvest this would pose significant challenges to future selection programs, as it may mean that size grading results in the removal of appreciable amounts of family-specific genetic diversity. Wang et al. (2008b) further highlighted the capricious nature of *L. calcarifer* spawns in commercial situations, where they found in a spawn involving 10 males and 10 females that 19 of the 20 parents genetically contributed (although individual broodstock contribution was highly skewed again), while in a second spawn with the same broodstock only five parents contributed.

Whilst the two studies highlighted above tracked the genetic contribution of broodstock early into production, two pieces of information essential to the design and conduct of selective breeding programs have until recently been lacking. These are; a) an understanding of when the apparent observed erosion of genetic diversity in a cohort stabilises (if at all) and, more importantly, b) how this genetic diversity is finally represented at harvest where differential growth of families occurs. Without this understanding, and under the limitation of having to apply DNA parentage techniques, it is impossible to know how many fish from the extremes of the distribution would need to be selected and held to ensure that sufficient familial genetic diversity is captured each generation. To address these gaps, Loughnan et al. (2013) and Domingos et al. (2013, unpublished data) used DNA parentage analyses to track progeny within a large mass spawn involving the induction of 12 dams and 21 sires. This spawning produced the highest number of Asian seabass families from one spawn reported to date, with DNA parentage analyses identifying > 66 half- and full-sib families in the resulting progeny cohort. As expected, familial contributions were highly skewed, though these studies contradicted that of Frost et al. (2006) in that family diversity and genetic contribution overall did not significantly change from that seen in the hatchery by the time fish were harvested, with a significant correlation observed between the contribution of individual sires and dams between the two sampling time points (Fig. 7.6). However, large differences in mean family growth performances were observed which meant that if a standard cut off value was used to select brood fish for the next generation many families would not be represented (Domingos et al. 2013, Fig. 7.7). For example, fish in the fastest growing family were on average 58.8% heavier than those represented in the slower growing families. Domingos et al. (unpublished data) used modelling to further show for this same cohort of seabass that if the objective was to retain 70% of the best families in a population comprising 66 families than up to 750 fish from the extreme of the distribution may need to be genotyped to capture this level of familial genetic diversity (Fig. 7.8). To capture 100% of familial diversity upwards of 5000 fish would have to be genotyped.

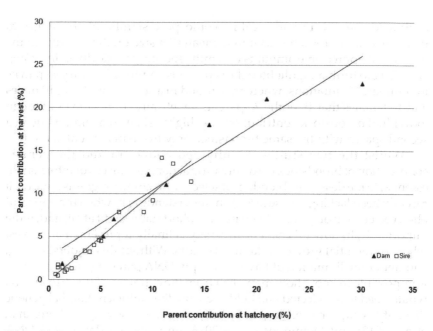

Figure 7.6. Percentage contribution of individual Asian seabass dams and sires at day 19 and then again at harvest (day 273) to a progeny cohort (Domingos et al. unpublished data).

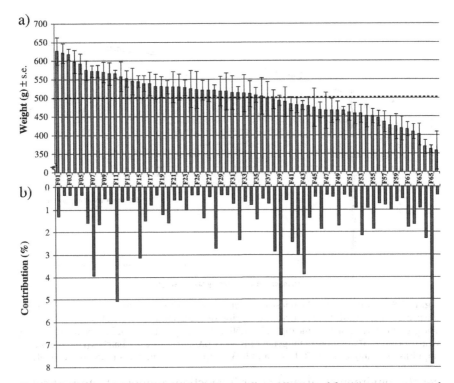

Figure 7.7. (a) Mean harvest weights and (b) overall contribution of the 66 most represented *Lates calcarifer* families (with at least five offspring assigned) commercially reared in intensive tanks (273 dph). Dashed line represents mean harvest weight (502 ± 7 g) of the 66 most represented families (Domingos et al. 2013).

7.5 Quantitative Genetics

Quantitative genetic theory predicts that phenotypic variation within a population is determined largely by both genetic and environmental factors according to the equation; $V_P = V_G + V_E + V_{GE}$, where V_P is the total phenotypic variation for a trait within a population, V_G is the amount of phenotypic variation attributable to genetic potential of an individual, V_E is the modification of genetic potential due to environmental variation, and V_{GE} is the variation attributable to environment specific interactions between the genotype of an animal and its local environment (G x E interactions) (Dunham 2004). Most selection programs aim to improve the contribution of the V_G component of phenotypic expression through targeted breeding of individuals possessing favourable genes for a trait of interest. Before the commencement of selective breeding programs, therefore, it is essential to understand the partitioning of phenotypic variance for a trait into its

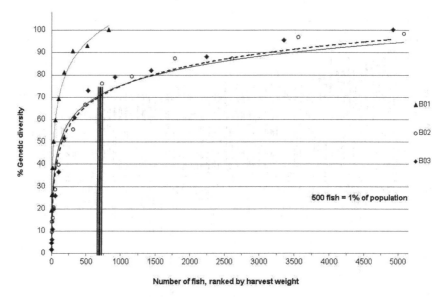

Figure 7.8. Number of the heaviest *L. calcarifer* needed to be genotyped in three seabass offspring batches (B01–B03) to include offspring from all families (100% of genetic diversity). The grey bar corresponds to the top 1.25 ~ 1.50% of the population (~ 625 to 750 heaviest fish), which includes 70% or more of the genetic diversity present in the three batches (Domingos et al. unpublished data). Cohorts B01, B02 and B03 comprises 7, 9 and 66 families, respectively.

genetic, environmental, and other interaction components. In particular, robust estimations of genetic parameters such as trait heritability (h^2), genetic correlations (r_g), and how genetic merit is realized in disparate environments (G x E), are all required before statistically informed breeding programs can be designed. Despite the intense interest in selective breeding of Asian seabass and the actual commencement of breeding programs in several countries, there have been few reports within the literature related to the estimation of genetic parameters in the species, with those that are reported primarily focusing on the heritability of growth traits during juvenile rearing (Kishore Chandra et al. 2000, Wang et al. 2008b), although Domingos et al. (2013) extended our understanding of additive genetic control of growth up to harvest. All of these studies have demonstrated that important growth traits such as weight gain, body length, and body depth are moderately heritable and therefore can be improved through targeted selection approaches. To date there have been no explorations into genetic parameters of carcass quality, or disease traits in *L. calcarifer*, as has been seen for other aquaculture fish species.

7.5.1 Heritability

Heritability is the proportion of phenotypic variation in a population that is due to additive genetic variation between individuals (i.e., V_A) (Falconer and Mackay 1996). Kishore Chandra et al. (2000) looked at heritability of early growth traits such as body length and body weight based on an experimental design involving four broodstock pair-matings. In the resulting cohort of progeny, coefficient of variation for body length was observed to range between 21.6 and 38.0, and 65.0 to 89.4 for body weight. Heritability estimates of these two traits were relatively high and for body length ranged between $h^2 = 0.15 \pm 0.16$ in 21 day old larvae, to $h^2 = 0.96 \pm 0.42$ in 80 day old juveniles. Heritabilities for body weight were $h^2 = 0.50 \pm 0.34$ at 50 days and 0.77 ± 0.40 at 80 days. However, due to the low numbers of families and progeny with phenotypic measurements involved in estimation of the genetic parameters their actual use in a selective breeding program deserves some caution as standard errors around these estimates were very large (Table 7.3).

Table 7.3. Heritability of body length, body weight, body depth and Fulton's condition factor for Asian seabass, *Lates calcarifer*, at varying ages.

Traits	Age (days)	h^2 ($\pm$ SE)	Source
Body length (mm)	21	0.15 ($\pm$ 0.16)	Kishore Chandra et al. (2000)
	50	0.88 ($\pm$ 0.14)	Kishore Chandra et al. (2000)
	80	0.96 ($\pm$ 0.42)	Kishore Chandra et al. (2000)
	90	0.31 ($\pm$ 0.14)	Wang et al. (2008b)
	90	0.24 ($\pm$ 0.21)	Wang et al. (2008b)
	62	0.25 ($\pm$0.14)	Domingos et al. (2013)
	273	0.42 ($\pm$0.11)	Domingos et al. (2013)
Weight (g)	50	0.53 ($\pm$ 0.33)	Kishore Chandra et al. (2000)
	80	0.77 ($\pm$ 0.40)	Kishore Chandra et al. (2000)
	90	0.22 ($\pm$ 0.16)	Wang et al. (2008b)
	90	0.25 ($\pm$ 0.18)	Wang et al. (2008b)
	62	0.21 ($\pm$0.11)	Domingos et al. (2013)
	273	0.42 ($\pm$0.11)	Domingos et al. (2013)
Fulton's condition (F)	90	0.22 ($\pm$ 0.22)	Wang et al. (2008b)
	90	0.15 ($\pm$ 0.09)	Wang et al. (2008b)
	62	0.13 ($\pm$ 0.10)	Domingos et al. (2013)
	273	0.20 ($\pm$ 0.07)	Domingos et al. (2013)
Body depth	62	0.13 ($\pm$ 0.09)	Domingos et al. (2013)
	273	0.13 ($\pm$ 0.09)	Domingos et al. (2013)

One of the major problems with attaining genetic parameters for *L. calcarifer*, as evidenced by the difficulties of Kishore Chandra et al. (2000), has been the determination of the progeny pedigree as a consequence of the mass spawning behaviour of the species. Due to a dependence on mass spawning it has been particularly difficult to obtain the large numbers of families necessary for genetic parameter estimation using single broodstock pair-matings. However, the development of DNA parentage techniques which can assign progeny back to their parents now allows researchers to decipher pedigree relationships among seabass individuals and to take advantage of the large numbers of full- and half-sib families commonly present within mass spawns (Wang et al. 2008b, Loughnan et al. 2013, Domingos et al. 2013, unpublished data). Utilization of this technology has allowed for larger numbers of families to be examined and conceivably more accurate genetic parameter estimates to be obtained. Accordingly, Wang et al. (2008b) and Domingos et al. (2013) refined heritability estimates for barramundi growth traits (Table 7.3). Important growth traits were shown to have a moderate additive genetic component offering potential for genetic gains with heritability for body length, body weight and Fulton's condition index estimated as 0.42 (±0.11), 0.42 (±0.11) and 0.20 (± 0.07) at 273 days of age, respectively (Domingos et al. 2013).

7.5.2 Genetic correlations

Estimation of the genetic correlation between traits gives an indication of how selection on one primary trait will influence the genetic progress on a secondary trait (Falconer and Mackay 1996). Positive genetic correlations between growth traits are usually advantageous, as it indicates that selection for one trait leads to improvement in secondary and other growth traits. In *L. calcarifer* this is indeed the case where all growth traits so far examined are moderate to highly genetically correlated (Table 7.4).

Table 7.4. Genetic and phenotypic correlations between growth traits in *Lates calcarifer*.

Traits	Age	r_G	r_P	Source
Body length-weight	50	0.93**	0.89**	(Kishore Chandra et al. 2000)
	80	0.88**	0.98**	(Kishore Chandra et al. 2000)
	273	0.98**	0.96**	(Domingos et al. 2013a)
Body length-depth	273	0.97**	0.91**	(Domingos et al. 2013a)
Body weight (due to age)	90–289	-	0.60	(Wang et al. 2008b)
Body weight-depth	273	0.99**	0.95**	(Domingos et al. 2013a)
Body depth-Fulton's K	273	0.38	0.33	(Domingos et al. 2013a)

**Significant at 1% level

7.5.3 Genotype-environment interactions

Genotype by environment (G x E) interactions occur when genotypes of the cultured species express their phenotypes differently when reared under diverse environments (i.e., the genetic potential of the animal is modified depending on the specific genotype of the animal and its response to the environmental influence exerted on it—most commonly as a consequence of differing selective pressures). There are two different types of G x E. The first type of interaction is observed when the relative performance rank of two or more genotypes changes when compared in two or more environments (i.e., ranking of family performance), while the other G x E interaction is where only the magnitude of phenotypic differences present amongst the genotypes varies (i.e., differences in mean trait performance increases/decreases dependent on the environment). As *L. calcarifer* can be farmed in many diverse culture environments, including marine and/or freshwaters, sea cage, pond, tank or raceways, in addition to the fact that fingerlings are commonly shipped between countries, it is of particular interest to future selective breeding programs whether barramundi selected under one particular environment will perform equally well in other environments. Given the importance of knowing how G x E may impact on traits like growth it is surprising that very little attention has been directed to date in elucidating possible impacts, with only one study reporting on G x E in Asian seabass within the literature. Here, Domingos et al. (2013) suggests that G x E for the growth traits body weight, length, depth and condition may be negligible due to the type of culture system, or salinity, in which fish are reared (Table 7.5). For example, the genetic correlation between family growth performance in saltwater *vs* freshwater was very high (0.98 ± 0.26), as was family performance in recirculation *vs* pond based culture systems (0.99 ± 0.23). However, this G x E analysis was based on a relatively small number of families (~ 10 families) and it will be imperative to obtain data from larger numbers of families before the impacts of G x E can be properly assessed. Additionally, due to the strong genetic population structure evident among Asian seabass populations (Section 7.3) it is highly likely

Table 7.5. Genotype by environment (r_G) estimates for *Lates calcarifer* traits measured in fish reared in fresh (FW) and salt water (SW) cages at 62 days post-hatch (dph) and at harvest in freshwater intensive tanks (343 dph) and a semi-intensive pond (469 dph). Ls = standard length, W = body weight, BD = body depth, K = Fulton's condition factor (Domingos et al. 2013).

Trait	FW v. SW	Tanks v. Pond
W	0.96 ± 0.34	0.97 ± 0.22
Ls	0.99 ± 0.22	0.98 ± 0.17
BD	0.95 ± 0.31	0.98 ± 0.16
K	0.56 ± 1.86	–0.18 ± 1.19

that population-specific G x E factors will be present when populations are reared outside their natal geographical range. Therefore, given the fact that hatchery produced seabass are increasingly being shipped around the globe for aquaculture production, quantifying population-specific G x E impacts is of urgent importance to the wider industry.

7.6 Genetics of Thermal Tolerance

Tropical latitudes are impacted upon by complex climate patterns, including monsoonal intra-year variability in rainfall and evaporation rate, mid-latitude seasonal oscillations, intra-annual fluctuations in the Madden Julian Oscillation, as well as longer global climate impacts such as the Interdecadal Pacific and El Niño Southern Oscillation Indices (Balston 2007). Predicted climate change up to 2070 is expected to further introduce an additional layer of stochastic variability onto this already complex climate pattern through average annual temperature rises of 0.3–5.2°C and more frequent and intense extreme rainfall events (i.e., –5%–15% in the wet tropics of Australia; Balston 2007, Meynecke et al. 2013—Chapter 3). As a consequence, tropical aquatic environments will be impacted through changes in water temperature, freshwater flow regimes, salinity levels driven by evaporation, precipitation and saltwater incursions, and nutrient pulses, all of which may cause changes in recruitment and migration patterns of fish, growth rates, and exposure to stress-induced disease epidemics (Putten and Rassam 2001).

Across the entire south-east Asian and Australian distribution of *L. calcarifer* this species forms the basis of important wild-harvest, aquaculture and recreational fisheries. Consequently, due to the importance of *L. calcarifer* to tropical communities understanding how climate change will impact on fishery access and the clear identification of adaptation options is of upmost importance to the future prosperity and resilience of many of these tropical communities. Of particular relevance, is elucidation of the capacity for *L. calcarifer* to adapt and deal with predicted temperature rises under climate change.

Studies examining thermal tolerance have been restricted to Australian *L. calcarifer*. In Australia, *L. calcarifer* populations exhibit strong neutral genetic structuring (see Section 7.3) and inhabit environments across 16 degrees of latitude. Consequently, northern populations experience different thermal regimes than those from southern populations (i.e., annual temperatures range from 23.2–32°C in Darwin, Northern Territory, to 18.5–27.7°C in Gladstone, central Queensland). Studies that have examined thermal tolerance in Australian *L. calcarifer* using various molecular and phenotype indicators show that this thermal gradient underpins adaptive variability in populations and that this variability is correlated with latitude.

For example, Newton et al. (2010) and later Jerry et al. (unpublished data) examined the correlation between the ratio of live/dead caudal fin cells and loss of swimming equilibrium (LOSE) of fish from genetically and geographically divergent populations and showed that lower latitude (i.e., northern) populations were more tolerant to upper thermal stress (Fig. 7.9). Thermally induced differences are also evident between populations in locomotor phenotypes (i.e., critical swimming speeds or U_{crit}) (Edmunds et al. 2010), lactate dehydrogenase gene expression pattern (Edmunds et al. 2010) and transcriptome regulation (Newton 2013). When *L. calcarifer* from northern (low latitude) and southern (high latitude) populations were swum, for instance, under native (30°C and 25°C) and non-native (20°C and 35°C) temperatures, distinct differences in the swimming performance of individuals were observed. More specifically, under cold-stress conditions (20°C) fish from cooler latitudes exhibited significantly faster swimming speeds (32.10 ± 0.33 cm·s^{-1}) than their northern counterparts (28.58 ± 0.64 cm·s^{-1}), while under heat-stress (35°C) northern fish from warmer latitudes performed significantly better (51.63 ± 2.1 cm·s^{-1}) than their southern counterparts (44.18 ± 3.11 cm·s^{-1}) (Fig. 7.10). Belying one possible explanation for the underlining molecular explanation of these differences, LDH-B gene expression patterns also differed between these same swum fish (Edmunds et al. 2012).

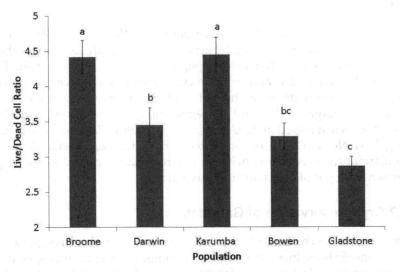

Figure 7.9. Relative upper thermal tolerance predicted by challenge of caudal fin cells to heat stress among five Australian *Lates calcarifer* populations representing different genetic stocks across a broad latitudinal range. The higher the live/dead cell ratio the lower the susceptibility of cells shocked under a heat stress (43°C) for 60 min. Different superscripts represent significant differences at $P < 0.05$. (Jerry et al. unpublished data).

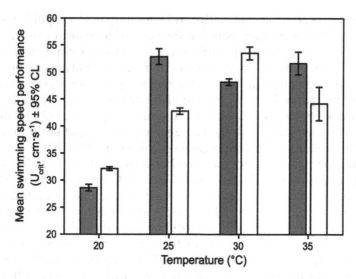

Figure 7.10. Mean critical swimming speed performance (U_{crit}) of *Lates calcarifer* from northern (shaded bars) and southern (open bars) Australian populations exposed to four thermal treatments during early growth ($n = 8$ individuals per bar). Note that northern (low latitude) fish exhibit a significantly slower U_{crit} at 20°C and a relatively greater increase in U_{crit} from 20°C to 25°C than southern (high latitude) fish. U_{crit} values are mass-adjusted means ($n = 8$) with 95% confidence limits (CL) calculated from a general linear mixed model (GLM) analysis (Edmunds et al. 2010).

As final evidence of adaptive genetic differences in Australian *L. calcarifer* to temperature, digital gene expression analyses from northern (Darwin) and southern (Gladstone) populations when fish were reared for 3.5 months under 22°C and 36°C conclusively showed variation in broad gene expression patterns, with fish from the Darwin population showing a massive reorganisation and up-regulation of genes compared to that seen both when reared at 22°C and in the southern Gladstone population (Fig. 7.11) (Newton 2013). Whilst evidence for underlying genetic adaptation to thermal stress is now conclusive in Australian *L. calcarifer*, nothing is known from populations outside Australia.

7.7 Cryopreservation of Gametes

One of the major challenges to the instigation of selective breeding programs in *L. calcarifer* is the inability to undertake single broodstock pair spawns on a commercial scale that allows for the creation of large numbers of families. This is because behaviorally *L. calcarifer* are a mass spawning species and best commercial production of eggs results when several males are mated with one or two females. Also handling stress of broodstock appears to reduce the quality of semen that can be expressed by abdominal pressure,

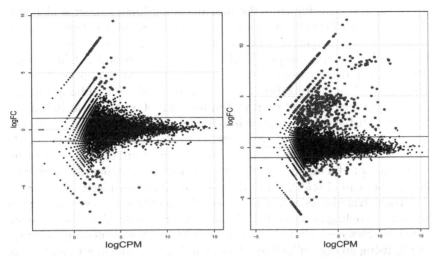

Figure 7.11. Genes found to be differentially expressed in *Lates calcarifer* from Darwin and Gladstone reared for 106 days at 22°C and 36°C. Grey stars represent gene contiqs which exhibit significant expression differences between fish from Darwin relative to Gladstone after correction for false discovery. Scales are x-axis—number of gene contiqs identified for that gene, y-axis—log-fold changes in the abundance (expression) of these gene contiqs in the Darwin population relative to that seen in the Gladstone population (Newton 2013).

while natural sex inversion from male to female creates uncertainty in the number of functional males in broodstock populations (Palmer et al. 1993). The ability to strip gametes, cryopreserve semen for later use, and artificially fertilize eggs would allow more complex mating designs to be implemented and inbreeding to be more effectively controlled, as it will be easier to determine whether males have contributed to a progeny cohort (Robinson and Jerry 2009, Robinson et al. 2010, Macbeth and Palmer 2011). Currently, due to use of tank-based mass spawns, DNA parentage techniques need to be used to sort out the pedigree of progeny from such mating events.

There are two reported studies on cryopreservation of sperm in *L. calcarifer*. Leung (1987) evaluated the effectiveness of three cryoprotectants, dimethylsulpoxide (DMSO), glycerol and methanol to cryopreserve *L. calcarifer* sperm. They found that the addition of milt to a 5% DMSO diluent (1:4 *v:v*) in freshwater teleost Ringer's solution (Kurokura et al. 1984) with either 15% (*w:v*) milk powder or 20% egg yolk (*v:v*) resulted in a post-thaw spermatozoa motility of between 70–100% upon activation in sea water. Methanol had the lowest protective ability. However, Leung (1987) did not conduct fertility trials to ascertain if cryopreserved semen could fertilize eggs. Subsequently, Palmer et al. (1993) extended the results of Leung (1987) by cryopreserving semen using either a 5% DMSO or 10% glycerol diluent in marine teleost Ringer's solution and then assessed post-thaw fertility by mixing semen with freshly stripped ova. Although motility post-activation was higher in the 10% glycerol diluent Palmer et al. (1993) showed that fertilisation success and subsequent viability of developmental stages was higher when 5% DMSO was used (Table 7.6). A semen to ova ratio of 1:100 (v/v) also provided high fertilisation and hatching rates. These two studies therefore show that it is possible to cryopreserve semen and use artificial fertilization techniques in *L. calcarifer*, although to our knowledge there has not been widespread adoption of this technique by industry. Further investigations into the feasibility of large-scale artificial fertilization techniques are encouraged in the species, as being able to reliably obtain unfertilized gametes allows opportunities such as triploidy induction, production of double haploids for whole genome sequencing, and refinement of selective breeding programs to take place.

There has also been limited research into the short term storage of semen by cooling between 0–5°C using different extenders (Hogan et al. 1987). Studies have shown that it is possible to store semen up to 9 days at 5°C using an egg yolk citrate diluent and still be able to activate sperm. However, it is unlikely that sperm will retain a high fertilization capacity after this time.

Table 7.6. Mean viability estimates (%) at three developmental stages of *Lates calcarifer* ova fertilised with spermatozoa diluted or cryopreserved using DMSO or glycerol. Values are means ±standard error (n=3). Within columns, values with different superscripts are significantly different ($P < 0.05$) (Palmer et al. 1993).

Insemination treatment	Developmental viability estimates		
	Blastula-gastrula	Late embryo	Hatched larvae
Ringer's solution only; unfrozen semen	87.8[a] ± 2.67	92.6[a] ± 3.73	80.7[a] ± 1.93
Ringer's + DMSO; unfrozen semen	87.3[a] ± 3.66	83.4[a] ± 4.96	83.7[a] ± 0.81
Ringer's + DMSO; frozen semen	86.1[a] ± 2.41	89.6[a] ± 1.96	84.1[a] ± 1.20
Ringer's + glycerol; unfrozen semen	71.1[a] ± 6.70	59.6[b] ± 7.30	42.0[c] ± 8.74
Ringer's + glycerol; frozen semen	75.4[a] ± 5.59	67.4[b] ± 2.33	60.9[b] ± 1.05

7.8 Cell Lines

The establishment of species-specific cell lines are an important tool in disease, toxicology, cellular physiology and genetic regulation studies. The first cell line for *L. calcarifer* was created by Chong et al. (1987) and was derived from whole seabass fry. This cell-line (termed SB cell-line) was developed to aid the identification of viral infections; a common method to determine if a virus is resident in a healthy population of fish. However, the SB cell-line was subsequently found to be chronically infected with an unidentified virus and became unusable for viral studies. Consequently, Chang et al. (2001) derived a new cell line based on day old larvae, which they subsequently sub-cultured 85 times. Both these cell cultures predominately comprised epithelial-like cells. Chang et al. (2001) subsequently went on to test their cell-line's susceptibility to 10 viruses, including grouper nodavirus, guppy reovirus and several iridoviruses. Cytopathic cellular effects were observed as the result of six viruses demonstrating utility of this new cell line in the isolation of various fish viruses.

Other authors have also created various tissue cell-lines of *L. calcarifer*, including kidney, spleen, and brain, to obtain culture cells specific to the preferred tissue of viral infection (Hameed et al. 2006, Parameswaran et al. 2006a, b, Parameswaran et al. 2007, Hasoon et al. 2011). For example, Hasoon et al. (2011) established a brain cell-line from Malayasian *L. calcarifer* as a tool to isolate and cultivate tissue-specific picine nodaviruses responsible for viral nervous necrosis (VNN). Their cell-line was shown to be susceptible to nodavirus infection with cytopathic effects visible (Fig. 7.12). As intensification of fish culture increases, particularly in the face of preceived climate change and translocation of aquaculture species, cell-lines will remain a useful tool in viral and bacterialogical research into the forseeable future.

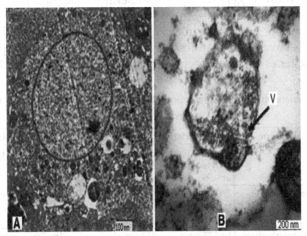

Figure 7.12. Transmission Electron Microscopy of infected *Lates calcarifer* brain cells with nodavirus. (A) vacuolation in the cytoplasm of infected 3 day post-infected cells, (B) virions distributed as viral aggregates in the cytoplasm or membrane bound organelles (Hasoon et al. 2011).

7.9 Genome Mapping and Genomics

7.9.1 DNA markers

Due to both the commercial aquaculture and wild fishery management importance of *L. calcarifer* in the Asia-Pacific there have been numerous publications reporting the development and use of novel genetic marker systems for this species. As genetic technologies have evolved we have likewise seen an evolution of the marker systems applied from the use of allozymes (Shaklee and Salini 1985, Salini and Shaklee 1988), mitochondrial DNA (Chenoweth et al. 1998a, b, Doupe et al. 1999, Lin et al. 2006, Ward et al. 2008, Norfatimah et al. 2009) and microsatellites (Yue et al. 2002, Sim and Othman 2005, Zhu et al. 2006, 2009), to now the adoption of next-generation sequencing technologies that allow easy identification of thousands of genes and tens to hundreds of thousands of single-nucleotide polymorphisms within the genome that may associate with traits of aquaculture and ecological importance. Consequently, there are now a large number of genetic markers for *L. calcarifer* that can be applied to address most population genetic and gene mapping purposes. Table 7.7 list publications reporting for the first time genetic marker systems for *L. calcarifer* and details the purpose for which they have been used.

In addition to genetic markers useful for population genetic and genome mapping purposes, there has also been focus on the isolation of specific genes of interest, primarily in trying to understand the influence these genes have in the manifestation of aquaculture relevant phenotypes. Important genes that have been characterized in *L. calcarifer* include insulin

Table 7.7. Genetic marker systems reported for *Lates calcarifer* and some of their potential applications.

Genetic marker type	Characteristics	Reference	Useful application
Allozymes	46 loci, 11 polymorphic at 0.99	Shaklee and Salini (1985)	population genetics
MtDNA control region	290 basepair fragment region 1	Chenoweth et al. (1998a, b)	population genetics, product traceability, forensics
Complete mitochondrial genome	16,535 base pairs	Lin et al. (2006)2006	mtDNA genomics, isolation of mitochondrial genes
Microsatellites	28 loci 7 loci 74 loci 51 loci 10 loci	Yue et al. (2002) Sim and Othman (2005) Zhu et al. (2006) Zhu et al. (2009) Zhu et al. (2010)	Population genetics, DNA parentage, product traceability, forensic, genome mapping

growth factor-I and II (Richardson et al. 1995, Drakenberg et al. 1997, Collet et al. 1998, Degger et al. 1999, Stahlbom et al. 1999, Degger et al. 2000, Nankervis et al. 2000, Yue et al. 2001), calreticulin (Bai et al. 2012), lactate dehydrogenase-B (Edmunds et al. 2009, 2010, 2012), myostatin (De Santis et al. 2008, De Santis and Jerry 2011, Newton et al. 2012), parvalbumin (Xu et al. 2006, Schiefenhovel and Rehbein 2011), prolactin (He et al. 2012), growth hormone (Yowe and Epping 1996, Yue et al. 2001) and amylase (Ma et al. 2001, 2004). Most of these genes have been characterized in the search for major genes linked to growth and/or disease tolerance. For example, Myostatin (MSTN) is a protein within the Transforming growth-factor B (TGF-β) gene superfamily that has been linked to muscle growth in vertebrates. The interest in characterizing this gene in *L. calcarifer* is related to the fact that MSTN inhibits muscle growth and affects both fiber hyperplasia and hypertrophy by preventing myoblast cell cycle progression. MSTN has also been shown to influence myogenesis from early into embryonic development (Amali et al. 2008, Rodgers and Garikipati 2008). In mammals where the gene has been made defective and the active protein is not produced, a "double-muscling" phenotype is observed. Similarly in zebrafish when this gene is knocked down using RNA interference increased muscle growth has been observed (Amali et al. 2008, Acosta et al. 2005, Lee et al. 2009). Thus the possibility of increasing Asian seabass growth through selecting for natural gene variants with lower expression, or directly inhibiting the activity of this gene/protein, has attracted interest. Unlike mammals, however, in teleosts there are two Mstn paralogs and there is uncertainty to whether these two paralogs have the same physiological role, or whether functional differentiation has occurred. Consequently, De Santis et al. (2008) characterized the two *L. calcarifer* Mstn paralogs and examined how their expression changes throughout embryogenesis and larval growth. They showed that Mstn paralogs underwent some level of differential expression during embryogenesis and, most significantly, that it was Mstn-2 that showed the highest negative correlation with muscle hypertrophy post-metamorphosis making this gene the likely target for targeted molecular intervention or selection.

Another gene of interest as a possible target for selection towards growth is that of parvalbumin (Xu et al. 2006). Parvalbumin is a calcium-binding albumin protein that among other physiological roles has been implicated in muscle fiber relaxation (Muntener et al. 1995). In vertebrates, proteins within the parvalbumin family form two distinct lineages: that of alpha and beta-parvalbumins. Characterization of cDNA of two beta-lineage parvalbumin genes in *L. calcarifer* identified the presence of a microsatellite (CT_{17}) in the 3'-untranslated region of parvalbumin1 which associated significantly with body weight and length ($P < 0.01$) of Singaporean *L. calcarifer* at 90 days post-hatch. In this population two microsatellite alleles (261 and 263 bp) were identified and segregation analyses resulting from mating a single male and female broodstock carrying these alleles showed that progeny homozygous for the 263/263 allele were heavier and longer than heterozygotes or homozygotes for 261/261 (Table 7.8) (Xu et al. 2006).

The gene that has received the most research attention in *L. calcarifer*, however, is that of insulin-like growth factor-I (IGF-I), and to a lesser extent IGF-II (Richardson et al. 1995, Collet et al. 1997, Drakenberg et al. 1997, Matthews et al. 1997, Collet et al. 1998, Degger et al. 1999, Stahlbom et al. 1999, Degger et al. 2000, Nankervis et al. 2000, Degger et al. 2001, Yue et al. 2001, Wang et al. 2008a). Insulin-like growth factors are proteins that play a critical role in the regulation of cellular growth and anabolism in mammals and, as such, these genes have been evaluated as potential markers to monitor and predict growth in fish. Liver is the major source of IGF-I and its mRNA transcript production is primarily regulated by growth hormone (GH) through the GH-IGF-I hormone axis. Consequently, growth-promoting actions of GH are surmised to be accomplished as a result of IGF-I levels (Stahlbom et al. 1999). Due to the potential of this gene as a marker of growth, regulation of IGF-I has been targeted as the gene of choice in *L. calcarifer* in studies evaluating the effects of dietary manipulations on growth parameters. For example, Matthews et al. (1997) examined how IGF-I mRNA expression in the liver and brain of juvenile barramundi varied as a consequence of either being fed to satiation, or starved for 6 weeks. They showed that hepatic IGF-I mRNA expression was significantly impacted by nutritional status, while the brain IGF-I expression was 10 times lower and un-influenced by feeding regimes. This suggests that in *L. calcarifer* the liver is the major site of IGF-I mRNA synthesis and is regulated by food availability. However, the exact control of IGF-I regulation still has not been fully elucidated, as both dietary lipid-carbohydrate ratio manipulation resulting in growth differences and addition of growth hormone failed to induce a change in IGF-I regulatory response (Stahlbom et al. 1999, Nankervis et al. 2000). In fact, injection of recombinant human IGF-I and recombinant human insulin did not have a stimulatory effect on the incorporation of glucose into muscle glycogen

Table 7.8. Growth-related traits of Asian seabass with different genotypes at a microsatellite locus (alleles 261 and 263) in parvalbumin1 (Xu et al. 2006). Different subscripts represent growth differences at $P < 0.01$).

Genotype	n	Traits	
		Body weight (g)	Body length (mm)
261/261	18	26.6 ± 3.2^b	96.6 ± 3.8^b
261/263	57	20.9 ± 1.0^b	89.0 ± 2.2^b
263/263	21	38.0 ± 3.5^a	110.1 ± 3.4^a

and protein in *L. calcarifer*, despite the contrary being observed in toadfish (*Opsanustaui*), hagfish (*Myxineglutinosa*) and northern pike (*Esoxlucius*) (Tashima and Cahill 1968, Ince and Thorpe 1976, Emdin 1982, Drakenberg et al. 1997). More work obviously needs to be done in this area to fully elucidate the role of IGF-I in barramundi growth.

7.9.2 Genome structure

Traditionally, animal improvement programs developed in conjunction with the evolution of modern quantitative genetic theory were based on "broad-brush" statistical estimates of the additive and covariance components of trait phenotypic variance. The central tenet of quantitative genetic theory assumes that the individual effects of genes that influence a quantitative trait are unknown and usually small in magnitude (the so-called infinitesimal model of trait variation) and that the cumulative effect of genes can be used to estimate an animal's breeding value for selection (Goddard 2003). Whilst the practice of selection under this genetic theory has led to the successful improvement of many animal production species, more recently on the back of the genomics revolution, has been recognition that quantifying the role individual genes play in trait expression can dramatically improve the accuracy of breeding value estimation, and therefore, genetic gains (Meuwissen et al. 2001). As a result there has been increased researcher focus into understanding the structure of animal genomes and in the detection of genes that may disproportionately influence the expression of an economic trait (so termed candidate genes or quantitative trait loci QTL—see Section 7.9.3).

The identification of loci and/or gene regions underlying significant variation in quantitative genetic traits is not an easy task and usually involves two major steps (Dekkers 2004). Firstly, some knowledge of genomic structure of the organism is required so that contributing gene, or gene regions, can be isolated. Usually this requires the construction of a moderately dense genetic linkage map comprising 100's of random markers spanning the breadth of the genome (Haley and Visscher 2000). The framework upon which these genetic linkage maps are constructed can be any type of polymorphic genetic marker, and traditionally have been based on restriction fragment length polymorphisms (RFLPs), amplified fragment length polymorphisms (AFLPs), single nucleotide polymorphisms (SNPs) and microsatellites. Once a genetic linkage map for the species is in place genome scans in conjunction with interval mapping can then be undertaken to test the association between markers on the genetic map and the occurrence of a particular phenotype. This genome scan approach identifies chromosomal regions, or QTL that co-segregate, and are in linkage disequilibrium with the trait of interest and significantly narrows down the search for economically important genes.

There has been significant research elucidating the genome structure of *L. calcarifer* and on preliminary QTL detection for several traits. Wang et al. (2007) produced a first-generation genetic linkage map for the species which was based on 240 microsatellite markers which segregated into 24 linkage groups ranging in size from 1.1 to 201.7 cM. Later this *L. calcarifer* genetic map was further refined to include 780 microsatellite and 10 SNP markers spanning a sex-averaged genetic length of 2411.5 cM, with an average intermarker distance of 3.0 cM or 1.1 Mb (Wang et al. 2011a). Between the two sexes differences were seen in the sex-specific length of the maps, with the length of the male map being 2674.6 cM and the female map spanned 2294.8 cM. Differences are also apparent in the recombination rates between the sexes, with females having much lower recombination rates in telomeric regions than the males, while males have higher recombination rates in regions proximal to the centromere. Average recombination rate across all linkage groups in *L. calcarifer* is approximately 3.4 cM/Mb, which is approximately double that observed for other fish species including catfish (1.65 cM/Mb) (Kucuktas et al. 2009), zebrafish (1.35 cM/Mb) (Shimoda et al. 1999) and tilapia (1.3 cM/Mb) (Lee et al. 2009). Comparative genome analysis with *Tetraodon nigroviridis* also found conserved synteny blocks in 16 of the 24 Asian seabass linkage groups, the largest of which was on LG5 that spanned 48 cM in seabass and matched 4 Mb of chromosome 18 in *T. nigroviridis* (Wang et al. 2011a).

Whilst genetic linkage maps are useful to understand genome structure and in the preliminary identification of QTL, having a physical map of the genome greatly increases our knowledge of gene order and facilitates fine-mapping of QTL and the positional cloning of genes located within QTL. Although there are apparent efforts to sequence and produce a draft *L. calcarifer* genome, to date the best knowledge on positional locations of many genes in the species has been based on sequencing of Bacterial Artificial Chromosomes (BACs) and subsequent integration of the sequence data with the second generation linkage map (Xia et al. 2010, Wang et al. 2011a). The current physical map was constructed based on fingerprinting 38,208 clones with a mean insert size of 98 kb. The longest contiq of the map currently spans 1,478 kb in length, with the average contiq size being ~232 kb (range 44 to 1,478 kb). The total physical length of the map sums up to ~665 Mb of sequence, indicating that given the *L. calcarifer* genome is approximately 700 Mb, that some genome regions are still poorly represented within the physical map. However, given the deficiency in the current physical map of the species, this physical map will still prove invaluable in validating and improving contiq positioning resulting from future next-generation whole genome sequencing efforts.

7.9.3 QTL and candidate gene positioning

It appears that most of the traits of interest to aquaculture production are under the control of many genes with small effect and subject to environmental modification. Such traits are traditionally called quantitative phenotypic traits. However, some of the genes controlling a quantitative character may have a disproportionate contribution to the overall trait variance in the population. Quantitative trait loci (QTL) are therefore loci that underlie detectable variation in the expression of a quantitative phenotypic character and are identified by examining the co-segregation of genetic markers on the whole genome with the phenotype of interest. QTL are not necessarily the actual functional gene influencing the phenotypic variation observed, but are markers which reside close enough within the genome to the actual functional gene to co-segregate (exhibit linkage disequilibrium) despite the presence of recombination as a result of meiotic chromosomal rearrangements. If the association between the marker and the phenotype is robust, then improved selection accuracy is predicted to be achieved by incorporating DNA marker information into the overall assessment of the genetic merit of an individual—so-called marker assisted selection.

The detection of QTL in aquaculture species has received a lot of attention, as many of the traits that we want to improve are hard to select for using traditional phenotypic approaches, either because they exhibit low levels of heritability within the population (i.e., some disease resistance and reproductive fitness traits), can only be selected on late into production (such as reproductive maturation, sex), or we would have to use destructive sampling to obtain the phenotypic measure of an individual's performance (flesh quality, body composition traits). The use of QTL therefore promises a way by which the genetic merit of an individual can be predicted simply by genotyping the individual. An added advantage is that this genotyping can be done at any stage of its life cycle negating the requirement to grow an individual for a long period of time before making a selection decision.

In *L. calcarifer*, the detection of QTL has been performed for growth-related traits (Wang et al. 2006, Wang et al. 2008a, Wang et al. 2011a), caudal fin length (Wang et al. 2011b), and cold tolerance (Bai et al. 2012). Five significant and 27 suggestive QTL for the growth traits body weight, standard body length, and total body length at three months post-hatch have been detected and interval and multiple QTL model mapped in a single F1 family to 10 linkage groups (Wang et al. 2006). Three of the five significant QTL detected (*qBW2-a*, *qTL2-a* and *qSL2-a*) putatively controlling body weight, total and standard length, respectively, were mapped to the same region of the LG2 near the microsatellite *Lca287* and accounted for 28.8%, 58.9%, 59.7% of the total phenotypic variance. Significant associations were also evident between these same traits and *Lca371*. Subsequent QTL mapping involving barramundi from two new F1 families originating from Singaporean and Thailand brooders confirmed the segregation of the detected QTL for body weight and length linked to the marker *Lca371* on LG2, but only linkage to *Lca287* in the Singaporean family. The fact that the *Lca371* marker has been mapped to the same QTL region in three distinct F1 families is good evidence that the marker is in strong linkage disequilibrium with the QTL for body weight and length. However, other QTL like that of *Lca287* that had been previously mapped were not universally detected suggesting that they may be linkage equilibrium markers and family-specific. More work is needed to fully evaluate the transferability of these growth-related QTL to other families and populations of *L. calcarifer*.

Water temperature is an important environmental factor that influences physiological variables in fishes such as growth, survival, stress and reproduction (Myrick and Cech 2000). Fishes under favorable ambient conditions typically have a functional immune system and, consequently, high rates of growth, disease resistance and reproductive activity are observed (Dominguez et al. 2004). *Lates calcarifer* is a tropical fish species with optimal growth around 32°C (Bermudes et al. 2010). At temperatures below 20°C seabass undergo increased thermal stress, growth rates decline, and fish become more susceptible to diseases like *Streptococcus iniae* (Bromage 2004). As a consequence it may be beneficial to identify seabass individuals possessing genes which provide increased cold tolerance and to select for this trait. In recognition of the importance of selecting for cold tolerant strains Bai et al. (2012) identified a SNP (C/G) within exon 3 of the calreticulin gene, a gene which may confer some tolerance to cool water conditions, and examined the association of two genotypes (C/C and G/C) with mortality when 200 fish were challenged at 16°C for 5 h. Forty-eight live and 48 dead individuals from the challenge were then genotyped. This analysis showed some association of cold tolerance to the SNP genotype (live fish: 39 showed the C/C genotype, and nine the G/C genotype; dead fish: 36 carried G/C genotype and 12 the C/C genotype, chi-square = 34.95, df =1 and P < 0.001). However, as for growth QTL, the transferability of this particular SNP to the wider *L. calcarifer* population has not been evaluated and its usefulness to select for cold tolerant strains not confirmed.

7.10 Future Directions and Conclusions

Over the past 30 years the genetics of *L. calcarifer* has received considerable attention. Large numbers of genetic markers have been developed, an understanding of both population and hatchery genetic variation has been established, and preliminary identification of gene regions influencing economically important traits have resulted. With the development of next-generation sequencing technologies the genome of *L. calcarifer* may soon also be sequenced aiding in the further identification of genes of importance and the placement of the species in the tree of life. Given the relatively unique biology of *L. calcarifer* having a complete genome sequence will help understand the genetic basis of protandry and thus the genes responsible for sex-change, for example. However, if the Asian seabass aquaculture industry is going to prosper and fully utilize the genetic gains that can come from selective breeding it will need to overcome several deficiencies in genetic knowledge. Firstly, the industry will need to take control of the reproductive biology of the fish through the commercial adoption of cryopreservation and strip-spawning techniques that will permit easy production of large numbers of single parent-pair families. The key to successful fish breeding programs is the production of large numbers of genetically diverse families and the current reliance on mass spawns which result in cohorts of progeny dominated by several highly skewed families is a hindrance to breeding programs. Related to this is a requirement to identify genes responsible for sex change and the development of intervention strategies to modify their effects. Ultimately, if seabass selective breeding programs are to become common, small, precocious, females that can be mated with males from the same generation will need to be produced. Small females will not only allow more rapid genetic gains to be achieved due to halving of the generation time of breeding candidates, but also will negate the industries current infrastructure intensive requirement to hold numerous large (10kg+) breeding females, or superior performing males for several years until they sex-change and can be used as breeding females. Thirdly, information on quantitative genetic parameters for traits other than growth (i.e., disease resistance, carcass quality, temperature tolerance) are vastly lacking, as is information on how seabass selected for one environment will perform in other disparate environments. This later aspect is particularly relevant given the diverse types of environments seabass can be farmed and the current shipping of fingerlings all over the globe. More attention than has been devoted to date is required to better understanding the quantitative genetic basis of important commercial traits.

Finally, numerous genetic studies have demonstrated the high levels of genetic structure and diversity evident throughout *L. calcarifer*'s distributional range. This diversity represents both a potential opportunity, as well as a threat. The high genetic diversity among populations may dictate that some genetic strains are better suited for aquaculture than others and may naturally exhibit attributes of benefit including faster growth, lower fat content, or tolerance to viral diseases like VNN. Systematic genetic strain comparison trials may therefore be of benefit to the industry instead of relying on local strains. However, given that several populations of what were formally considered *L. calcarifer* have recently been found to be a distinct species, coupled with large genetic differences between Asian and Australian stocks of the recognized *L. calcarifer*, the whole-scale movement of seabass fingerlings for aquaculture and stocking purposes poses a serious threat to natural populations. Further population genetic work is urgently required to identify unique evolutionary units of *L. calcarifer* through-out the species broader geographical range and this should be undertaken along with the development of a regional inter-governmental translocation policy for the species, a necessary policy to conserve residual genetic variability before it is lost.

References

Acosta, J., Y. Carpio, I. Borroto, O. Gonzalez and M.P. Estrada. 2005. Myostatin gene silenced by RNAi show a zebrafish giant phenotype. J. Biotechnol. 119: 324–331.

Amali, A.A, C.J.F. Lin, Y.H. Chen, W.L. Wang, H.Y. Gong, R.D. Rekha, J.K. Lu, T.T. Chen and J.L.Wu. 2008. Overexpression of Myostatin2 in zebrafish reduces the expression of dystrophin associated protein complex (DAPC) which leads to muscle dystrophy. J. Biomed. Sci. 15: 595–604.

Bai, Z.Y., Z.Y. Zhu, C.M. Wang, J.H. Xia, X.P. He and G.H. Yue. 2012. Cloning andcharacterization of the calreticulin gene in Asian seabass (*Lates calcarifer*). Animal 6: 887–893.

Balston, J.M. 2007. An examination of the impacts of climate variability and climate change on the wild barramundi (*Lates calcarifer*): Atropical estuarine fishery of north-eastern Queensland, Australia. Ph.D. Thesis, James Cook University, Queensland.

Bermudes, M., B. Glencross, K. Austen and W. Hawkins. 2010. The effects of temperature and size on the growth, energy budget and waste outputs of barramundi (*Lates calcarifer*). Aquaculture 306: 160–166.

Bromage, E. 2004. The humoral immune response of *Lates calcarifer* to *Streptococcus iniae*. Ph.D. Thesis, James Cook University, Queensland.

Carey, G. and P. Mather. 1999. Karyotypes of four Australian fish species *Melanotaenia duboulayi*, *Bidyanus bidyanus*, *Macquaria novemaculeata* and *Lates calcarifer*. Cytobios 100: 137–146.

Chang, S., G.H. Ngoh, L.F.S. Kueh, Q.W. Qin, C.L. Chen, T.J. Lam and Y.M. Sin. 2001. Development of a tropical marine fish cell line from Asian seabass (*Lates calcarifer*) for virus isolation. Aquaculture 192: 133–145.

Chenoweth, S.F., J.M. Hughes, C.P. Keenan and S. Lavery. 1998a. Concordance between dispersal and mitochondrial gene flow: isolation by distance in a tropical teleost, *Lates calcarifer* (Australian barramundi). Heredity 80: 187–197.

Chenoweth, S.F., J.M. Hughes, C.P. Keenan and S. Lavery. 1998b. When oceans meet: a teleost shows secondary intergradation at an Indian-Pacific interface. Proc. Royal Soc. Lond. B Bio. 265: 415–420.

Chong, S.Y., G.H. Ngoh, M.K. Ng and K.T. Chu. 1987. Growth of Lymphocystis virus in a seabass, *Lates calcarifer* Bloch, cell line. Singapore Vet. J. 11: 78–85.

Collet, C., J. Candy, N. Richardson andV. Sara. 1997. Organization, sequence, and expression of the gene encoding IGFII from barramundi (Teleosteii; *Lates calcarifer*). Biochem. Genet. 35: 211–224.

Collet, C., J. Candy and V. Sara. 1998. Tyrosine hydroxylase and insulin-like growth factor II but not insulin are adjacent in the teleost species barramundi, *Lates calcarifer*. Anim. Genet. 29: 30–32.

Cross, T.F. 2000. Genetic implications of translocation and stocking of fish species, with particular reference to Western Australia. Aquacult. Res. 31: 83–94.

De Santis, C. and D.R. Jerry. 2011. Differential tissue-regulation of myostatin genes in the teleost fish *Lates calcarifer* in response to fasting. Evidence for functional differentiation. Mol. Cell. Endocrinol. 335: 158–165.

De Santis, C., B.S. Evans, C. Smith-Keune and D.R. Jerry. 2008. Molecular characterization, tissue expression and sequence variability of the barramundi (*Lates calcarifer*) myostatin gene. BMC Genomics 9: 82.

Degger, B.G., N. Richardson, C. Collet, F.J. Ballard and Z. Upton. 1999. In vitro characterization and *in vivo* clearance of recombinant barramundi (*Lates calcarifer*) IGF-I. Aquaculture 177: 153–160.

Degger, B., Z. Upton, K. Soole, C. Collet and N. Richardson. 2000. Comparison of recombinant barramundi and human insulin-like growth factor (IGF)-I in juvenile barramundi (*Lates calcarifer*): *in vivo* metabolic effects, association with circulating IGF-binding proteins, and tissue localisation. Gen. Comp. Endocrinol. 117: 395–403.

Degger, B., N. Richardson, C. Collet and Z. Upton. 2001. Production, *in vitro* characterisation, *in vivo* clearance, and tissue localisation of recombinant barramundi (*Lates calcarifer*) insulin-like growth factor II. Gen. Comp. Endocrinol. 123: 38–50.

Dekkers, J. 2004. Commercial application of marker- and gene-assisted selectionin livestock: strategies and lessons. J. Anim. Sci. 82: 313–328.

Domingos, J.A., C. Smith-Keune, N. Robinson, S. Loughnan and D.R. Jerry. 2013. Heritability of harvest growth traits and genotype-environment interactions in barramundi *Lates calcarifer* (Bloch). Aquaculture 402–403: 66–75.

Dominguez, M., A. Takemura, M. Tsuchiya and S. Nakamura. 2004. Impact of different environmental factors on the circulating immunoglobulin levels in the Nile tilapia, *Oreochromis niloticus*. Aquaculture 241: 491–500.

Doupe, R.G., P. Horwitz and A.J. Lymbery. 1999. Mitochondrial genealogy of Western Australian barramundi: applications of inbreeding coefficients and coalescent analysis for separating temporal population processes. J. Fish Biol. 54: 1197–1209.

Drakenberg, K., G. Carey, P. Mather, A. Anderson and V.R. Sara. 1997. Characterization of an insulin-like growth factor (IGF) receptor and the insulin-like effects of IGF-1 in the bony fish, *Lates calcarifer*. Regul. Pept. 69: 41–45.

Dunham, R.A. 2004. Aquaculture and Fisheries Biotechnology—Genetic Approaches. CAB International, Oxfordshire.

Edmunds, R.C., L. van Herwerden, C. Smith-Keune and D.R. Jerry. 2009. Comparative characterization of a temperature responsive gene (lactate dehydrogenase-B, ldh-b) in two congeneric tropical fish, *Lates calcarifer* and *Lates niloticus*. Int. J. Biol. Sci. 5: 558–569.

Edmunds, R.C., L. van Herwerden and C.J. Fulton. 2010. Population-specific locomotor phenotypes are displayed by barramundi, *Lates calcarifer*, in response to thermal stress. Can. J. Fish. Aquat. Sci. 67: 1068–1074.

Edmunds, R.C., C. Smith-Keune, L. van Herwerden, C.J. Fulton and D.R. Jerry. 2012. Exposing local adaptation: synergistic stressors elicit population-specific lactate dehydrogenase-B (ldh-b) expression profiles in Australian barramundi, *Lates calcarifer*. Aquat. Sci. 74: 171–178.

Emdin, S.O. 1982. Effects of hagfish insulin in the Atlantic hagfish, *Myxineglutinosa*—the *in vivo* metabolism of c-14-labeled glucose and c-14-labeled leucine and studies on starvation and glucose-loading. Gen. Comp. Endocrinol. 47: 414–425.

Evanno, G., S. Regnaut and J. Goudet. 2005. Detecting the number of clusters of individuals using the software structure: a simulation study. Mol. Ecol. 14: 2611–2620.

Falconer, D.S. and T.F.C. Mackay. 1996. Introduction to Quantitative Genetics. Pearson Education Ltd., Edinburgh Gate, Essex.

Frost, L.A., B.S. Evans and D.R. Jerry. 2006. Loss of genetic diversity due to hatchery culture practices in barramundi (*Lates calcarifer*). Aquaculture 261: 1056–1064.

Goddard, M.E. 2003. Animal breeding in the (post-) genomic era. Anim. Sci. 76: 353–365.

Haley, C. and P. Visscher. 2000. DNA markers and genetic testing in farm animal improvement: current applications and future prospects. Roslin Institute Annual Report 1998–1999: 28–39.

Hameed, A.S.S., V. Parameswaran, R. Shukla, I.S.B. Singh, A.R. Thirunavukkarasu and R.R. Bhonde. 2006. Establishment and characterization of India's first marine fish cell line (SISK) from the kidney of seabass (*Lates calcarifer*). Aquaculture 257: 92–103.

Hasoon, M.F., H.M. Daud, A.A. Abdullah, S.S. Arshad and H.M. Bejo. 2011. Development and partial characterization of new marine cell line from brain of Asian seabass *Lates calcarifer* for virus isolation. *In vitro* Cell. Dev. Biol. Anim. 47: 16–25.

He, X.P., J.H. Xia, C.M. Wang, H.Y. Pang and G.H. Yue. 2012. Significant associations of polymorphisms in the prolactin gene with growth traits in Asian seabass (*Lates calcarifer*). Anim. Genet. 43: 233–236.

Hogan, A.E., C.G. Barlow and P.J. Palmer. 1987. Short-term storage of barramundi sperm. Aust. Fish. 46: 18–19.

Ince, B.W. and A. Thorpe. 1976. *In vivo* metabolism of c-14 glucose and c-14 glycine in insulin-treated northern pike (*Esoxlucius*). Gen. Comp. Endocrinol. 28: 481–486.

Jerry, D.R. and C. Smith-Keune. 2010. Molecular discrimination of imported product from Australian barramundi *Lates calcarifer*. Project 2008/758 Final Report to the SEAFOOD CRC, Flinders University, Adelaide.

Keenan, C.P. 1994. Recent evolution of population structure in Australian barramundi, *Lates calcarifer* (Bloch): An example of isolation by distance in one dimension. Aust. J. Mar. Freshwater Res. 45: 1123–1148.

Keenan, C.P. 2000. Should we allow human-induced migration of the Indo-West Pacific fish, barramundi *Lates calcarifer* (Bloch) within Australia? Aquacult. Res. 31: 121–131.

Kishore Chandra, P., M. Kailasam, A.R. Thirunavukkarasu and M. Abraham. 2000. Genetic parameters for early growth traits in *Lates calcarifer* (Bloch). J. Mar. Biol. Assoc. India 42: 194–199.

Kucuktas, H., S.L. Wang, P. Li, C.B. He, P. Xu, Z.X. Sha, H. Liu, Y.L. Jiang, P. Baoprasertkul, B. Somridhivej, Y.P. Wang, J. Abernathy, X.M. Guo, L. Liu, W. Muir and Z.J. Liu. 2009. Construction of genetic linkage maps and comparative genome analysis of catfish using gene-associated markers. Genetics 181: 1649–1660.

Kurokura, H., R. Hirano, M. Tomita and M. Iwahashi. 1984. Cryopreservation of carp sperm. Aquaculture 37: 267–273.

Lee, B.Y., W.J. Lee, J.T. Streelman, K.L. Carleton, A.E. Howe, G. Hulata, A.Slettan, J.E. Stern, Y. Terai and T.D. Kocher. 2009. A second-generation genetic linkage map of tilapia (*Oreochromis* spp.). Genetics 170: 237–244.

Leung, L.P. 1987. Cryopreservation of spermatozoa of the barramundi, *Lates calcarifer* (Teleostei, Centropomidae). Aquaculture 64: 243–247.

Lin, G., L. Lo, Z.Y. Zhu, F. Feng, R. Chou and G.H. Yue. 2006. The complete mitochondrial genome sequence and characterization of single-nucleotide polymorphisms in the control region of the Asian seabass (*Lates calcarifer*). Mar. Biotechnol. 8: 71–79.

Loughnan, S.R., J.A. Domingos, C. Smith-Keune, J. Forrestor, D.R. Jerry, L.B. Beheregaryand L.A. Robinson. 2013. Broodstock contribution after mass spawning and size grading to prevent cannibalism in barramundi (*Lates calcarifer*, Bloch). Aquaculture 404–405: 139–149.

Ma, P., B. Sivaloganathan, P.K. Reddy, W.K. Chan and T.L. Lam. 2001. Ontogeny of alpha-amylase gene expression in seabass larvae (*Lates calcarifer*). Mar. Biotechnol. 3: 463–469.

Ma, P., K.P. Reddy, W.K. Chan and T.J. Lam. 2004. Hormonal influence on amylase gene expression during Seabass (*Lates calcarifer*) larval development. Gen. Comp. Endocrinol. 138: 14–19.

Macbeth, G.M. and P.J. Palmer. 2011. A novel breeding programme for improved growth in barramundi *Lates calcarifer* (Bloch) using foundation stock from progeny-tested parents. Aquaculture 318: 325–334.

Marshall, C.R.E. 2005. Evolutionary genetics of barramundi (*Lates calcarifer*) in the Australian Region. Ph.D. Thesis, Murdoch University, Western Australia.
Matthews, S.J., A.K. Kinhult, P. Hoeben, V.R. Sara and T. Anderson. 1997. Nutritional regulation of insulin-like growth factor activities in barramundi, *Lates calcarifer*. J. Mol. Endocrinol. 273–276.
Meuwissen, T.H.E., B.J. Hayes and M.E. Goddard. 2001. Prediction of total genetic value using genome-wide dense marker maps. Genetics 157: 1819–1829.
Meynecke, J.-O., J. Robins and J. Balston. 2013. Climate effects on recruitment and catch effort of *Lates calcarifer*. *In:* D.R. Jerry and F.G. Ayson (eds.). Ecology and Culture of Asian Seabass *Lates calcarifer*. CRC Press
Muntener, M., L. Kaser, J. Weber and M.W. Berchtold. 1995. Increase of skeletal-muscle relaxation speed by direct-injection of parvalbumin cDNA. PNAS 92: 6504–6508.
Myrick, C.A. and J.J. Cech. 2000. Temperature influences on California rainbow trout physiological performance. Fish Physiol. Biochem. 22: 245–254.
Nankervis, L., S.J. Matthews and P. Appleford. 2000. Effect of dietary non-protein energy source on growth, nutrient retention and circulating insulin-like growth factor I and triiodothyronine levels in juvenile barramundi, *Lates calcarifer*. Aquaculture 191: 323–335.
Newton, J.R. 2013. Investigating the genetics of thermal tolerance and adaptation to temperature amongst populations of Australian barramundi (*Lates calcarifer*). Ph.D. Thesis, James Cook University, Queensland.
Newton, J.R., C. De Santis and D.R. Jerry. 2012. The gene expression response of the catadromous perciform barramundi *Lates calcarifer* to an acute heat stress. J. Fish Biol. 81: 81–93.
Newton, J.R., C. Smith-Keune and D.R. Jerry. 2010. Thermal tolerance varies in tropical and sub-tropical populations of barramundi (*Lates calcarifer*) consistent with local adaptation. Aquaculture 308: S128–S132.
Noble, T.H. 2012. Impacts of escaped farm barramundi (*Lates calcarifer*) on the genetic integrity and long-term fitness of wild populations. Honours Thesis, James Cook University, Queensland.
Norfatimah, M.Y., M.N.S. Azizah, A.S. Othman, I. Patimah and A.F.J. Jamsari. 2009. Genetic variation of *Lates calcarifer* in Peninsular Malaysia based on the cytochrome b gene. Aquacult. Res. 40: 1742–1749.
Palmer, P.J., A.W. Blackshaw and R.N. Garrett. 1993. Successful fertility experiments with cryopreserved spermatozoa of barramundi, *Lates calcarifer* (Bloch), using dimethylsulfoxide and glycerol as cryoprotectants.Reprod. Fertil. Dev. 5: 285–293.
Parameswaran, V., R. Shukla, R. Bhonde and A.S.S. Hameed. 2006a. Establishment of embryonic cell line from seabass (*Lates calcarifer*) for virus isolation. J. Virol. Methods 137: 309–316.
Parameswaran, V., R. Shukla, R.R. Bhonde and A.S.S. Hameed. 2006b. Splenic cell line from seabass, *Lates calcarifer*: Establishment and characterization. Aquaculture 261: 43–53.
Parameswaran, V., R. Shukla, R. Bhonde and A.S.S. Hameed. 2007. Development of a pluripotent ES-like cell line from Asian seabass (*Lates calcarifer*)—an oviparous stem cell line mimicking viviparous ES cells. Mar. Biotechnol. 9: 766–775.
Pethiyagoda, R. and A.C. Gill. 2012. Description of two new species of seabass (Teleostei: Latidae: *Lates*) from Myanmar and Sri Lanka. Zootaxa 3314: 1–16.
Putten, M.V. and G. Rassam. 2001. Columns—Director's line fisheries, climate change, and working together—The American Fisheries Society and National Wildlife Federation are working together to help address global climate change. Fisheries—American Fisheries Society 26: 4–5.
Richardson, N.A., A.J. Anderson, M.A. Rimmer and V.R. Sara. 1995. Localization of insulin-like growth factor-I immunoreactivity in larval and juvenile barramundi (*Lates calcarifer*). Gen. Comp. Endocrinol. 100: 282–292.

Robinson, N. and D.R. Jerry. 2009. Development of a genetic management and improvement strategy for Australian cultured barramundi. 2008/758 Final report to the Australian Seafood CRC. Flinders University, Adelaide.

Robinson, N.A., G. Schipp, J. Bosmans and D.R. Jerry. 2010. Modelling selective breeding in protandrous, batch-reared Asian seabass (*Lates calcarifer*, Bloch) using walkback selection. Aquacult. Res. 41: 643–655.

Rodgers, B.D. and D.K. Garikipati. 2008. Clinical, agricultural, and evolutionary biology of myostatin: A comparative review. Endocr. Rev. 29: 513–534.

Salini, J. and J.B. Shaklee. 1988. Genetic structure of barramundi (*Lates calcarifer*) stocks from northern Australia. Aust. J. Mar. Freshwater Res. 39: 317–329.

Schiefenhovel, K. and H. Rehbein. 2011. Identification of barramundi (*Lates calcarifer*) and tilapia (*Oreochromis* spp.) fillets by DNA- and protein-analytical methods. J. Verbraucherschutz Lebensmittelsicherh. 6: 203–214.

Shaklee, J.B. and J.P. Salini. 1985. Genetic variation and population subdivision in Australian barramundi, *Lates calcarifer* (Bloch). Aust. J. Mar. Freshwater Res. 36: 203–218.

Shaklee, J.B., J. Salini and R.N. Garrett. 1993. Electrophoretic characterization of multiple genetic stocks of barramundi perch in Queensland, Australia. Trans. Am. Fish. Soc. 122: 685–701.

Shimoda, N., E.W. Knapik, J. Ziniti, C. Sim, E. Yamada, S. Kaplan, D. Jackson, F. de Sauvage, H. Jacob and M.C. Fishman. 1999. Zebrafish genetic map with 2000 microsatellite markers. Genomics 58: 219–232.

Sim, M.P. and A.S. Othman. 2005. Isolation and characterization of microsatellite DNA loci in seabass, *Lates calcarifer* Bloch. Mol. Ecol. Notes 5: 873–875.

Stahlbom, A.K., V.R. Sara and P. Hoeben. 1999. Insulin-like growth factor mRNA in barramundi (*Lates calcarifer*): Alternative splicing and nonresponsiveness to growth hormone. Biochem. Genet. 37: 69–93.

Sudhesh, P.S., G. John and I.D. Gupta. 1992. Chromosomes of *Lates calcarifer*. J. Inland Fish. Soc. India 24: 26–29.

Tashima, L. and G.F. Cahill. 1968. Effects of insulin in the toadfish *Opsanus tau*. Gen. Comp. Endocrinol. 11: 262–71.

Wang, C.M., L.C. Lo, Z.Y. Zhu and G.H. Yue. 2006. A genome scan for quantitative trait loci affecting growth-related traits in an F1 family of Asian seabass (*Lates calcarifer*). BMC Genomics 7: 274.

Wang, C.M., Z.Y. Zhu, L.C. Lo, F. Feng, G. Lin, W.T. Yang, J. Li and G.H. Yue. 2007. A microsatellite linkage map of Barramundi, *Lates calcarifer*. Genetics 175: 907–915.

Wang, C.M., L.C. Lo, F. Feng, Z.Y. Zhu and G.H. Yue. 2008a. Identification and verification of QTL associated with growth traits in two genetic backgrounds of Barramundi (*Lates calcarifer*). Anim. Genet. 39: 34–39.

Wang, C.M., L.C. Lo, Z.Y. Zhu, G. Lin, F. Feng, J. Li, W.T. Yang, J. Tan, R. Chou, H.S. Lim, L. Orban and G.H. Yue. 2008b. Estimating reproductive success of brooders and heritability of growth traits in Asian seabass (*Lates calcarifer*) using microsatellites. Aquacult. Res. 39: 1612–1619.

Wang, C.M., Z.Y. Bai, X.P. He, G. Lin, J.H. Xia, F. Sun, L.C. Lo, F. Feng, Z.Y. Zhu and G.H. Yue. 2011a. A high-resolution linkage map for comparative genome analysis and QTL fine mapping in Asian seabass, *Lates calcarifer*. BMC Genomics 12: 174.

Wang, C.M., L.C. Lo, Z.Y. Zhu, H.Y. Pang, H.M. Liu, J.Tan, H.S. Lim, R. Chou, L. Orban and G.H. Yue. 2011b. Mapping QTL for an adaptive trait: the length of caudal fin in *Lates calcarifer*. Mar. Biotechnol. 13: 74–82.

Ward, R.D., B.H. Holmes and G.K. Yearsley. 2008. DNA barcoding reveals a likely second species of Asian seabass (barramundi) (*Lates calcarifer*). J. Fish Biol. 72: 458–463.

Xia, J.H., F. Feng, G. Lin, C.M. Wang and G.H. Yue. 2010. A first generation BAC-based physical map of the Asian seabass (*Lates calcarifer*). PLoS ONE5: e11974. doi: 10.1371/journal.pone.0011974.

Xu, Y.X., Z.Y. Zhu, L.C. Lo, C.M. Wang, G. Lin, F. Feng and G.H. Yue. 2006. Characterization of two parvalbumin genes and their association with growth traits in Asian seabass (*Lates calcarifer*). Anim. Genet. 37: 266–268.

Yowe, D.L. and R.J. Epping. 1996. A minisatellite polymorphism in intron III of the barramundi (*Lates calcarifer*) growth hormone gene. Genome 39: 934–940.

Yue, G., Y. Li and L. Orban. 2001. Characterization of microsatellites in the IGF-2 and GH genes of Asian seabass (*Lates calcarifer*). Mar. Biotechnol. 3: 1–3.

Yue, G.H., Y. Li, T.M. Chao, R. Chou and L. Orban. 2002. Novel microsatellites from Asian seabass (*Lates calcarifer*) and their application to broodstock analysis. Mar. Biotechnol. 4: 503–511.

Yue, G.H., Z.Y. Zhu, L.C. Lo, C.M. Wang, G. Lin, F. Fenf, H.Y. Pang, J. Li, P. Gong, H.M. Liu, J.Tan, R.Chou,H.Lim and L. Orban. 2009. Genetic variation and population structure of Asian seabass (*Lates calcarifer*) in the Asia-Pacific region. Aquaculture 293: 22–28.

Yue, G.H., J.H. Xia, F. Liu and G. Lin. 2012. Evidence for female-biased dispersal in the protandrous hermaphroditic Asian seabass, *Lates calcarifer*. PLoS ONE 7: e37976 doi: 10.1371/journal.pone.0037976.

Zhu, Z.Y., C.M. Wang, L.C. Lo, F. Feng, G. Lin and G.H. Yue. 2006. Isolation, characterization, and linkage analyses of 74 novel microsatellites in barramundi (*Lates calcarifer*). Genome 49: 969–976.

Zhu, Z.Y., C.M. Wang, F. Feng and G.H. Yue. 2009. Isolation and characterization of 51 microsatellites from BAC clones in Asian seabass, *Lates calcarifer*. Anim. Genet. 40: 125–126.

Zhu, Z.Y., C.M. Wang, L.C. Lo, G. Lin, F. Feng, J. Tan, R. Chou, H.S. Lim, L. Orban and G.H. Yue. 2010. A standard panel of microsatellites for Asian seabass (*Lates calcarifer*). Anim. Genet. 41: 208–212.

8

Lates calcarifer Nutrition and Feeding Practices

Brett Glencross,* Nick Wade and **Katherine Morton**

8.1 Introduction

Feed development for *L. calcarifer* has undergone considerable evolution over the past 30 years from that of feeding bait-fish or trash-fish, the use of simple pellet-pressed diets, to the point where modern extrusion technologies are now applied to produce floating and sinking pellets with a range of nutrient densities depending on farming country and practices. Considerable effort has been made in Australia, Thailand and the Philippines in defining the nutritional requirements of *L. calcarifer*, with much of these efforts being reviewed in Boonyaratpalin and Williams (2001) and Glencross (2006). This chapter builds on these early reviews and further synthesises the latest research related to nutrition and feeding of *L. calcarifer*. For ease of presentation the chapter is divided into three sections covering essential nutrient and energy requirements, raw material applications and feeding management options for this species.

CSIRO Food Futures Flagship and CSIRO Marine and Atmospheric Research, GPO Box 2583, Brisbane, QLD 4001, Australia.
Email: Brett.Glencross@csiro.au
*Corresponding author

8.2 Essential Nutrient Requirements

Growth in *L. calcarifer*, like all animals, is underpinned by the dietary intake of certain essential nutrients. The optimal concentration of these essential nutrients in any diet is largely driven by the energy content of the feed eaten by the fish. As such, the optimal levels of most nutrients can largely be considered to be energy intake dependent, though in many cases energy independent assessments have been determined. Where possible both will be presented.

8.2.1 Protein and amino acids

Protein and amino acids constitute the key group of essential nutrients required by all animals for synthesis of protein, which tends to be the key driver of growth. There have been numerous studies in *L. calcarifer* undertaken to define protein requirements. However, by comparison there have been few studies examining specific requirements for essential amino acids.

8.2.2 Protein requirements

Most studies examining the requirements for protein by *L. calcarifer* suggest a relatively high protein demand, consistent with the carnivorous/piscivorous nature of the fish (Glencross 2006). These recommendations vary somewhat and are influenced by a range of factors including dietary energy density and fish size. An overview of these assessments is presented in Table 8.1. The earliest estimates of protein requirements for *L. calcarifer* were those by Sakaras et al. (1988, 1989). The series of studies undertaken by Sakaras et al. (1988, 1989) examined protein requirements for small fish using a factorial study with three protein levels (45%, 50% and 55%) and two lipid levels (10% and 15%). In the initial study the best performance was observed with a diet of 50% protein and 15% lipid (Sakaras et al. 1988), though in these authors later work performance was improved with a diet of 45% protein and 18% lipid (Sakaras et al. 1989) (Table 8.1).

Catacutan and Coloso (1995) also undertook a factorial study with three protein levels (35%, 42.5% and 50%), but with three lipid levels (5%, 10% and 15%) with very small fish of an initial weight of 1.5 g. Highest growth rates were observed with fish fed a diet of 50% protein and 15% lipid. However, the growth rate of the fish fed the 50% protein and 15% lipid diet was not significantly better than those fed a diet of 42.5% protein and 10% lipid.

Table 8.1. Summary of protein requirement estimates for Lates calcarifer.

Crude Protein levels examined (% to %)	Optimal Level (%)	Gross Energy level at Optima (MJ/kg)	Initial Fish Size (g)	Temp (°C)	Authors
35–55	45–55	13.4–16.4	n/d	n/d	Cuzon and Fuchs 1988
45–55	50	n/d	7.5	n/d	Sakaras et al. 1988
45–55	45	n/d	n/d	n/d	Sakaras et al. 1989
n/d	40–45	n/d	n/d	n/d	Wong and Chou 1989
35–50	50	50	1.3	29	Catacutan and Coloso 1995
29–55	46–55	18.4–18.7	76	28	Williams and Barlow 1999
38–52	52	17.8–21.0	230	28	Williams et al. 2003a
44–65	60	20.9–22.8	80	28	Williams et al. 2003a
43–56	46	20.2–29.6	1180	30	Glencross et al. 2007
46–65	65	21.3–22.4	2.9	30	Glencross et al. (unpublished)

n/d: not defined

Williams et al. (2003a) also examined the protein requirements of *L. calcarifer* and did so under a relatively constant energy regime (18.4–18.7 MJ/kg GE). These researchers also made a significant advancement over earlier studies in that this work was performed on a digestible nutrient and energy basis. Additionally, the authors also began to present the requirement data on a g/MJ basis. To do this a series of diets containing digestible protein from 29.0% to 55.5% with varying lipid levels to maintain iso-energetic diets (Fig. 8.1) were fed to juvenile (76 g) *L. calcarifer*. The optimal performance in this study was observed at the higher protein levels (25 to 30 g/MJ). Although attempts were made to maintain each of the diets as iso-energetic with each other on a digestible basis, the higher protein diets were substantially more energy dense on a digestible basis. Consequently, it was suggested by the authors that the performance of the fish at the highest protein levels may have been a response to dietary energy and that marginally lower protein levels could satisfy the protein requirement, provided total energy supply was maintained.

In a second series of studies by Williams et al. (2003a), the requirement for protein and energy were examined in a factorial array over two experiments. The dietary energy levels in each experiment were largely manipulated by altering diet lipid content and, as such, also allowed some reflection on the effects of lipid inclusion in the diet (see section 8.2.2). In the first experiment using larger fish of 230 g, four protein levels from

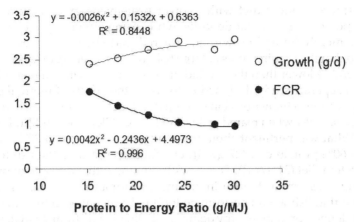

Figure 8.1. Influence of digestible protein to digestible energy ratio on growth and feed conversion by juvenile (76 g) *Lates calcarifer*. Data from Williams and Barlow (1999).

38% to 52% were examined at three energy levels. Growth was best at the highest protein and energy levels (52% protein, 21.0 MJ/kg), and there was a clear response by the fish to the graded protein levels, suggesting that it would be worth examining higher protein and energy levels. A second experiment was undertaken with smaller fish (80 g), and using a range of higher protein levels from 44% to 65% across three energy levels. As with the first study with larger fish, the best growth was again seen with higher protein levels, although at the higher energy densities (highest lipid inclusion levels) a marginally reduced growth performance was also noted. Across both experiments though a consistent increase in growth was seen with increasing protein to energy ratio.

Building on the work of Williams et al. (2003a), a series of studies were undertaken by Glencross (2008), Glencross et al. (2008), and Glencross and Bermudes (2010, 2011, 2012) that used a bioenergetic modelling approach to define not only demands for energy, but also for protein and amino acids (see sections 8.2.1.4; 8.2). From these studies it was demonstrated that there was a critical relationship between fish size and protein requirements and that much of the vagaries around the different protein requirement data reported could be explained by the wide range of fish sizes used in the various studies.

To demonstrate this and to also act as an independent validation of the developed model, Glencross et al. (2008) undertook a study with large (1180 g) *L. calcarifer* fed diets with digestible protein levels of 43% to 50% (on a dry matter basis) and digestible protein to digestible energy ratios of 15.8 g/MJ to 29.7 g/MJ (on a dry matter basis). Here it was demonstrated that with large fish the most efficient growth and protein retention was with a diet of 20 g/MJ equating to a diet with 46% protein and 30% lipid.

Furthermore, using a diet with a lower protein:energy density resulted in fish possessing more fat deposits, while using a higher protein:energy density merely limited growth and/or reduced efficiency. These results showed that the requirements of large fish (relative to energy density) were considerably lower than the myriad of other studies with smaller fish.

In a separate study designed to focus on the effects of thermal stress, the application of higher protein levels (60% cf. 52%) was examined in small (18 g) *L. calcarifer* when reared at 30°C and 37 °C (Glencross and Rutherford 2010). What was pertinent about this study, was that even at 30°C the fish fed the 60% protein diet (27 g/MJ) clearly outperformed the fish fed the 52% protein diet (21 g/MJ). This work highlighted that the ratio of protein to energy was a critical factor in driving performance (especially in small fish) and that fish would perform better if fed a diet more closely matched to their defined protein : energy demands. Following on from this, Glencross et al. (unpub. data) fed three diets of the same lipid content (10%), but with protein at 46%, 56% or 65%, equating to protein energy ratios of 22 g/MJ, 26 g/MJ and 29 g/MJ, to small *L. calcarifer* (2.5 g) and found that growth was again best at the highest protein:energy ratio, though not significantly more so than the 26 g/MJ diet. However, there was a clear curvilinear response to protein : energy ratio with the estimate peak at around 29 g/MJ.

In combination these results clearly define that there is a curvilinear relationship in *L. calcarifer* between fish size and protein : energy demands, which are discussed more fully in section 8.2.1.4.

8.2.1.1 Amino Acid Requirements

It is generally considered that the qualitative requirements for essential amino acid by *L. calcarifer* are for the same 10 amino acids as that of all other fish. However, there is no specific data to verify or refute this hypothesis. Estimations of essential amino acid (EAA) demands can be derived based on the amino acid composition of the body tissues (Table 8.2). To do this, the relative ratios of each of the EAA relative to lysine, usually regarded as the first limiting amino acid in most formulated diets, are defined. The demands for lysine are then determined empirically and the requirements for the remaining nine EAA determined based on that "ideal" ratio. This concept is often referred to as the "ideal protein" concept and works on the assumption that the ideal amino acid composition for a diet is one that reflects the amino acid composition of the animal's body proteins. However, this method can potentially underestimate those EAA with high metabolic turnover, or overestimate those EAA that are preferentially utilised in protein synthesis. Comparison of 'ideal protein' data relative to the experimentally determined requirement for lysine by Millamena (1994) shows some degree of homology (Glencross 2006, Table 8.2).

Table 8.2. Essential amino acid composition of whole *Lates calcarifer* and key organs (g/16g N, unless otherwise stated). From Glencross (2006).

Amino Acid	Muscle	Liver	GIT	Whole-body	Whole-body relative to Lysine (%)
Arginine	6.2	5.5	7.7	6.9	111
Histidine	2.1	2.0	1.2	1.5	24
Isoleucine	4.5	4.4	3.5	3.6	59
Leucine	8.6	9.6	6.5	7.1	115
Lysine	8.3	3.9	5.1	6.2	100
Methionine	3.3	3.0	2.7	3.0	48
Phenylalanine	4.2	5.1	4.3	4.2	67
Threonine	4.6	6.4	4.7	4.5	73
Valine	4.8	6.3	4.3	4.4	71

GIT: Gastrointestinal tract. Data from Glencross (2006). Tryptophan data not reported.

There have been some estimates of *L. calcarifer* dietary requirements for a couple of amino acids. Colosso et al. (1993) estimated that the requirement for tryptophan was about 0.5% of dietary protein, which is similar to the estimates for other species (NRC 2011). The requirements for methionine, lysine and arginine have been determined to be about 2.2%, 4.9% and 3.8% of dietary protein respectively (Millamena 1994). These estimates are also largely consistent with those amino acid requirements reported for most other carnivorous fish species (NRC 2011). Interestingly, it has also been reported that excessive dietary tyrosine can cause kidney malfunction in *L. calcarifer* (Boonyaratpalin 1997).

Williams et al. (2001) undertook a series of experiments that examined the capacity of *L. calcarifer* to use either crystalline or protein-bound amino acids. The first study added crystalline lysine to a wheat gluten based, high-protein diet, and compared its utilisation to complementary diets that had been modified to have an equivalent level of lysine enrichment, but with protein-bound amino acids. The second study similarly examined the addition of crystalline amino acids, but this time into a lower protein diet. Both studies showed that the utilisation of the crystalline amino acids was as effective as that of protein-bound amino acids. This work was important in that it demonstrated the potential to utilise crystalline amino acids to balance essential amino acid composition of diets as the need arises.

8.2.1.3 Protein and amino acid utilisation

Williams and Barlow (1999) determined that protein utilisation by *L. calcarifer* was about 46% efficient in a diet with 47% protein, and reduced to 34% when a 37% protein diet was fed to the same fish. This suggested that with differential protein intake that there were differences in the utilisation efficiency of protein by the animal. Glencross (2008) subsequently identified that the partial efficiencies of protein utilisation by *L. calcarifer* were about 50% efficient. This is similar to that observed in other fish species (Lupatsch et al. 2003). Presently, there are no published data for specific amino acid utilisation efficiencies, and it is largely assumed that they will be similar to that of protein. Certainly an amino acid utilisation efficiency of this magnitude would be similar to that observed in other fish species (Hauler and Carter 2001). However, further work in this area is required as future reliance on grain and other non-fish protein sources will begin impinging on amino acid requirements of the animal and more accurate data are required to ensure minimum requirements are met.

8.2.1.4 Iterative protein and amino acid requirements

Most studies estimating protein and energy requirements of *L. calcarifer* have been based on empirical dose-response studies. However, in recent years nutritional models have evolved to the point where considerable progress has been made in estimating nutritional requirements using iterative (reverse engineering) methods. This strategy is based on the notion that an animal's dietary needs can be estimated by developing an integrative approach to defining what the animal needs for deposition and maintenance processes, as well as the inefficiencies associated with each of those processes (Glencross 2008). This integrative approach relies on several assumptions and defined parameters/algorithms, one of which is the determination of the partial efficiency of protein and/or amino acid deposition relative to dietary intake (see Fig. 8.2). For a more comprehensive description of this process see Glencross (2008).

One of the advantages of the iteratively determined requirements for protein and lysine is that they clearly demonstrate the multi-faceted nature of the requirement for nutrients. This approach also clearly demonstrates the varying requirement for these nutrients as live-weight of the fish changes (Table 8.3 and Fig. 8.3). Furthermore, these findings are consistent with the outcomes of a wide range of empirical studies (see sections 8.2.1.1 and 8.2.1.2). This change in nutrient requirements is largely driven by the relative change in energy demands of the fish as it grows.

8.2.2 Lipid and essential fatty acids

Lipids are an important dietary energy source for *L. calcarifer* and are also the source of essential fatty acids. In concert with the work on protein requirements there has also been a considerable amount of work devoted

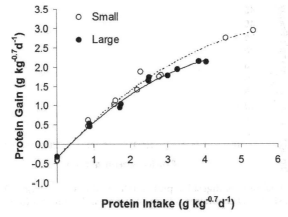

Figure 8.2. Protein gain with varying digestible protein intake levels by *Lates calcarifer* of two different size classes. Non-linear regression equation for each fish size is: Protein gain (small-fish: 15g) = –0.40 + 11.55 • (Protein intake) • (15.45 + (Protein intake))$^{-1}$, (R^2 = 0.993). Protein gain (large-fish: 400g) = –0.40 + 3.98 • (Protein intake) • (3.49 + (Protein intake))$^{-1}$, (R^2 = 0.944). Linear regression equation below an intake of 3 g kg$^{-0.7}$ d^{-1} is: Protein gain = 0.822 • (Protein intake) –0.370, (R^2 = 0.967). Maintenance protein intake level as estimated based on this linear regression is 0.45 g kg$^{-0.7}$ d^{-1}. Linear regression equations above an intake of 2.5 g kg$^{-0.7}$ d^{-1} are: Protein gain (small-fish) = 0.484 • (Protein intake) = 0.423, (R^2 = 0.986) and Protein gain (large-fish) = 0.496 • (Protein intake) = 0.756, (R^2 = 0.939).

Table 8.3. Estimated protein and lysine requirements of *Lates calcarifer* over varying live-weights as determined iteratively based on energetic demands, nutrient utilisation efficiencies and deposition demands (Glencross 2006).

Fish weight (g)	10	50	100	500	1000	2000
15 MJ Digestible Energy diet						
%BW intake/day	4.9%	2.7%	2.1%	1.2%	1.0%	0.8%
Digestible Protein (% diet)	56%	46%	42%	34%	31%	28%
Digestible Lysine (% diet)	3.5%	2.8%	2.6%	2.1%	1.9%	1.8%
17 MJ Digestible Energy diet						
%BW intake/day	4.3%	2.4%	1.9%	1.1%	0.8%	0.7%
Digestible Protein (% diet)	63%	52%	48%	39%	35%	32%
Digestible Lysine (% diet)	3.9%	3.2%	3.0%	2.4%	2.2%	2.0%

to exploring the inclusion of lipids in *L. calcarifer* diets to increase their energy density. There has also been some work on examining various facets of essential fatty acid requirements.

186 Biology and Culture of Asian Seabass

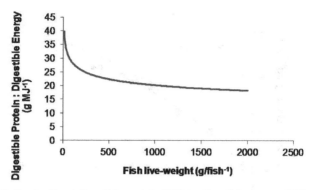

Figure 8.3. Optimal ratio of digestible protein (DP) to digestible energy (DE) with varying *Lates calcarifer* fish size. Data is derived from Glencross and Bermudes 2012.

8.2.2.1 Total lipids

Lipids are the most energy dense of all dietary nutrients, providing almost double the energy of protein and more than double that of carbohydrates. There have been several studies that have examined the variable inclusion of lipids, primarily for energetic reasons (Sakaras et al. 1988, Catacutan and Coloso 1995, Williams et al. 2003a, Glencross et al. 2008). The best growth was observed from *L. calcarifer* fed diets with 15% to 18% lipid content when protein levels were 45% to 50% (Sakaras et al. 1988, 1989). Tucker et al. (1988) found little effect on growth from *L. calcarifer* fed diets with either 9% or 13% lipids, but noted that feed conversion ratio was lower with the higher lipid levels.

Catacutan and Coloso (1995) examined inclusion levels of 5%, 10% and 15% lipids with three protein levels (35%, 42.5% and 50%). The 15% lipid level sustained the highest growth rate, provided protein was also at the highest levels (50%). Growth was similar with those fish fed diets containing 10% lipids and 42.5% protein. However, increased levels of somatic fat deposition were observed with increasing dietary fat levels.

Williams et al. (2003a) examined the influences of varying dietary lipid levels on the utilisation of varying protein levels. The authors conducted two experiments with different sized fish using different levels of both lipid and protein. For the first experiment, larger (230 g) fish were fed diets with protein levels varying from 38% to 53% and lipid levels from 7%, 13% or 19%. In that experiment the fish performed better when fed the higher lipid levels, though this was dependent on a high protein level in the diet. In a second experiment, smaller (80 g) fish were fed a similar array of diets with lipid levels of 13%, 18% or 23% and protein levels varying from 44% to 65%. In this second experiment there was an improvement in FCR with the use of higher lipid levels, though not necessarily an improvement in growth rate. The improvement in FCR/feed efficiency at higher lipid levels suggests an energy dependent effect of growth. The best performance from the small fish was seen with a diet containing 18% lipid and 60% protein.

In a study of some extruded commercial diets, Glencross et al. (2003) found that two diets of similar protein levels, but differing substantially in lipid levels (16% vs. 22%) sustained equivalent growth when fed to 555 g fish, and that the higher lipid level diets resulted in a significantly lower feed conversion ratio. This was primarily a response by the fish to the energy density of the diets. Subsequently, Glencross et al. (2008) found that increasing the lipid content of diets fed to large (1180 g) fish resulted in both improvements in FCR as well as growth. In this study, lipid levels were varied from 10% to 34%, with the best growth resulting from diets containing about 30% lipid. A clear relationship between energy intake and fish performance was observed, and in this case it was the addition of lipid as fish oil that was the primary driver of that energy density.

8.2.2.2 Essential fatty acids

The long-chain polyunsaturated fatty acids have been shown to provide some essential fatty acid (EFA) value to *L. calcarifer* (Borlongan and Parazo 1991, Boonyaratpalin 1997). Early studies by Buranapanidgit et al. (1988, 1989), suggested that total n-3 EFA levels, primarily as a mix of 20:5n-3 Eicosapentaenoic acid (EPA) and 22:6n-3 Docosahexaenoic acid (DHA) of 1.0 to 1.7% of the diet were optimal for growth of *L. calcarifer*. These estimates were later confirmed by studies by Wanakowat et al. (1993). The demands for lc-PUFA (EPA + DHA) were also explored in a study examining gross inclusion levels as manipulated by the addition of various blends of fish oil and soybean oil (Williams et al. 2006). This study also examined the effect of manipulating the n-3 to n-6 fatty acid ratio on fish performance (Williams et al. 2006). Optimal performance of *L. calcarifer* was observed at inclusion levels of 18:3n-3 linolenic acid (LNA), 20:5n-3 and 22:6n-3 of 0.45%, 0.75% and 1.15% respectively (a total lc-PUFA content of 1.9%). An optimal n-3 to n-6 fatty acid ratio of 1.5–1.8:1 was suggested.

More recently, the clear requirements for combined inclusion of EPA and DHA were identified through specific studies on addition of each of 20:4n-6 arachidonic acid (ARA) and EPA to diets containing DHA, and also the incremental inclusion of DHA from 0.1% to 1.9% (Glencross and Rutherford 2011). Limited effect of DHA on growth was observed in this study (Fig. 8.4a), though increasing the level of DHA had a profound effect on fish health with an increasing degree of ionic dysfunction occurring. This ionic dysfunction was ameliorated by the addition of EPA, but not by the addition of ARA. The known role of these C20 fatty acids in the production

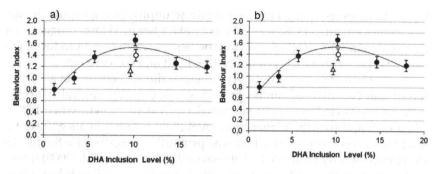

Figure 8.4. (a) Growth in *Lates calcarifer* response to varying inclusion levels of DHA (●) and 1% DHA + 1% EPA (○) or 1% DHA + 1% ARA (Δ). (b) Behaviour responses from fish in response to varying inclusion levels of DHA (●) and 1% DHA + 1% EPA (○) or 1% DHA + 1% ARA (Δ). Data is derived from Glencross and Rutherford (2011).

of prostaglandins and leukotrienes in other fish was implicated in *L. calcarifer* in terms of their involvement in ionic regulation (Glencross 2009). Another effect of the incremental inclusion of DHA was the effect on fish behaviour. As DHA increased in the diet the incidence of positive feeding behaviour also increased, suggesting a strong link between DHA and either brain or visual development (Fig. 8.4b). Interestingly, the increasing inclusion of DHA also increased the "effective" retention efficacy of EPA, whilst DHA retention declined with increasing inclusion (Fig. 8.5). The retention of ARA was also exacerbated by increasing inclusion of DHA, but not to the same extent as seen for EPA.

Studies on the expression of elongase and desaturase genes (including functional characterisation) in *L. calcarifer* have shown some capacity to chain elongate and desaturate shorter-chain PUFA (18:2n-6 (LOA) and LNA) to make 18:3 and 18:4 fatty acids (Mohd-Yusof et al. 2010). Specifically, activities for a fatty acid desaturase (FADS) Δ6 and an elongase (Elovl5) were identified. However, no ability to make 20:5 from 20:4 or 20:4 from 20:3 was identified, indicating an inability in *L. calcarifer* it appears to generate lc-PUFA like EPA and DHA *de novo* (Mohd-Yusof et al. 2010).

Alhazzaa et al. (2011a, b) also demonstrated that *L. calcarifer* had limited ability to elongate and desaturate PUFA to make lc-PUFA. The authors postulated that it was the FADS Δ6 that limited this capacity and therefore the addition of stearidonic acid (SDA; 18:4n-3) to diets should circumvent this problem. Diets containing Echium oil (rich in SDA) fed to juvenile (5 g) fish demonstrated that there was little to no elongation and desaturation relative to a diet provided with LNA and LOA as the predominant PUFA. Whilst the fish were being fed these diets it was noted that there was an up-regulation in expression of genes for FADS Δ6, but that this made little

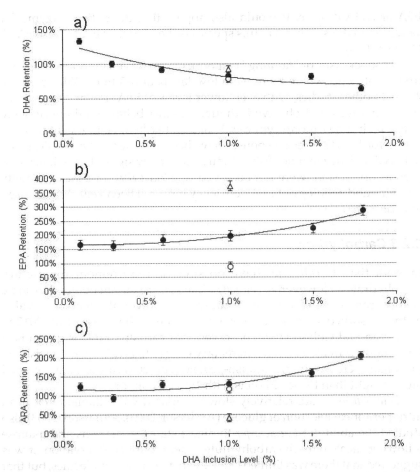

Figure 8.5. (a) Retention of DHA in *Lates calcarifer* in response to varying inclusion levels of DHA (●) and 1% DHA + 1% EPA (○) or 1% DHA + 1% ARA (Δ). (b) Retention of EPA in response to varying inclusion levels of DHA (●) and 1% DHA + 1% EPA (○) or 1% DHA + 1% ARA (Δ). (c) Retention of ARA in response to varying inclusion levels of DHA (●) and 1% DHA + 1% EPA (○) or 1% DHA + 1% ARA (Δ). Data is derived from Glencross and Rutherford (2011).

difference to the total accumulation of EPA or DHA. It was hypothesised that this limitation must be at some other point of the elongation and desaturation pathway.

Tu et al. (2012a, b) also hypothesised that it was the Δ6 FADS that limits the ability of a fish to produce lc-PUFA. In a series of two studies it was demonstrated that the FADS from *L. calcarifer* had both Δ6 and Δ8 desaturase activities and that *L. calcarifer* can make 18:4 from 18:3 and 20:4 from 20:3, but there was no evidence for synthesis of 20:5 from 20:4 (usually performed by a Δ5 FADS). Therefore, it appears that *L. calcarifer* are unable to generate

EPA or DHA *de novo*. It would also appear that the critical enzyme for *de novo* lc-PUFA synthesis in this species is the $\Delta 5$ rather than the $\Delta 6$ FADS (Tocher 2010).

A "shock-like" or "fainting" response was observed in some *L. calcarifer* from treatments where there were low levels of n-3 EFA (Williams et al. 2006). In other studies, with even lower levels of EFA similar "shock-like" symptoms were not observed, though distinct behavioural effects were noted, including a more cryptic and panicky like behaviour (Glencross and Rutherford 2011). These responses have been reported to be a typical sign of EFA deficiency in other fish, although other key signs of EFA deficiency, such as fin erosion, were not observed (Castell 1979). In general though, the signs and aetiology of EFA deficiency have not been well characterised in *L. calcarifer*.

8.2.3 Carbohydrates

Like all fish, *L. calcarifer* have no specific requirement for dietary carbohydrates. However, *L. calcarifer*, like some fish, can obtain digestible energy from some carbohydrate sources. However, the metabolic value of this absorbed glucose has been questioned (Glencross et al. 2012b). Catacutan and Coloso (1997) observed that carbohydrate, as gelatinised bread flour, provided significant dietary energy to *L. calcarifer*. The fish fed diets that were iso-lipidic and iso-proteic with 20% carbohydrate gained more weight than those fed diets with 15% carbohydrates.

Lates calcarifer are relatively slow, compared with tilapia, to clear an intra-peritoneal injection of glucose (Anderson 2003). Anderson (2003) also studied the ability of *L. calcarifer* to digest starches, dextrins and maltose following meals of each carbohydrate type. From his observation, it was suggested that there was limited capacity to digest these molecules, but that *L. calcarifer* did absorb free glucose quickly. *Lates calcarifer* are also unable to metabolise galactose or xylose at any appreciable rate (Stone 2003).

Starch digestibility at 15% and 30% inclusion levels was examined by McMeniman (2003). The digestibility of glucose, maltose and different starch sources were examined, as were the effects of varying degrees of starch gelatinisation (McMeniman 2003). Increasing starch inclusion in the diet reduced the apparent digestibility of the energy content of the diets, but not that of the protein content. The component digestibilities for the energy content of starch at 15% and 30% inclusion were calculated at 29% and 19% respectively. Increasing gelatinisation of starch, from 0% to 80%, decreased its energy digestibility from 20% digestible at 0% to 11% digestible at 80% gelatinisation. Substantial differences in the component digestibilities of maltose, glucose and various starch sources were also identified. Interestingly, in contrast to previous reports, pre-gelled wheat starch was poorly digested and had an energy value of essentially 0% (McMeniman 2003). The digestibility of ungelatinised wheat starch was unclear. Pea starch energy digestibility was poor at about 5%, and the energy digestibility of glucose and maltose was 40%. The dextrin content of starch was also shown to not influence its energy digestibility. It was clear from this work, that further research into the digestibility of starch is required before definitive conclusions can be drawn regarding the effects of different starch sources and different forms of starch on *L. calcarifer* digestibility.

In a more recent study, Glencross et al. (2012a) examined the protein, energy and starch digestibility of a wide range of cereal varieties. Included among these were wheat, barley, sorghum, tapioca and faba beans. From this work Glencross et al. (2012a) were able to define a clear relationship between starch content and the digestibility of starch (Fig. 8.6). More specifically it was the amylopectin content that was the defining parameter of the response to digestibility, with higher levels of amylopectin correlating with reduced starch digestibility. The relationship with amylose digestibility, while more dramatic, was not as strong. When pre-gelatinised starch was added as the sole carbohydrate source and at different inclusion levels, Irvin et al.

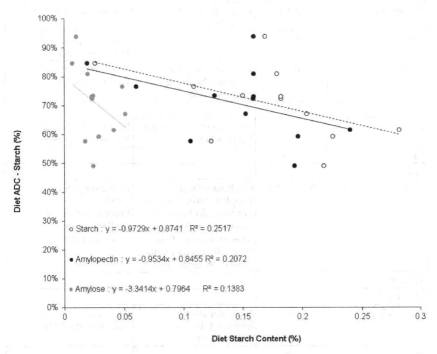

Figure 8.6. Apparent digestibility coefficient (ADC) of starch by *Lates calcarifer* when examined from a broad range of starch sources when each was included in test diets at 30%. Derived from Glencross et al. (2012).

(2012) observed that this form of starch was well absorbed and digested as an energy source and had little effect on the digestion of protein. Other carbohydrates, like lignin, pectin and cellulose, presented contrasting challenges to protein and energy digestion when included in the diet.

8.2.4 Vitamins

Like most other fish species, *L. calcarifer* also have certain essential dietary vitamin requirements (Table 8.4). However, the list of those vitamins that have been examined is not complete, and those where requirements have not

Table 8.4. Summary of vitamin requirements (mg/kg of diet) for *Lates calcarifer*.

Vitamin	Requirement (mg/kg diet)	Deficiency Signs	Authors
Thiamine	R	Poor growth, High mortality, Stress susceptible	Boonyaratpalin and Wanakowat (1993)
Riboflavin	R	Erratic swimming, Cataracts	Boonyaratpalin and Wanakowat (1993)
Pyridoxine	5–10	Erratic swimming, High mortality, Convulsions	Wanakowat et al. (1989)
Pantothenic acid	15–90	High mortality	Boonyaratpalin and Williams (2001)
Nicotinic acid	n/a	Fin hemorrhaging and erosion, Clubbed gills, High mortality	Boonyaratpalin and Williams (2001)
Biotin	n/a		
Inositol	R	Poor growth, Abnormal bone formation	Boonyaratpalin and Wanakowat (1993)
Choline	n/a		
Folic acid	n/a		
Ascorbic acid (Vit C)	25–30[a] (700[b])	Gill hemorrhages, Exophthalmia, Scoliosis, Lordosis, Broken back syndrome, Fatty liver, Muscle degeneration, Poor gill development, Bone deformations	Boonyaratpalin et al. (1989) Phromkunthong et al. (1997) Fraser and deNys (2011)
Vitamin A	n/a		
Vitamin D	n/a		
Vitamin E	R	Muscular atrophy, Increased disease susceptibility	Boonyaratpalin and Wanakowat (1993)
Vitamin K	n/a		

R: required, but not quantitatively defined. n/a: no data identified. [a] Based on ascorbyl-2-monophosphate-Mg or ascorbic acid glucose. [b] Based on crystalline ascorbic acid.

been specifically identified are still thought to be required, consistent with findings from other species. Further work on defining the requirement for all vitamins is needed, though it is acknowledged that this has been difficult because of poor acceptance by *L. calcarifer* of purified diets. Application of different forms of some vitamins (e.g., stabilised versus unstablised forms of ascorbic acid) also need to be evaluated more thoroughly.

There have been several studies on the requirement for vitamin C. Early studies using crystalline ascorbic acid suggested that an inclusion level of 500 to 700 mg/kg in the diet supported normal growth. Diets without added vitamin C tended to have fish that had normal growth for two weeks before growth declined and deficiency signs developed (Boonyaratpalin et al. 1989, 1994). Addition of vitamin C has also been shown to reduce the incidence of spine and jaw deformities in small fish (Fraser and de Nys 2011). An alternative form of vitamin C is ascorbyl-2-monophosphate-magnesium. In a study using levels of 0, 30, 60 and 100 mg/kg of dietary inclusion, *L. calcarifer* demonstrated better growth, survival and feed conversion at all inclusion levels of vitamin C compared to the 0 mg/kg control, although no differences were noted between any of the test levels (Phromkunthong et al. 1997). Clinical signs of vitamin C deficiency included fin erosion, erratic swimming behaviour and an increase in the incidence of spinal deformities.

For the other vitamins, few studies exist. Boonyaratpalin et al. (1989) found no effects on growth, feed conversion or survival when inositol, niacin, choline or vitamin E were absent from a practical diet formulation. However, poor weight gain and feed conversion were observed from fish fed similar diets without added thiamine and riboflavin (Pimoljinda and Boonyaratpalin 1989). In a series of studies using semi-purified diets, Boonyaratpalin and Wanakowat (1993) later demonstrated that thiamine, pantothenic acid, inositol and vitamin E were each required for normal growth by *L. calcarifer*.

In studies using semi-purified diets, 5 mg/kg diet of pyridoxine was required for normal growth and, at inclusion levels of 10 mg/kg of diet, increased levels of lymphocytes were observed (Wanakowat et al. 1989). The requirement for pantothenic acid (supplied as calcium D-pantothenate) has been shown to be 15 mg/kg of diet. Without a minimum supplementation level of pantothenic acid, total mortality occurred within six weeks (Boonyaratpalin et al. 1994).

8.2.5 Minerals

There are few studies on any of the mineral requirements of *L. calcarifer*. The exception to this is the requirement for phosphorous, which was examined based on the incremental inclusion of monosodium phosphate to a fish meal based diet (Boonyaratpalin and Phongmaneerat 1990). Best growth was observed with the 0.5% supplementation level, but the feed conversion and protein retention of the fish was better with 1.0% supplementation. The authors estimated the dietary phosphorus requirement for juvenile *L. calcarifer* as 0.55% to 0.65%, which is marginally lower than that estimated for salmonids (NRC 2011). However, given the higher level of bone content in *L. calcarifer* relative to salmonids, it is suspected that the endogenous phosphorus in the experimental diet was not accounted for in this study and therefore the requirement in reality is probably substantially higher (e.g., 0.8 to 1.0 g/kg). Supporting this notion, were the results from another study which examined the supplementation of monosodium phosphate at 0, 0.25, 0.5 and 1.0% inclusion to diets based on low-ash and high-ash fish meal. This therefore allowed the examination of diets with low and high levels of fish meal phosphorus content, respectively. The results of this work showed that addition of 0.5% and 1.0% monosodium phosphate significantly increased total phosphorus deposition in the fish (Chaimongkol and Boonyaratpalin 2001). Irrespective, the requirements for phosphorus by *L. calcarifer* require clarification.

Ironically, despite its close relationship with phosphorus metabolism, there are no reported studies on the nutritional requirements for calcium in *L. calcarifer*. This also needs some rectification.

Studies on requirements for other trace elements like iron, copper, manganese, magnesium and selenium are largely absent. In most commercial diet formulations the mineral mix used is largely consistent with that used for salmonids (NRC 2011).

8.2.6 Energy requirements

In *L. calcarifer*, as with other fish, the energy density of the feed has been shown to directly influence the amount of feed consumed (Glencross et al. 2003). With higher energy dense diets there is a reduction in the amount of food consumed and this has direct implications on the concentration of essential nutrients required in daily intake to satisfy growth demands (Glencross 2006, 2008).

The energetic content of the three macro-nutrient classes; protein, lipids and carbohydrates is the source of this dietary energy, though the capacity of *L. calcarifer* to utilise each of these nutrients is characteristic of a high-order carnivore (Glencross et al. 2012b). In this sense, dietary energy demands are largely met through the metabolism of dietary protein and lipid intake, and to only a limited extent through the carbohydrate content of any food consumed.

8.2.6.1 Energy demand

Dietary energy demands are generally assumed to be the sum of the needs of a growing fish to satisfy its maintenance and growth energy demands. Within the energy demands for maintenance, is also energy utilised for activity and heat losses. However, being a poikilothermic animal, the energy demands by *L. calcarifer* directly reflect the temperature of their environment. Across normal thermal regimes (20°C to 32°C) for *L. calcarifer*, the energy demands for growth and maintenance follow a Gompertz relationship (Bermudes et al. 2010). However, above and below critical thermal ranges there will be deterioration in the nature of this balance between maintenance and growth energy demands, and also the efficiency with which dietary energy is utilised (Bermudes et al. 2010, Glencross and Bermudes 2010, 2011, 2012).

Classical nutrition defines that from the moment a fish is fed the utilisation of energy can be partitioned according to its different roles and inefficiencies (Fig. 8.7, NRC 2011). The net outcome of this is the deposition of energy as tissue growth or 'recovered energy' (RE). However, assessment of energy partitioning for each of these parameters has been fraught with error, so a more simplistic approach of defining energy partitioning has been developed. In this simpler format (factorial bioenergetic approach), the flow of energy is ascribed into either deposition or non-deposition formats and parameters can be measured from empirical studies to describe the efficiencies of those utilisation processes (Glencross 2008).

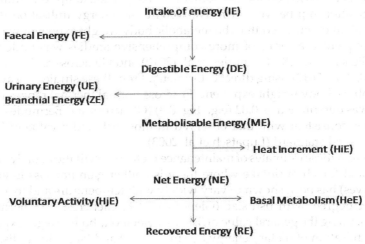

Figure 8.7. Schematic of energy flow through a fish. Derived from NRC (2011).

8.2.6.2 Energy utilisation and demands for maintenance

Energy demand by all animals is a mass-specific relationship (Withers 1992). Therefore, to understand the relationship between a fish's live-weight and its energy demand, an equation needs to be defined. Typically these relationships are described by an equation such as $y=ax^b$. In this relationship the exponent 'b' describes the relationship between the animal's live-weight and its energy demand for maintenance (often referred to as the metabolic body weight exponent), while the coefficient 'a' describes the species and temperature specific nature of that energy demand (Glencross and Bermudes 2011).

The relationship between body size and energy demand has been estimated using several methods in *L. calcarifer*. Both direct and indirect calorimetry have been used to examine the energetics of *L. calcarifer* during starvation, this is usually referred to as the standard metabolic rate (SMR). Using direct calorimetry the loss of somatic energy reserves from starved fish over a period of time (weeks) can be estimated (Glencross and Bermudes 2011). By contrast, indirect calorimetry typically bases its assessment of energy use on the measurement of an indirect indicator of energy consumption such as oxygen use or carbon dioxide production by fish 'fasted' for a minimum of 24 h (Glencross and Felsing 2006). It should be noted that there is often a difference in the amount of energy loss measured when fish are either 'fasted' or 'starved'.

Glencross and Felsing (2006) studied *L. calcarifer* using indirect calorimetry based on the assessment of oxygen consumption. From this, the relationship between fish live-weight and energy utilisation during starvation determined that the metabolic body weight exponent was 0.73. Subsequently, a series of more comprehensive studies was undertaken by Glencross (2008), Bermudes et al. (2010) and Glencross and Bermudes (2010, 2011, 2012) using direct calorimetry. From these studies the average metabolic body weight exponent (b) across normal thermal ranges (23–32 °C) was determined as 0.82 (e.g., Fig. 8.8) (Glencross and Bermudes 2011). This is consistent with that observed in most fish and a value of 0.80 is considered standard (Lupatsch et al. 2003).

From these estimates of maintenance metabolic activities a daily energy demand for HEm (intake where there is neither gain nor loss in energy reserves) has been shown to vary widely with temperature and to a much lesser degree, with fish size (Glencross 2008, Glencross and Bermudes 2010). Using the general value of 0.80 as a metabolic body weight exponent the utilisation of dietary energy by small (15 g) and large (400 g) fish was transformed (Glencross 2008). From this a saturation kinetic function was considered the most appropriate function to describe this relationship (Fig. 8.9). Here; Energy Gain (kJ/ $kg^{0.8}$/d) = –37.0 + 403 * (Energy intake) /

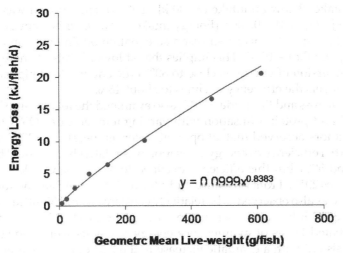

Figure 8.8. Energy loss through starvation by juvenile *Lates calcarifer* of varying live-weights when held at 32°C. Data derived from Glencross and Bermudes (2011).

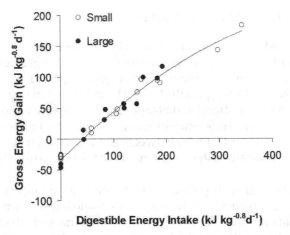

Figure 8.9. Energy gain with varying digestible energy intake by *Lates calcarifer* of two different size classes. From Glencross (2008).

(376 + (Energy intake). However, traditionally this relationship is described using a linear equation such that regression over all intake levels in Fig. 8.9 is Energy Gain (kJ/ kg$^{0.8}$/d) = 0.684 * (Energy intake) − 26.80. This equation indicates that energy utilisation over this energy intake range is about 68% efficient. Energy intake is also based on units of kJ/kg$^{0.80}$/d. From this the intercept at the X-axis can also estimate the maintenance energy intake level (HEm) as 42.6 kJ/kg$^{0.80}$/d. However, given that the line was clearly not linear, linear regression was applied at two different levels of

feed intake. Below an intake of 100 kJ/ $kg^{0.8}$/d, energy gain was; Energy Gain (kJ/ $kg^{0.8}$/d) = 0.806 * (Energy intake) – 34.39, while above an intake of 150 kJ/ $kg^{0.8}$/d) linear regression an equation of: Energy gain = 0.463 • (Energy intake) – 15.67. This implies that at lower levels of energy intake the conversion of energy is close to 80% efficient, while at higher energy intake levels the efficiency declines to about 46%.

Glencross and Bermudes (2010) also examined the relationship between energy and protein utilisation with varying temperatures (23°C to 37°C). The authors observed that, at optimal water temperatures (25°C to 32°C), the partial efficiency of energy utilisation was relatively consistent between 53% and 59%, but this efficiency declined to 42% at 36°C (Glencross and Bermudes 2010). From this same study, the maintenance demand for energy (HEm) was also observed to be relatively constant around 35–40 kJ/ $kg^{0.80}$/d, but at 36°C it began to increase to an estimate of 110 kJ/$kg^{0.80}$/d. These exacerbated levels of maintenance energy demands were also shown to be consistent with a dramatic increase in the loss of protein as an energy reserve above 32°C.

8.2.6.3 Energy demands for growth

Energy demands for growth are dictated by a series of inter-related factors. These factors are generally considered as; the amount of growth potential at a given water temperature and fish live-weight (Fig. 8.10), the composition (energy density) of that growth (Fig. 8.11) and the partial efficiency with which consumed and digested dietary energy is converted into growth. For *L. calcarifer* the partial efficiency of energy conversion has been estimated to be between 0.42 and 0.68 (or between 42% and 68% efficient) subject to water temperature and fish size (Glencross 2008, Glencross and Bermudes 2010).

Estimates of growth potential are dependent on empirical data and clearly the most practical data for such an estimate is from farm production. However, data reported from a combination of farm and laboratory based data provides one of the most robust assessments of growth for *L. calcarifer* over the temperature range of 16°C to 39°C and for a wide range of fish sizes (10 g–3000 g) (Glencross and Bermudes 2012). These authors reported a functional growth equation/model depicted as Fig 8.10 equation available in Glencross and Bermudes (2012).

The composition of whole *L. calcarifer* over a wide live-weight range (2 g–2400 g) has been reported by Glencross and Bermudes (2012) (Fig. 8.11). This data showed a clear exponential relationship between live-weight, lipid and dry matter density of the fish. This also translated to a similar exponential relationship between live-weight and energy density; Energy density (kJ/g) = 3.82 * (live-weight)$^{0.12}$. The inflection point of this energy

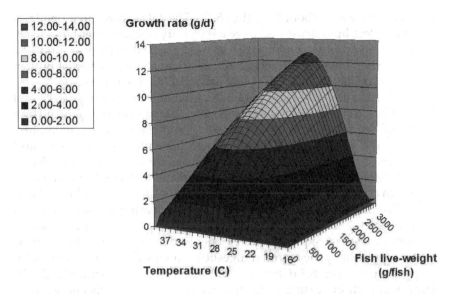

Figure 8.10. Growth model for *Lates calcarifer* growth (g d⁻¹) from 16°C to 39°C and up to fish of 3000 g in weight. Model includes variance in optimal temperatures between small and large fish, with larger fish reaching peak performance at 29°C and smaller fish at 31°C. Derived from Glencross and Bermudes (2012).

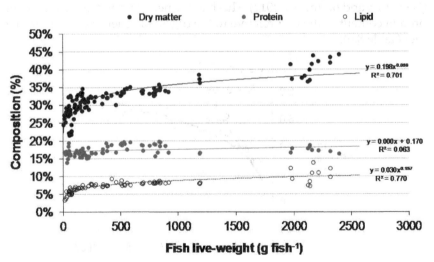

Figure 8.11. Body composition of *Lates calcarifer* at different size classes. Derived from Glencross and Bermudes (2012).

density curve was at about 200 g (Fig. 8.12). The data showed that small fish are clearly less lipid dense and have a lower dry matter content (inversely a higher water content) and that fish grow to a lipid density of around 10% subject to diets being effectively managed. It can also be noted that protein content is constant across fish size at around 17% of live-weight.

By integrating all of the parameters defined in sections 8.2.6.2 and 8.2.6.3 it becomes possible to predict the total digestible energy demand by a *L. calcarifer* of any size (from 10 g to 3000 g) and at any water temperature (within 16°C to 39°C) and this has been presented by Glencross and Bermudes (2012). This factorial bioenergetic modelling process can also be used to define the total digestible protein demand and together an iterative approach can be used to define ideal dietary specifications (Tables 8.5 and 8.6). What this approach showed is that with increasing fish size there was a clear increase in demand for energy and protein, but that the relative amount of protein required against energy demand diminishes. This is best exemplified by the required digestible protein: digestible energy ratio as presented in Fig. 8.3 (Glencross 2008, Glencross and Bermudes 2012). However, with increasing water temperature there was also an increase in energy demand and typically this can be managed by increasing the ration allocation to the fish (see Table 8.6). Nonetheless, under conditions of thermal stress (> 32°C) the relationship between digestible protein:digestible energy ratio changed and there was an increase in the demand for protein (Glencross and Bermudes 2012). This had the net effect of dictating a higher protein content in the diet relative to the digestible energy density of that diet (Table 8.6).

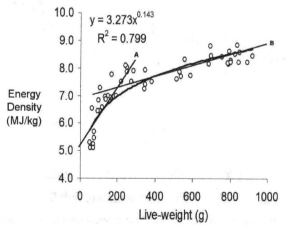

Figure 8.12. The prescription of diet protein and energy specifications based on *Lates calcarifer* somatic energy demands for growth. Derived from Glencross (2006). Represented on the figure as tangential lines are theoretical assignments of diets A and B. Diet A would be a lower energy, higher-protein diet, while diet B would be a lower-protein, medium-energy diet.

Table 8.5. Derived energy and protein demands in growing *Lates calcarifer* at 30°C based on the growth, energy and protein utilisation models from Glencross (2008).

Fish live-weight (g fish⁻¹)	50	100	500	1000	2000
Expected growth (g d⁻¹) [a]	2.24	3.01	5.95	7.98	10.70
Energy requirement					
Metabolic BW (kg)$^{0.80}$ [b]	0.09	0.16	0.57	1.00	1.74
DEmaint (kJ fish⁻¹ d⁻¹) [c]	5.05	8.79	31.86	55.48	96.59
Energy gain (kJ fish⁻¹ d⁻¹) [d]	12.83	19.00	47.35	70.15	103.94
DEgrowth (kJ fish⁻¹ d⁻¹) [e]	18.86	27.95	69.63	103.16	152.85
DEtotal (kJ fish⁻¹ d⁻¹) [f]	23.91	36.74	101.49	158.64	249.45
Protein requirement					
Protein BW (kg)$^{0.70}$ [g]	0.123	0.200	0.616	1.000	1.625
DProt-maint (g fish⁻¹ d⁻¹) [h]	0.06	0.09	0.28	0.45	0.73
Protein gain (g fish⁻¹ d⁻¹) [i]	0.37	0.50	0.99	1.33	1.78
DProt-growth (g fish⁻¹ d⁻¹) [j]	0.76	1.02	2.02	2.71	3.63
DProt-total (g fish⁻¹ d⁻¹) [k]	0.82	1.11	2.29	3.16	4.36
Optimal DProt:DE (g MJ⁻¹) [l]	34.1	30.2	22.6	19.9	17.5

[a] Daily weight gain determined using growth equation and fish live-weight and a temperature of 30°C.
[b] Live-weight in kilograms transformed by metabolic body weight exponent of 0.80.
[c] Maintenance digestible energy requirement determined using maintenance energy equation and fish live-weight and a temperature of 30°C per metabolic body weight.
[d] Energy content of live-weight gain.
[e] Digestible energy demand based on energy gain through growth divided by G_E (0.68).
[f] Combined digestible energy demand for maintenance and growth.
[g] Live-weight in kilograms transformed by protein body weight exponent of 0.70.
[h] Maintenance digestible protein requirement based on 0.45 g of digestible protein fish⁻¹ d⁻¹ per protein body weight.
[i] Protein content of live-weight gain (based on gain * 0.166).
[j] Digestible protein demand based on protein gain through growth divided by G_P (0.49).
[k] Combined digestible protein demand for maintenance and growth.
[l] Based on the equation for protein:energy ratio with varying fish live-weight.

8.3 Feed Management

Irrespective of how well a feed is specified or formulated, if it is not managed effectively then its value is severely diminished. There are two aspects to the management of feed, the choice of what feed to use at a particular stage of production and also how to allocate the appropriate ration of that feed to the fish to ensure its best use.

Table 8.6. Derived energy and protein demands in growing *Lates calcarifer* at 25°C, 30°C and 35°C based on the growth, energy and protein utilisation models from Glencross and Bermudes 2012.

Temperature	25°C	30°C	30°C	30°C	35 °C
Fish weight (g/fish)	100	10	100	1000	100
Expected growth (g/d)	2.11	1.06	2.88	7.85	2.27
Energy requirement					
Metabolic BW Exponent (x)	0.80	0.80	0.80	0.80	0.87
Metabolic BW (kg)^x	0.16	0.02	0.16	1.00	0.14
DEmaint (kJ/fish/d)	4.72	1.16	7.40	47.10	10.16
Energy gain (kJ/fish/d)	14.00	5.33	19.14	68.71	15.09
Energy utilisation efficiency (%)	53.50	61.20	61.20	61.20	49.50
DEgrowth (kJ/fish/d)	26.20	8.71	31.27	112.25	30.50
DEtotal (kJ/fish/d)	30.92	9.88	38.67	159.35	40.66
Protein requirement					
Protein BW Exponent (x)	0.68	0.70	0.70	0.70	0.83
Protein BW (kg)^x	0.21	0.04	0.20	1.00	0.15
DProt-maint (g/fish/d)	0.10	0.02	0.10	0.48	0.17
Protein gain (g/fish/d)	0.36	0.18	0.49	1.33	0.39
Protein utilisation efficiency (%)	46.70	48.70	48.70	48.70	31.20
DProt-growth (g/fish/d)	0.77	0.37	1.01	2.74	1.24
DProt-total (g/fish/d)	0.87	0.39	1.10	3.22	1.41
Optimal DProt:DE (g/MJ)	28.0	39.4	28.5	20.2	34.7

8.3.1 Feed choice

The choice of what feed type and/or specification to feed to fish at any particular point of their growth cycle will influence not only the efficiency with which they use that feed, but also the nutrient composition of their growth. Feeding a diet that is too energy dense, relative to the level of provided nutrients within the diet, is likely to result in a higher level of fat deposition within the fish. Feeding a low energy density will cause the fish to eat more as it seeks to satisfy its energy demands, thus result in a poorer feed conversion.

Maintaining growth efficiency as high as possible is more important with larger fish than smaller fish because of the implications that this has on the overall volumes of feed used. Because of this it is more important to use high-energy diets with larger fish (>500 g) than with smaller fish (<500 g). The point at which the diet should be changed from a low-energy to high-energy density can be estimated based on the changes in somatic composition of energy in the fish (Fig. 8.12). In this representation, tangential lines are fitted to the fish's energy density curve. The lines intercept where a change in diet is considered appropriate. The format of the tangential lines on the figure also provides additional information about optimal feed choice. The higher the gradient of the line the higher the projected requirement for digestible protein and other key nutrients in that diet. The higher the Y-intercept of the line, then a greater energy density of the diet that should be considered. In the example given (Fig. 8.12) diet A would be a lower energy, higher-protein diet (15 MJ DE/kg, 50% crude protein, 14% lipid), while diet B would be a lower-protein, medium-energy diet (17 MJDE/kg, 46% crude protein, 20% lipid). The change of diets would be appropriate at around a fish live-weight of 200 g.

Table 8.7. Typical feed specifications used by the Australian aquafeed manufacturing sector for *Lates calcarifer* of varying sizes.

	Fish Size (g/fish)		
	< 200	200 - 1000	> 1000
Dry Matter (%)	90	90	90
Crude Protein (%)	53	46	42
Digestible Protein (%)	48	41	38
Crude Fat (%)	10	20	30
Starch (%)	10	10	10
Ash/Filler (%)	17	14	8
Gross Energy (MJ/kg)	19.0	21.5	23.0
Digestible Energy (MJ/kg)	16.0	18.0	20.0
DP:DE (g/MJ)	30	23	19

Source: www.ridleyaquafeed.com.au

8.3.2 Feed ration

It is generally considered that feed intake in fish is largely regulated by energy demand and the digestible energy density of the feed (Bureau et al. 2002). This demand for energy by the fish is influenced by both water temperature and also the energy demands to sustain growth at a specific phase of the fish's growth cycle (Glencross 2008, Dumas et al. 2010). By examining the relationship between temperature and energy deposition demand, it becomes possible to estimate the total daily energy demand. This relationship can then be linked to total daily feed ration for any diet based on its digestible energy density (Glencross 2008, Glencross and Bermudes 2012). This approach to feed ration determination is termed a prescriptive approach.

From Table 8.8, it can be seen that, based on the data of Glencross and Bermudes (2012), the feeding demand of *L. calcarifer* will respond strongly to both increases in water temperature and fish size. Table 8.8 also demonstrates the effects of digestible energy density on feed ration allocation, with a decrease in feed fed for the higher energy density diet relative to the lower digestible energy density diet.

Although these models can define the amount of digestible energy, and therefore food that a fish needs to eat on any given day, the number of feeding sessions that is appropriate needs to be considered empirically. Ideally, it could be argued that the number of feeding sessions required, and the feeding rate used to distribute the feed, should be simply the number

Table 8.8. Feeding ration table for diets of either 16 MJ/kg digestible energy (16 MJDE) or 18 MJ/kg digestible energy (18 MJDE) fed to *Lates calcarifer* of various sizes at various temperatures. Derived from Glencross and Bermudes (2012).

16 MJDE					18 MJDE				
	Temperature°C					Temperature°C			
	20	25	30	35		20	25	30	35
Fish weight	g/fish				Fish weight	g/fish			
10	0.23	0.46	0.62	0.68	10	0.21	0.41	0.55	0.60
50	0.67	1.25	1.60	1.69	50	0.60	1.11	1.42	1.50
100	1.06	1.93	2.42	2.54	100	0.94	1.72	2.15	2.26
500	3.07	5.38	6.46	6.88	500	2.73	4.78	5.74	6.12
1000	4.86	8.41	9.96	10.82	1000	4.32	7.47	8.85	9.62
2000	7.70	13.17	15.45	17.28	2000	6.85	11.71	13.73	15.36
3000	10.08	17.15	20.03	22.89	3000	8.96	15.25	17.80	20.35
Fish weight	%BW				Fish weight	%BW			
10	2.32	4.56	6.17	6.78	10	2.06	4.05	5.49	6.03
50	1.34	2.50	3.19	3.38	50	1.19	2.22	2.84	3.00
100	1.06	1.93	2.42	2.54	100	0.94	1.72	2.15	2.26
500	0.61	1.08	1.29	1.38	500	0.55	0.96	1.15	1.22
1000	0.49	0.84	1.00	1.08	1000	0.43	0.75	0.89	0.96
2000	0.39	0.66	0.77	0.86	2000	0.34	0.59	0.69	0.77
3000	0.34	0.57	0.67	0.76	3000	0.30	0.51	0.59	0.68

required to ensure that the fish eat the entire specified ration. Williams and Barlow (1999) examined the influence of water temperature and ration frequency on performance of *L. calcarifer* in a series of empirical studies over a range of fish sizes (Table 8.9 and 8.10).

It has been suggested that for small fish that there is little benefit in feeding more than twice daily, and that for large fish this could be reduced to once daily without significant effects (Williams and Barlow 1999). However, despite this it can be seen in many instances there is additional growth resulting from extra feeding.

With temperature though, irrespective of fish size, there is a clear increase in feed intake and growth with increasing water temperature and this has been noted in multiple studies (Williams and Barlow 1999, Bermudes et al. 2010) (Tables 8.8, 8.9 and 8.10 and Fig. 8.13). The feed conversion ratio is lowest for fish at their point of optimal growth. The study by Williams and Barlow (1999) suggested that optimal growth occurred at around 26°C to 29 °C, though there were no treatments conducted above 29°C. More recent studies by Bermudes et al. (2010) showed that optimal growth occurred at closer to 32°C and this also corresponds with the optimal/lowest FCR

Table 8.9. Effects of varying temperature and feeding regimes on fish production of small (40 g) *Lates calcarifer*. Data derived from Williams and Barlow (1999).

	Regime	20°C	23°C	26°C	29°C
Growth rate (g/day)	2/d	0.75	1.01	1.71	1.79
	3/d	0.77	1.24	1.58	1.85
Feed conversion (g/g)	2/d	1.32	1.18	1.04	1.08
	3/d	1.39	1.18	0.98	1.14
Feed intake (g/day)	2/d	0.99	1.19	1.82	1.93
	3/d	1.06	1.46	1.61	1.86

Table 8.10. Effects of varying temperature and feeding regimes on fish production of larger (270 g) *L. calcarifer*. Data derived from Williams and Barlow (1999).

	Regime	20°C	23°C	26°C	29°C
Growth rate (g/day)	1/d*	1.92	2.34	2.97	3.43
	2/d	2.07	2.80	3.94	4.27
Feed conversion (g/g)	1/d*	1.15	1.13	1.08	1.01
	2/d	1.12	1.05	0.98	1.02
Feed intake (g/day)	1/d*	2.17	2.63	3.20	3.47
	2/d	2.31	2.93	3.86	4.40

*1/d values are averages of two treatments where fish were fed either in the morning or afternoon.

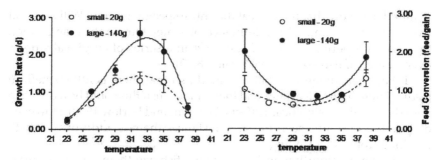

Figure 8.13. Effects of temperature on *Lates calcarifer* weight gain rate (g/d) and feed conversion ratio (feed/gain) from 23°C to 38°C. Indicated are also the initial weights of the fish used. Data derived from Bermudes et al. (2010).

(Fig. 8.13). Recent models have also noted that the optimal temperature varies slightly between large and small fish, with small fish generally being more tolerant of warmer waters (Glencross and Bermudes 2012).

The influences of photoperiod on growth and feeding of larval and juvenile *L. calcarifer* were examined by Barlow et al. (1995). Extended light periods were observed to promote faster growth of larval *L. calcarifer*. However, varying photoperiods (light-dark) of 12L-12D, 16L-8D, 24L-0D were not shown to influence growth of small (11–12 mm total length) juveniles (Barlow et al. 1995). During the 12L-12D regime fish were noted to cease feeding during periods of darkness, while during the 24L-0D regime fish were observed to feed during what would have been the normal light period, but then ceased feeding during what would have been the onset of the dark period, only to recommence feeding around midnight. For the larger fish, food consumption was about 40% greater in a continuously lit regime than in a 12L-12D regime though performance not significantly different. Despite these apparent anomalies, the authors concluded that extended light had little benefit beyond the larval phases of growth. Further work in this area would be useful.

The compensatory growth responses of *L. calcarifer* deprived of food for a one, two or three week period was examined by Tian and Qin (2002). The authors observed that fish deprived of food for one week displayed compensatory growth, obtaining a similar overall weight gain to fish that had not been starved, after a three-week re-feeding period. However, fish that had been starved for two or three weeks did not display the compensatory response that allowed them to recover lost growth. This work was followed by a later study that examined the compensatory growth responses of *L. calcarifer* deprived of various levels of food (0%, 25%, 50% or 75%) for two weeks (Tian and Qin 2004). Following these various levels of restriction, satiation feeding was restored for the following five weeks for all treatments. A control group was also fed continuously throughout the study. Fish fed at 75% and 50% satiation exhibited compensatory growth and caught up with the control fed fish within 5 weeks after resumption of satiation feeding. No other treatments managed to recover lost growth performance. Furthermore, all fish fed restrictively showed signs of hyperphagia, suggesting that the efficiency of food utilisation was no better in the restrictedly fed fish than the control group.

8.4 Ingredient Utilization

The manufacture of diets for *L. calcarifer* requires the consideration of a range of factors. The formulation of the diet has to provide adequate nutrient and energy levels for the fish, but it must also encompass certain requirements that allow it to be manufactured through extrusion processing. The use of ingredients that the fish finds palatable is also crucial as poor feed intake can completely override any optimised diet specifications. Therefore the choice of ingredients for effective diet manufacture is critical. To underpin this process of ingredient evaluation, attempts have been made to standardise the ingredient assessment process, recently reviewed by Glencross et al. (2007). Therefore much of the following section will follow that recommended format of information required of; characterisation, digestibility, palatability and utilisation.

8.4.1 Ingredient characterisation and digestibility

Studies on the digestibility of feed ingredients for *L. calcarifer* are among the most useful of the ingredient evaluation studies available as the information can be adapted to influence formulation databases. Moreover, the findings can affect a broad range of outcomes, both commercial and scientific. However, as a foundation to achieve this, valid methods for assessing nutrient and energy digestibility from ingredients needs to be established. Williams and McMeniman (1998) studied the process of faecal collection from *L. calcarifer*, concluding that stripping of digesta was the only appropriate way to collect samples from *L. calcarifer*. The authors concluded that the high dissolution rates of nutrients from faeces voided into the water column rendered settlement methods invalid. However, despite the influences of digesta collection method on the digestibility assessment, variations in product quality should not be overlooked for these differences (Glencross et al. 2007). In addition, the calculation of ingredient digestibilities in most early studies did not take into account the relative nutrient contributions of the ingredient and reference diet components and, as such, may not be entirely accurate (Glencross et al. 2007). More recently Blyth et al. (2012) examined the effects of collection method (stripping or settlement) and time since introduction to the diet as variables on the digestibility of diets fed to *L. calcarifer*. From this work it was noted that stripping produced a more conservative and reliable digesta/faecal sample. Furthermore, it was also observed acclimation to a new diet needs to be at least four days, as prior to this the digestibility values observed have not stabilised.

McMeniman (1998) using stripping methods for collection of digesta provided one of the earliest assessments of a range of ingredients (Table 8.11). In that study protein digestibility was typically higher in the plant protein ingredients than the animal meal ingredients evaluated. Energy digestibility was variable among the ingredients studied and appeared to reflect the fat and protein content of each ingredient. Glencross (2011), Blyth et al. (2012) and Glencross et al. (2012a) went on to further benchmark the digestibility of a suite of protein and starch based ingredients (Table 8.11).

In an attempt to validate if data from other species could be applied to *L. calcarifer*, Glencross (2011) conducted a digestibility assessment with rainbow trout and *L. calcarifer* fed the same diets and both fish species stripped to collect digesta samples (Fig. 8.14). The findings demonstrated that the digestibility values were interchangeable, but more so for some parameters than others (e.g., energy digestibilities were good, but protein digestibilities less so). In addition, when calculating digestibility values across species a correction factor is needed.

8.4.2 Ingredient palatability and nutrient utilisation

Traditionally pelleted diets for fish were made largely from the compounding of fish meals and oils with a binder, like wheat flour. This proved relatively effective and the high inclusion of fish meal stimulated a good feed intake, or palatability response. However, with limited stocks of fishery resources available for use as feed resources, other raw materials/ingredients have had to become relied upon to provide the necessary dietary protein and energy (Tacon and Metian 2008). It has been demonstrated that there is a critical factor in fishmeal that induces a palatability response in *L. calcarifer* and below a certain inclusion of fishmeal in the diet then the feed intake declines (Fig. 8.15) (Glencross et al. 2011b). Furthermore, when diets are formulated on an equivalent digestible energy and digestible protein basis from a wide range of alternatives ingredients, then any vagaries in growth response can largely be linked back to vagaries in feed intake (Fig. 8.16) (Glencross et al. 2011b).

Table 8.11. Ingredient protein and energy digestibility (%) in *Lates calcarifer* from a range of feed ingredients.

Ingredient	Ingredient Composition (g/kg DM)					Digestibility		Data Source
	Protein	Lipid	Ash	Carbohydrates		Protein	Energy	
Meat meal A	n/a	n/a	340	n/a		53.9	58.2	McMeniman 1998
Meat meal B	n/a	n/a	240	n/a		63.5	66.5	"
Poultry offal meal	n/a	n/a	n/a	n/a		78.8	76.7	"
Soybean meal (solvent-extracted)	n/a	n/a	n/a	n/a		86.0	69.4	"
Soybean meal (full-fat)	n/a	n/a	n/a	n/a		84.8	75.9	"
Peanut meal	n/a	n/a	n/a	n/a		91.9	68.7	"
Canola meal (solvent-extracted)	n/a	n/a	n/a	n/a		81.0	56.1	"
Lupin (*L. angustifolius*) kernel meal	n/a	n/a	n/a	n/a		98.1	61.5	"
Wheat gluten	n/a	n/a	n/a	n/a		101.9	98.8	"
Yellow lupin (*L. luteus*) kernel meal	567	67	39	327		81.2	82.7	Glencross 2011
Lupin (*L. angustifolius*) kernel meal	412	64	35	489		96.1	73.4	"
Yellow lupin protein concentrate	819	112	29	40		98.6	113.0	"
Lupin (*L. angustifolius*) protein concentrate	754	153	23	70		86.0	100.4	"
Soybean meal (solvent extracted)	500	17	86	397		103.4	65.5	"
Canola meal (expeller extracted)	388	133	53	559		63.3	59.7	"
Poultry offal meal	608	119	160	113		39.7	52.5	"
Hydrolysed feather meal	802	144	17	37		74.8	67.9	"
Wheat	196	31	15	758		100.2	65.2	Glencross et al. 2012a
Barley	151	44	21	784		152.7	54.7	"
Oats	135	91	25	749		98.0	52.4	"

Table 8.11. contd....

Table 8.11. contd.

Ingredient	Ingredient Composition (g/kg DM)				Digestibility		Data Source
	Protein	Lipid	Ash	Carbohydrates	Protein	Energy	
Sorghum	138	39	15	808	109.9	53.7	Glencross et al. 2012a
Tapioca	7	3	4	986	n/a	58.0	"
Triticale	205	26	20	749	110.7	57.3	"
Corn	52	26	18	905	149.9	43.2	"
Faba beans	380	63	36	521	104.3	61.6	"
Fishmeal A	744	75	162	19	89.1	80.7	Blyth et al. 2012
Fishmeal B	721	85	158	36	96.1	82.0	"
White lupin kernel meal	482	86	37	395	98.1	83.3	"
Lupin (*L. angustifolius*) kernel meal	383	54	34	530	99.0	83.9	"
Blood meal	953	1	20	26	84.1	70.8	"
Gluten	710	46	8	236	121.9	107.3	"

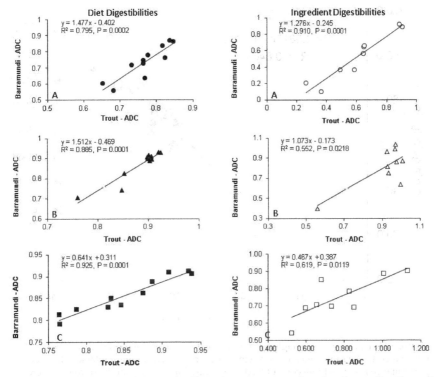

Figure 8.14. Comparison of *Lates calcarifer* and rainbow trout *Oncorhynchus mykiss* diet and ingredient apparent digestibility coefficients (ADC) for dry matter (A), protein (B) and energy (C). From Glencross 2011.

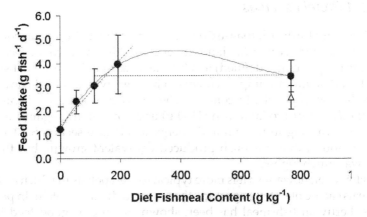

Figure 8.15. Variability in feed intake by *Lates calcarifer* when fed diets formulated the same digestible protein and energy specifications, but varying in the concentration of fishmeal. Figure derived from Glencross et al. (2011b).

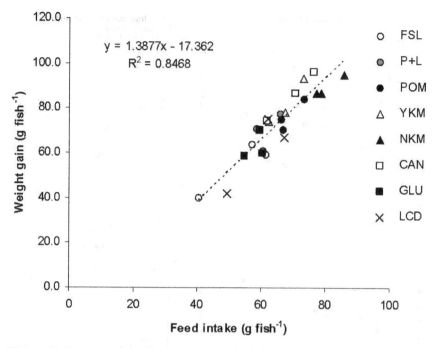

Figure 8.16. *Lates calcarifer* weight gain as a function of feed intake among a series of diets formulated to the same digestible protein and energy specifications. Notable is how growth performance is a direct reflection of feed intake variability. Figure derived from Glencross et al. (2011b).

8.4.2.1 Protein sources

Traditionally trash-fish (bait-fish) were used as the main protein source in diets for farmed *L. calcarifer*. Tantikitti et al. (2005) compared the feeding of trash-fish (species not defined) to a fishmeal based pellet in small fish (1 g) and found significant improvements in both growth and feed conversion with the use of pellets. In contrast, Glencross et al. (2008) fed trash-fish (*Sardinella lemuru*) to large fish (1180 g) and compared it against a series of pellets varying in protein and energy levels (see section 8.2.1.1). The authors found that trash-fish produced equivalent growth, but that the FCR was much higher.

However, nowadays it is more typical to feed pellets in which a variety of meals can be blended to provide the required dietary protein. In pelleted diets, Peruvian fish meal has been shown to support good feed intake and growth up to a 60% inclusion level. Regardless of inclusion level, no palatability problems were observed, as defined by levels of feed intake and growth consistent with fast growing fish (Glencross et al. 2011b). There are few studies comparing fish meal sources or qualities in diets for *L. calcarifer*. Studies aimed at defining threshold effects for minimum inclusion levels and/or maximum threshold levels for biogenic amines often found in fishmeal need to be undertaken.

Studies utilising meat meal (ovine or bovine) have shown that both high (60% protein) and medium quality (52% protein) meat meals can be used effectively as ingredients in *L. calcarifer* feeds (Williams 1998, Williams et al. 2003b). A series of experiments were undertaken using extruded diets where meat meal was shown to provide good nutritional value at inclusion levels of 40% and greater in diets fed to *L. calcarifer* reared in commercial farm trials (Table 8.12). Whilst growth rates were sub-optimal in this study, meat meals were shown to support growth equivalent to that of the high fish meal formulation (used as a reference). Moreover, meat meal diets did not exhibit any feed intake/palatability problems at the inclusion levels studied (Williams 1998). Further studies by Williams (1998) using a summit-dilution approach examined the inclusion of up to 70% of the diet as a rendered meat meal and observed no decline in growth or feed intake even at the highest inclusion levels (Table 8.13).

The inclusion of poultry offal meals has also shown good acceptance and utilisation by *L. calcarifer* (Glencross et al. 2011b). The authors examined diets with either 30% or 40% inclusion of poultry offal meal and demonstrated that both supported growth equal or superior to diets containing only fish meal as the protein source. This work also showed that there was considerable complementarity in using both a rendered animal protein (poultry offal meal) and grain protein sources (lupin kernel meal) together in an extruded diet for *L. calcarifer*.

The inclusion of soybean meals in *L. calcarifer* diets was initially evaluated by Boonyaratpalin et al. (1998). In this work, each of four different types of soybean meal (solvent-extracted, extruded-full-fat, steamed-full-fat, soaked-raw) was used to replace 15% of the fish meal of a reference diet. Despite such small levels of substitution, growth was poorer in each of the soybean treatments than the control, but not significantly so for the solvent-extracted soybean meal. In this study all diet specifications were maintained at a constant crude protein and gross energy levels. Owing to the suggested high digestibilities of most soybean meals it was suggested that it was a palatability problem that caused the poorer growth for most of the soybean meals.

Solvent extracted soybean meal was shown to support good growth up to a 30% inclusion level, not exhibiting any palatability problems until a 60% inclusion level (Table 8.14, Williams 1998). Based on data on the composition of the diets it is quite possible that dietary energy density effects may have influenced on the findings of this study. In addition, there was

Table 8.12. Influence of meat meal inclusion in two studies on growth performance of juvenile *Lates calcarifer*. Data derived from Williams et al. (2003b).

Formulation	Experiment 1			Experiment 3		
	CTL	M1	M2	CTL	M3	M4
Wheat flour	304	181	299		161	104
Fish meal	350	100	100			
Meat meal 1 (52% protein)		450			500	500
Meat meal 2 (60% protein)	100		400			
Blood meal (ring-dried)					70	90
Soybean (full-fat)	160	160	50		150	100
Soybean (solvent-extracted)			50			
Wheat gluten	50	50	50		50	100
Crystalline amino acids	1.5	9	9		9.5	11
Fish oil	25	40	32.5		50	60
Tallow						25
Common ingredients	9.5	10	9.5		9.5	10
Diet specifications						
Dry matter	924	943	943		892	901
Crude protein (g/kg DM)	474	456	456		476	531
Crude fat (g/kg DM)	102	135	134		172	142
Ash (g/kg DM)	103	158	101		164	156
Energy (MJ/kg DM)	21.0	20.4	21.5		21.3	22.3
Estimated DE (MJ/kg DM)	16.2	15.9	16.1		16.8	18.0
Fish performance						
Initial weight (g)	275	277	285	224	213	232
Final weight (g)	440	439	459	421	418	449
Gain rate (g/d)	2.50	2.45	2.64	2.98	3.11	3.29
DGC (%/d)	1.67	1.64	1.72	2.15	2.28	2.29
Feed intake (g/fish/d)	3.60	3.51	3.88	3.64	4.47	4.31
FCR (g fed / g gain)	1.44	1.43	1.47	1.22	1.44	1.31

CTL: Control diet. DGC: Daily Growth Coefficient (Final weight$^{0.33}$–Initial weight$^{0.33}$)/time x 100. FCR: Feed Conversion Ratio. DM: Dry Matter.

Table 8.13. Influence of meat meal inclusion in a summit-dilution study on growth performance of *Lates calcarifer*. Data derived from Williams (1998).

Parameter	Meat meal substitution								Diatomaceous earth substitution			
	Reference	10%	20%	30%	40%	50%	60%	70%	10%	20%	30%	40%
Diet specifications												
Dry matter	939	939	939	939	939	939	938	938	945	951	957	963
Crude protein (g/kg DM)	521	525	528	532	535	539	543	546	469	417	365	313
Crude fat (g/kg DM)	151	146	142	137	132	128	123	118	136	121	106	91
Ash (g/kg DM)	78	103	127	152	177	202	226	251	170	262	354	446
Energy (MJ/kg DM)	21.18	20.75	20.31	19.88	19.44	19.01	18.57	18.14	19.06	16.94	14.83	12.71
Fish performance												
Initial weight (g)	84.8	87.1	82.0	84.1	87.1	87.5	84.6	83.8	83.7	84.7	82.1	81.6
Final weight (g)	155.6	181.4	170.2	179.6	188.2	175.0	163.1	160.8	161.6	150.3	138.7	131.6
DGC (%/d)	2.01	2.51	2.44	2.57	2.65	2.35	2.19	2.17	2.19	1.89	1.69	1.53
Feed intake (g/fish/d)	1.40	1.81	1.66	1.70	1.80	1.71	1.62	1.65	1.56	1.62	1.56	1.69
FCR (g fed / g gain)	0.97	0.94	0.92	0.87	0.87	0.96	1.01	1.05	0.98	1.21	1.35	1.66

Table 8.14. Influence of solvent-extracted soybean meal inclusion in a summit-dilution study on growth performance of *Lates calcarifer*. Data derived from Williams (1998).

Parameter	Soybean meal substitution									Diatomaceous earth substitution			
	Reference	10%	20%	30%	40%	50%	60%	70%		10%	20%	30%	40%
Diet specifications													
Dry matter	927	925	923	921	919	918	916	914		933	940	946	952
Crude protein (g/kg DM)	521	521	520	520	519	519	518	518		469	417	365	313
Crude fat (g/kg DM)	151	137	124	110	96	83	69	55		136	121	106	91
Ash (g/kg DM)	78	77	77	76	76	75	74	74		169	260	352	443
Energy (MJ/kg DM)	22.18	21.90	21.62	21.35	21.07	20.79	20.51	20.23		20	18	16	13
Fish performance													
Initial weight (g)	151.5	152.3	159.2	149.1	153.9	153.1	150.2	148.5		155.0	152.3	147.9	154.6
Final weight (g)	219.0	223.5	231.8	209.6	212.1	206.4	191.0	182.3		229.6	213.0	196.6	173.2
DGC (%/d)	1.66	1.73	1.72	1.52	1.44	1.33	1.06	0.89		1.79	1.50	1.25	0.49
Feed intake (g/fish/d)	1.53	1.67	1.83	1.65	1.61	1.66	1.46	1.45		1.88	1.76	1.83	1.59
FCR (g fed/g gain)	0.95	0.99	1.06	1.16	1.16	1.32	1.50	1.80		1.06	1.22	1.58	3.60
Nitrogen retention (%)	41	42	43	34	36	32	29	19		42	42	36	24

no compensation made for diet protein or energy levels in this design, and this is a confounding factor with the summit-dilution experiment approach (Glencross et al. 2007).

The inclusion of soybean meals in *L. calcarifer* diets was also evaluated by Tantikitti et al. (2005). In this work each of six different inclusion levels (0% to 38%) of soybean meal (solvent-extracted) were used to replace the fish meal and rice flour content of a reference diet. There was a significant decline in growth of fish with every inclusion level (based on regression analysis) of soybean meal in this study. However, the formulation of diets on a crude specification basis in this work does question whether the effects observed are ingredients or diet specification effects? In this study, a palatability problem was observed that caused the poorer growth for most of the soybean meals, though differences in digestible protein and energy values of the diets cannot be ruled out either. Further studies on the use of soy products based on diets formulated to equivalent digestible protein and energy are required.

In a summit-dilution study, lupin (*Lupinus angustifolius*) kernel meal was also shown to support good growth up to a 40% inclusion level and did not exhibit any palatability problems until a 70% inclusion level in the diet (Williams 1998). Similar to the findings from the soybean study, it is quite possible that dietary digestible energy density effects may have had an influence on the findings of this study. No compensation was made for the substitution of lupin kernel meal with the control diet in this study, though the effect of ingredient substitution was not as marked as that with the soybean meal. A study by Katersky and Carter (2009) used a simplistic design with a single inclusion level (24%) of a lupin kernel meal compared against a fishmeal based diet. In this study the two diets were formulated on a crude basis only, and fed to 4 g fish for 15 days. No effects of diet on either growth or feed intake were noted. In a more thorough study Glencross et al. (2011b) first established the digestible nutrient and energy values of several lupin species (narrowleaf and yellow) and other feed grain resources like canola and wheat gluten (Glencross 2011). The authors then formulated diets on equivalent digestible protein and energy basis, and each diet contained a 30% inclusion level of narrowleaf or yellow lupin kernel meal, or canola meal. Feed intake of these diets and subsequent growth was among the best of the treatments in that study and exceeded the performance seen with the fishmeal based diet. This finding that an alternative protein source could exceed performance achieved with a diet comprising fishmeal as the protein source was consistent with a more recently finding, again using diets formulated to equivalent digestible specifications (Glencross et al. 2011a).

8.4.2.2 Lipid sources

The use of a series of plant lipid resources alternative to fish oil has been evaluated by Raso and Anderson (2002), Williams et al. (2006) and more recently by Alhazzaa et al. (2011a, b). In the work by Raso and Anderson (2002), a series of diets with 15% added oil (total diet lipid content of ~20%) were made with partial substitution of fish oil with canola, linseed or soybean oils (Raso and Anderson 2002). With increasing substitution of the alternatives for the fish oil, deterioration in both feed intake and growth was noted for small fish (19 g) fed the canola and linseed oil treatments, but not for those fish fed the soybean oil treatments. The use of soybean oil was also explored in a study of six blends of fish oil with soybean oil (Williams et al. 2006). The authors reported optimal performance of *L. calcarifer* at a blend of 23% soybean and 77% fish oil as the added dietary lipid sources, as well as a significant improvement in FCR with this blend. None of the diets, however, affected growth. Williams et al. (2006) hypothesised that this effect was due to a superior n-3:n-6 ratio in the diet (see section 8.2.2.2). Alhazzaa et al. (2011a, b) examined the use of Echium oil (rich in SDA) in addition to rapeseed oil when fed to juvenile fish (5 g). Alhazzaa et al. (2011a, b) demonstrated that there was no effect of using rapeseed oil compared to fish oil, but that the use of Echium oil resulted in poorer performance (growth and feed conversion) than either of the other two oils. Alhazzaa et al. (2011a, b) also examined the use of these oils in either fresh or marine water and found no influence of water salinity on the nutritional results.

More recently the complete substitution of fish oil with a rendered animal meal (poultry by-product oil) in diets with 25% fishmeal was observed to have no significant effects on growth or feed intake of larger fish (142 g–312 g; Glencross, unpub. data). In fact growth was numerically better than that of fish fed the control diet, which were fed solely fish oil as the added lipid source (Glencross, unpub. data). However, fatty acid composition of the flesh was affected by oil type. These results demonstrate that the potential for wholesale replacement of fish oil with alternatives is highly feasible; however, there were implications for reduced levels.

8.5 Flesh Quality

The flesh quality characteristics of fish produced for human food are an important consideration in the nutrition of any fish. It is well acknowledged that feed management can play a central role in managing flesh quality of many aquaculture species, and *L. calcarifer* are no different in this regard.

8.5.1 Nutrient and energy effects on flesh quality

The effects of diet specifications on flesh quality have been examined in several studies and these vary from physical, chemical and sensory quality measurements. The dress-out yield, a physical indication of the relative portion of gut and gill content of the carcass has been used as one indicator of product quality (Williams and Barlow 1999, Glencross et al. 2008). The fat content, a chemical indicator, of the whole fish and also the fillet has also been used as an indicator of quality (Williams and Barlow 1999, Glencross et al. 2008). In addition, there are also studies where taste testing has been conducted to assess the sensory qualities of the fish (Glencross et al. 2008).

The effects of varying dietary protein and energy ratios on the sensory qualities of *L. calcarifer* were first examined by Williams and Barlow (1999). Fish fed the diet with the highest protein:energy ratio had significantly better sensory scores for flavour than those fish fed a comparatively higher-energy or lower-protein diet. In another study comparing several commercial diets with similar protein content, but differing fat levels and energy densities, no differences were observed in the dress-out percentages. However, the fish from the two higher energy density diets had slightly higher fat levels in their carcass (Glencross et al. 2003). Glencross et al. (2003) concluded that this higher level of fat was related to the larger size of these fish, rather than an excess of fat deposition. It is now well established that there is a clear exponential relationship between fish live-weight and its fat density (Fig. 8.11). A subsequent study by Glencross et al. (2008) with large fish (1180 g) compared a range of four protein and energy densities and found significant effects on a range of sensory parameters (e.g., oiliness, dryness) among the different dietary treatments. Here it was suggested that the dryness was indicative of a low level of fat deposition in the flesh, while oiliness was considered indicative of excessive deposition of fat in the flesh and that these parameters could be related to over specification or under-specification of protein to energy ratios.

The effects of feed restriction and varying water temperatures on body composition were examined by Williams and Barlow (1999). With increasing water temperature and varying feed ration there was an improvement in the dress-out percentage from 85.0% at 20°C to 88.9% at 29°C. With decreasing levels of feed intake, a decrease in the deposition of fat was also observed.

Comparisons of lipid content between wild fish and cultured fish have been difficult because of well recognised size and diet effects on expected lipid content and composition of the animal. However, a study reporting a comparison of freshwater and saltwater *L. calcarifer* suggested that fat content of the fillet from freshwater *L. calcarifer* (0.9% wet-weight) was more than twice that from saltwater *L. calcarifer* (0.4%) (Nichols et al. 1998). However, there were no indications as to the source, size or even sample cut parameters of the fish and, as such, it is difficult to interpret this data. It is also recognised that there is substantial variability in the fat content of various sections of the *L. calcarifer* fillet (Percival et al. 2008). The highest levels of fat deposition are observed to occur in the belly-flap region (Fig. 8.17). It is now considered more standard to analyse quality parameters, like total fat or fatty acid content from the region referred to as the NQC cut (Fig. 8.17).

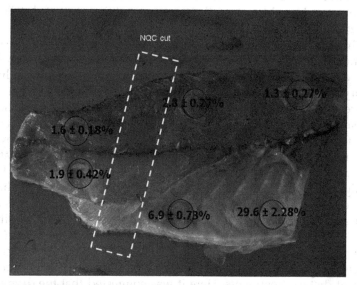

Figure 8.17. Variability in fat content (% live-weight) within the fillet of a large harvest size *Lates calcarifer*. Shown are the fat levels (%) determined from within the sample taken from the area marked by the respective ring. Also indicated is the region considered as the NQC cut in *L. calcarifer*.

8.5.2 Ingredient utilization effects on flesh quality

There are few studies examining the influence of ingredients on the flesh quality of *L. calcarifer*. One of those, however, was that of Williams et al. (2003c) who examined the effects of meat-meal inclusion on a range of sensory parameters. Fish that had been fed a series of diets containing high-levels of meat-meals and a fishmeal control were provided to a sensory panel for evaluation. The panellists rated the flesh colour, odour, flavour, texture and overall liking of each sample provided. Fish that had been fed diets with high meat-meal content ranked higher for flavour (sweetness) and texture compared with those fish fed diets with high levels of fish meal. Ironically, scores for 'fishy' flavour were also highest for those fish fed the meat meal diets rather than the fishmeal diets. This work also demonstrated that the complete exclusion of fish meal from a diet fed to *L. calcarifer* did not detract from the sensory value of the fish flesh.

There is also evidence to suggest that there is a distinct colour difference between the fillets of wild and farmed fish (Percival et al. 2008). A notable greyish tone is observed in farmed *L. calcarifer* in the dorsal and upper lateral muscle groups above the spine. This difference is even notable compared to ventral muscle groups in the animal. This phenomenon has been observed from fish produced in both freshwater and marine-water situations (Percival et al. 2008) and fish fed a range of protein and energy densities (Glencross et al. 2008) and also with varying protein sources (Williams et al. 2003c).

8.6 Future Directions

Substantial progress in the area of *L. calcarifer* nutritional research has been achieved over the past decade. Over time, changes to formulation strategies and also raw material developments have contributed to a changing 'landscape' in terms of research needs. To further optimise diets for the production of *L. calcarifer* a series of priorities are suggested in the following sections based on the information reviewed to date.

8.6.1 Nutrient and energy requirements

Most of the published dose-response studies on *L. calcarifer* nutrient requirements utilised fish less than 100 g in live-weight, though in recent years there has been some effort to work with larger fish (Glencross et al. 2008). While it is impractical to suggest that all key nutrients need to be re-examined in large fish, validation of some of these estimates would be appropriate, particularly as this is where the majority of feed use occurs. Nutritional models to arrive at iterative estimations have been shown to be useful in estimating the demands of larger fish (Glencross et al. 2008). It would therefore be beneficial to follow this modelling approach with an assessment of some key amino acid requirements for a range of fish sizes. For the formulation criteria to become more efficient, they also need to move towards being based on essential amino acid demands rather than on digestible protein demands.

As both fishmeal and fish oil continue to become constrained in feed formulations there will also be increasing pressure on essential fatty acid requirements for this species. To support this evolution in feeds there is a looming need to better define the specific requirements of both 20:5n-3

(EPA) and 22:6n-3 (DHA), both singly and in combination. Work has already progressed on requirements for DHA, though demands for EPA remain to be elucidated (Glencross and Rutherford 2010). This is particularly important given *L. calcarifer*'s poor capacity to elongate and desaturate short-chain PUFA (Mohd-Yusof et al. 2010).

Data on requirement for key minerals still remains poorly studied; notably those demands for both calcium and phosphorus need to be more fully defined. Moreover, the requirements for some vitamins need to be revisited in light of changes to diet specifications and also the forms in which the vitamins are provided (e.g., soluble or stabilised forms of vitamin C).

To assist the development of diets better suited to the long-term health of the fish, as is often required in grow-out to larger fish sizes, there is the need for the development of a series of standard blood chemistry responses to balanced and nutritionally incomplete diets. This would assist assessment of problems at a sub-clinical scale.

8.6.2 Ingredient evaluation

Traditionally there has been a lack of reliable ingredient digestibility data for most *L. calcarifer* energy and nutrient parameters, though this has recently begun to change. In particular, data for *L. calcarifer* on amino acid and phosphorus digestibilities are still lacking and further work is needed. However, a key constraint in this area, as with most other species, has been a reliable and coherent approach to the ingredient assessment process (Glencross et al. 2007). Undertaking future work using a standardised approach will add substantial value to the field. To address this a more systematic approach to characterisation of the digestible energy, phosphorus, protein and amino acid values of key industrial ingredients (and multiple sources of these) should be seen as a priority.

There is still limited information on palatability issues of specific ingredient and identification of tolerance thresholds. The early summit-dilution work of Williams (1998) examined this to a degree. The fact that there were some classical feed intake deterioration issues with those studies somewhat limits their intended value in interpreting nutrition value of the ingredients. However, the results do provide some useful information regarding palatability issues. The focus of much of this work was with rendered mammalian meals and there has been limited work on vegetable protein sources when fed to *L. calcarifer*. Studies on soybean are the most comprehensive (Boonyaratpalin et al. 1998, Glencross 2011), but systematic studies on others grains like lupins, rapeseed, peas and glutens are lacking. Furthermore, the implication of ingredient quality variability, on the nutritional value of feed ingredients to *L. calcarifer* is another key research issue that needs to be addressed.

8.6.3 Product quality

The development of a comprehensive standard characterisation of flesh composition is required to provide a reference for product quality variation within farmed *L. calcarifer*. Such a characterisation of standard quality characteristics also needs to consider what to refer to as a "standard" fish, given the substantial changes in recent years in use of both protein and lipid ingredients. This is known to have effects on parameters like fatty acid composition and also sensory qualities of the flesh (Glencross et al. 2003). Factors like live-weight and diet protein to energy ratios are already recognised to influence somatic energy density (lipid deposition) and the sensory evaluation of those fish (Glencross et al. 2008). Therefore it is likely that this and other factors will need to be considered so that standardisation criteria can be developed to ensure comparisons are relevant.

The occurrence of grey fillets has been observed in farmed *L. calcarifer*, especially compared to wild caught product (Percival et al. 2008). The nature and cause of this grey colouration has not been resolved, but it is suspected to be melanin based. Given that this condition occurs only in farmed fish suggests that it may be a nutritional issue, although this remains to be confirmed. To address this problem, identification of the grey pigment is the obvious starting place, followed by identifying potential nutritional mechanisms of alleviating the problem based on an understanding of the underlying biochemistry that induces the problem in the first place.

In some farming situations a darkening of the skin colour has also been noted as a problem. Anecdotal observations suggest that this melanisation of the skin can be remedied by changing the light and/or background regime (Percival et al. 2008). Despite the potential opportunities to resolve this through production management there may also be opportunities to examine nutritional strategies and promote such diets accordingly.

In some freshwater farming systems the occurrence of a muddy taint, through accumulation of the algal metabolites of geosmin and 2-methyl-iso-borneol (MIB), have been reported (Percival et al. 2008). While purging regimes have been shown to be useful in reducing taint problems (Percival et al. 2008), there may be opportunities to examine nutritional strategies to resolve this issue and promote such diets accordingly.

8.7 References

Alhazzaa, R., A.R. Bridle, P.D. Nichols and C.G. Carter. 2011a. Replacing dietary fish oil with *Echium* oil enriched barramundi with C18 PUFA rather than long-chain PUFA. Aquaculture 312: 162–171.

Alhazzaa, R., A.R. Bridle, P.D. Nichols and C.G. Carter. 2011b. Upregulated desaturase and elongase gene expression promoted accumulation of polyunsaturated fatty acid (PUFA) but not long-chain PUFA in *Lates calcarifer*, a tropical euryhaline fish, fed a stearidonic acid and γ-linoleic acid enriched diet. J. Agric. Food Chem. 59: 8423–8434.

Anderson, A.J. 2003. Metabolic studies on carbohydrate utilisation by barramundi and tilapia. *In*: G.L. Allan, M.A. Booth, D.A.J. Stone and A.J. Anderson (eds.). Aquaculture Diet Development Subprogram: Ingredient Evaluation. Project 1996-391, Final Report to the Fisheries R&D Corporation, Canberra, Australia, pp. 120–134.

Barlow, C.G., M.G. Pearce, L.J. Rodgers and P. Clayton. 1995. Effects of photoperiod on growth, survival and feeding periodicity of larval and juvenile barramundi *Lates calcarifer* (Bloch). Aquaculture 138: 159–168.

Bermudes, M., B.D. Glencross, K. Austen and W. Hawkins. 2010. Effect of high water temperatures on nutrient and energy retention in barramundi (*Lates calcarifer*). Aquaculture 306: 160–166.

Blyth, D., S. Irvin, N. Bourne and B.D. Glencross. 2012. Comparison of faecal collection methods, and diet acclimation times for the measurement of digestibility coefficients in barramundi (*Lates calcarifer*). *In*: International Society for Fish Nutrition and Feeding Conference. 4th–7th June 2012, Molde, Norway.

Boonyaratpalin, M. 1997. Nutrient requirements of marine food fish cultured in Southeast Asia. Aquaculture 151: 283–313.

Boonyaratpalin, M. and J. Phongmaneerat. 1990. Requirement of seabass for dietary phosphorus. Technical Paper No. 4, National Institute of Coastal Aquaculture, Department of Fisheries, Thailand 20 pp.

Boonyaratpalin, M. and J. Wanakowat. 1993. Effect of thiamine, riboflavin, pantothenic acid and inositol on growth, feed efficiency and mortality of juvenile seabass. *In*: S.J. Kaushik and P. Luget (eds.). Fish Nutrition in Practice. Biarritz, France, pp. 819–828.

Boonyaratpalin, M. and K.C. Williams. 2001. Asian sea bass, *Lates calcarifer*. *In*: C.D. Webster and C.E. Lim (eds.). Nutrient Requirements and Feeding of Finfish for Aquaculture. CABI Publishing, Wallingford, UK, pp 40–50.

Boonyaratpalin, M., N. Unprasert and J. Buranapanidgit. 1989. Optimal supplementary vitamin C level in seabass fingerling diet. *In*: M. Takeda and T. Watanabe (eds.). The Current Status of Fish Nutrition in Aquaculture. Tokyo University of Fisheries, Tokyo, Japan, pp. 149–157.

Boonyaratpalin, M., S. Boonyaratpalin and K. Supamataya. 1994. Ascorbyl-phosphate-Mg as a dietary vitamin C sources for seabass (*Lates calcarifer*). *In*: L.M. Chou, A.D. Munro, T.J. Lam, T.W. Chen, L.K.K. Cheong, J.K. Ding, K.K. Hooi, V.P.E. Phang, K.F. Shim and C.H. Tan (eds.). The Third Asian Fisheries Forum. Asian Fisheries Society, Manila, Philippines, pp. 725–728.

Boonyaratpalin, M., P. Suraneiranat and T. Tunpibal. 1998. Replacement of fish meal with various types of soybean products in diets for the Asian sea bass, *Lates calcarifer*. Aquaculture 161: 67–78.

Borlongan, I.G. and M.M. Parazo. 1991. Effect of dietary lipid sources on growth, survival and fatty acid composition of seabass (*Lates calcarifer*, Bloch) fry. Bamidgeh 43: 95–102.

Buranapanidgit, J., M. Boonyaratpalin, T. Watanabe, T. Pechmanee and R. Yashiro. 1988. Essential fatty acid requirement of juvenile seabass *Lates calcarifer*. Technical paper No. 3. National Institute of Coastal Aquaculture. Department of Fisheries, Thailand 21 pp.

Buranapanidgit, J., M. Boonyaratpalin, T. Watanabe, T. Pechmanee and R. Yashiro. 1989. Optimum level of ω3HUFA on juvenile seabass *Lates calcarifer*. *In*: IDRC Fish Nutrition Project Annual Report. Department of Fisheries, Thailand 23 pp.

Bureau, D.P., S.J. Kaushik and C.Y. Cho. 2002. Bioenergetics. *In*: Fish Nutrition, Third Edition. Elsevier Science, USA, pp 2–61.

Catacutan, M.R. and R.M. Coloso. 1995. Effect of dietary protein to energy ratios on growth, survival, and body composition of juvenile Asian seabass, *Lates calcarifer*. Aquaculture 131: 125–133.

Catacutan, M.R. and R.M. Coloso. 1997. Growth of juvenile Asian seabass, *Lates calcarifer*, fed varying carbohydrate and lipid levels. Aquaculture 149: 137–144.

Chaimongkol, A. and M. Boonyaratpalin. 2001. Effects of ash and inorganic phosphorus in diets on growth and mineral composition of sea bass *Lates calcarifer* (Bloch). Aquacult. Res. 32: 53–59.

Castell, J.D. 1979. Review of lipid requirement of finfish. In: J.E. Halver and K. Tiews (eds.). World Symposium on Finfish Nutrition and Fish Feed Technology. Hamburg, Germany, Heenemann, Berlin, Vol. 1, pp. 59–84.

Colosso, R.M., D.P. Murillo, I.G. Borlongan and M.K. Catacutan. 1993. Requirement of juvenile seabass *Lates calcarifer* Bloch, for tryptophan. In: Program and Abstracts of the VI International Symposium on Fish Nutrition and Feeding, 4–7 October 1993, Hobart, Australia.

Cuzon, G. and J. Fuchs. 1988. Preliminary nutritional studies of seabass *Lates calcarifer* (Bloch) protein and lipid requirements. In: Program and Abstracts, 19th Annual Conference and Exposition World Aquaculture Society, Hawaii pp. 15–16.

Dumas, A., J. France and D.P. Bureau. 2010. Modelling growth and body composition in fish nutrition: Where have we been and where are we going? Aquacult. Res. 41: 161–181.

Fraser, M.R. and R. de Nys. 2011. A quantitative determination of deformities in barramundi (*Lates calcarifer*; Bloch) fed a vitamin deficient diet. Aquacult. Nutr. 17: 235–243.

Glencross, B.D. 2006. Nutritional management of barramundi, *Lates calcarifer*—A review. Aquacult. Nutr. 12: 291–309.

Glencross, B.D. 2008. A factorial growth and feed utilisation model for barramundi, *Lates calcarifer* based on Australian production conditions. Aquacult. Nutr. 14: 360–373.

Glencross, B.D. 2009. Exploring the nutritional demand for essential fatty acids by aquaculture species. Reviews in Aquaculture 1: 71–124.

Glencross, B.D. 2011. A comparison of the diet and raw material digestibilities between rainbow trout (*Oncorhynchus mykiss*) and barramundi (*Lates calcarifer*)—Implications for inferences of digestibility among species. Aquacult. Nutr. 17: e207–e215.

Glencross, B.D. and M. Felsing. 2006. Influence of fish size and water temperature on the metabolic demand for oxygen by barramundi, *Lates calcarifer*, in freshwater. Aquacult. Res. 37: 1055–1062.

Glencross, B.D. and M. Bermudes. 2010. Effect of high water temperatures on the utilisation efficiencies of energy and protein by juvenile barramundi, *Lates calcarifer*. FAJ 14: 1–11.

Glencross, B.D. and M. Bermudes. 2011. Effect of high water temperatures on energetic allometric scaling in barramundi (*Lates calcarifer*). Comp. Biochem. Physiol., Part A: Mol. Integr. Physiol. 159: 167–174.

Glencross, B.D. and M. Bermudes. 2012. Using a bioenergetic modelling approach to understand the implications of heat stress on barramundi (*Lates calcarifer*) growth, feed utilisation and optimal protein and energy requirements—Options for adapting to climate change? Aquacult. Nutr. 18: 411–422.

Glencross, B.D. and N.R. Rutherford. 2010. Dietary strategies to reduce the impact of temperature stress on barramundi (*Lates calcarifer*) growth. Aquacult. Nutr. 16: 343–350.

Glencross, B.D. and N.R. Rutherford. 2011. The docosahexaenoic acid (DHA) requirements of juvenile barramundi (*Lates calcarifer*). Aquacult. Nutr. 17: e536–e548.

Glencross, B.D., N. Rutherford and W.E. Hawkins. 2003. Determining waste excretion parameters from barramundi aquaculture. Fisheries Contract Report Series No. 4. Department of Fisheries, Perth, Western Australia. pp 48.

Glencross, B.D., M. Booth and G.L. Allan. 2007. A feed is only as good as its ingredients—A review of ingredient evaluation for aquaculture feeds. Aquacult. Nutr. 13: 17–34.

Glencross, B.D., R. Michael, K. Austen and R. Hauler. 2008. Productivity, carcass composition, waste output and sensory characteristics of large barramundi *Lates calcarifer* fed high-nutrient density diets. Aquaculture 284: 167–173.

Glencross, B.D., N.R. Rutherford and W.E. Hawkins. 2011a. A comparison of the growth performance of rainbow trout (*Oncorhynchus mykiss*) when fed soybean, narrow-leaf or yellow lupin kernel meals in extruded diets. Aquacult. Nutr. 17: e317–e325.

Glencross, B.D., N.R. Rutherford and J.B. Jones. 2011b. Fishmeal replacement options for juvenile barramundi (*Lates calcarifer*). Aquacult. Nutr. 17: e722–e732.

Glencross, B.D., D. Blyth, S.J. Tabrett, N. Bourne, S. Irvin, T. Fox-Smith and R.P. Smullen. 2012a. An examination of digestibility and technical qualities of a range of cereal grains when fed to juvenile barramundi (*Lates calcarifer*) in extruded diets. Aquacult. Nutr. 18: 388–399.

Glencross, B.D., N. Wade, D. Blyth, S. Irvin and N. Bourne. 2012b. Examining the consequences of different macronutrient energy sources on growth, feed utilisation, energy partitioning and gene expression by barramundi, *Lates calcarifer*. *In*: International Society for Fish Nutrition and Feeding Conference. 4th–7th June 2012, Molde, Norway.

Hauler, R.C. and C.G. Carter. 2001. Re-evaluation of the quantitative dietary lysine requirements of fish. Rev. Fish. Sci. 9: 133–163.

Irvin, S., D. Blyth, N. Bourne and B.D. Glencross. 2012. Examining the discrete and interactive effect of different NSP non-starch polysaccharide (NSP) sources on feed digestibility by barramundi, *Lates calcarifer*. *In*: International Society for Fish Nutrition and Feeding Conference. 4th–7th June 2012, Molde, Norway.

Katersky, R.S. and C.G. Carter. 2009. Growth and protein synthesis of barramundi, *Lates calcarifer*, fed lupin as a partial protein replacement. Comp. Biochem. Physiol., Part A: Mol. Integr. Physiol. 152(4): 513–517.

Lupatsch, I., G.W. Kissil and D. Sklan. 2003. Comparison of energy and protein efficiency among three fish species *Sparus aurata*, *Dicentrarchus labrax* and *Epinephelus aeneus*: energy expenditure for protein and lipid deposition. Aquaculture 225: 175–189.

McMeniman, N. 1998. The apparent digestibility of feed ingredients based on stripping methods. In: Fishmeal Replacement in Aquaculture Feeds for Barramundi (K.C. Williams Ed.). Project 93/120-04. Final Report to Fisheries R&D Corporation. Canberra, Australia, pp. 46–70.

McMeniman, N. 2003. Digestibility and utilisation of starch by barramundi. *In*: Aquaculture Diet Development Subprogram: Ingredient Evaluation. G.L. Allan, M.A. Booth, D.A.J. Stone and A.J. Anderson (eds.). Project 1996-391, Final Report to the Fisheries R&D Corporation, Canberra, Australia, pp. 135–139.

Millamena, O.M. 1994. Review of SEAFDEC/AQD fish nutrition and feed development research. *In*: Feeds for Small-Scale Aquaculture, Proceedings of the National Seminar-Workshop on Fish Nutrition and Feeds. C.B. Santiago, R.M. Coloso, O.M. Millamena and I.G. Borlongan (eds.). SEAFDEC Aquaculture Department, Iloilo, Philippines, pp. 52–63.

Mohd-Yusof, N.Y., O. Monroig, A. Mohd-Adnan, K.L Wan and D.R. Tocher. 2010. Investigation of highly unsaturated fatty acid metabolism in the Asian seabass (*Lates calcarifer*). Fish Physiol. Biochem. 36: 827–844.

Nichols, P.D., P. Virtue, B.D. Mooney, N.G. Elliot and G.K Yeardsley. 1998. Seafood the good food: the oil (fat) content and composition of Australian commercial fishes, shellfishes and crustaceans. CSIRO Division of Marine Research, Hobart, Australia 201 pp.

NRC (National Research Council). 2011. Nutrient Requirements of Fish and Shrimp. National Academy Press, Washington, DC.

Percival, S., P. Drabsch and B.D. Glencross. 2008. Determining factors affecting muddy-flavour taint in farmed barramundi, *Lates calcarifer*. Aquaculture 284: 136–143.

Phromkunthong, W., M. Boonyaratpalin and V. Storch. 1997. Different concentrations of ascorbyl-2-monophosphate-magnesium as dietary sources of vitamin C for seabass, *Lates calcarifer*. Aquaculture 151: 225–243.

Pimoljinda, T. and M. Boonyaratpalin. 1989. Study on vitamin requirements of seabass *Lates calcarifer* Bloch, in seawater. Technical Paper No. 3. Phuket Brackishwater Fisheries Station, Department of Fisheries, Thailand, 24 pp.

Raso, S. and T. Anderson. 2002. Effects of dietary fish oil replacement on growth and carcass proximate composition of juvenile barramundi (*Lates calcarifer*). Aquacult. Res. 34: 813–819.

Sakaras, W., M. Boonyaratpalin, N. Unpraser and P. Kumpang. 1988. Optimum dietary protein energy ratio in seabass feed I. Technical Paper No. 7. Rayong Brackishwater Fisheries Station, Thailand 20 pp.

Sakaras, W., M. Boonyaratpalin, N. Unpraser and P. Kumpang. 1989. Optimum dietary protein energy ratio in seabass feed II. Technical Paper No. 8. Rayong Brackishwater Fisheries Station, Thailand 22 pp.

Stone, D.A.J. 2003. Dietary carbohydrate utilisation by fish. Reviews in Fisheries Science 11: 337–369.

Tacon, A.G.J. and M. Metian. 2008. Global overview on the use of fish meal and fish oil in industrially compounded aquafeeds: Trends and future prospects. Aquaculture 285: 146–158.

Tantikitti, C., W. Sangpong and S. Chiavareesajja. 2005. Effects of defatted soybean protein levels on growth performance and nitrogen and phosphorus excretion in Asian seabass (*Lates calcarifer*). Aquaculture 248: 41–50.

Tian, X. and J.G. Qin. 2002. A single phase of food deprivation provoked compensatory growth in barramundi *Lates calcarifer*. Aquaculture 224: 169–179.

Tian, X. and J.G. Qin. 2004. Effects of previous ration restriction on compensatory growth in barramundi *Lates calcarifer*. Aquaculture 235: 273–283.

Tocher, D.R. 2010. Fatty acid requirements in ontogeny of marine and freshwater fish. Aquacult. Res. 41: 717–732.

Tu, W.C., R.J. Cook-Johnson, M.J. James, B.S. Mühlhäusler, D.A.J. Stone and R.A. Gibson. 2012a. Barramundi (*Lates calcarifer*) desaturase with $\Delta 6/\Delta 8$ dual activities. Biotechnol. Lett. 34: 1283–1296.

Tu, W.C., B.S. Mühlhäusler, M.J. James, D.A.J. Stone and R.A. Gibson. 2012b. An alternative n-3 fatty acid elongation pathway utilising 18:3n-3 in barramundi (*Lates calcarifer*). Biochem. Biophys. Res. Commun. 423: 176–182.

Tucker, J.W., M.R. MacKinnon, D.J. Russell, J.J. O'Brien and E. Cazzola. 1988. Growth of juvenile barramundi (*Lates calcarifer*) on dry feeds. The Progressive Fish Culturist 50: 81–85.

Wanakowat, J., M. Boonyaratpalin, T. Pimolindja and M. Assavaaree. 1989. Vitamin B6 requirement of juvenile seabass *Lates calcarifer*. *In*: The Current Status of Fish Nutrition in Aquaculture. M. Takeda and T. Watanabe (eds.). Tokyo University of Fisheries, Tokyo, Japan, pp. 141–147.

Wanakowat, J., M. Boonyaratpalin and T. Watanabe. 1993. Essential fatty acid requirement of juvenile seabass. *In*: Fish Nutrition in Practice. S.J. Kaushik and P. Luquet (eds.). Paris, France, pp. 807–817.

Williams, K.C. 1998. Fishmeal Replacement in Aquaculture Feeds for barramundi. Project 93/120-04. Final Report to Fisheries R&D Corporation. Canberra, Australia, pp. 17–21.

Williams, K.C. and C.G. Barlow. 1999. Dietary requirement and optimal feeding practices for barramundi (Lates calcarifer). Project 92/63, Final Report to Fisheries R&D Corporation, Canberra, Australia 95 pp.

Williams, K.C., C.G. Barlow and L. Rodgers. 2001. Efficacy of crystalline and protein-bound amino acids for amino acid enrichment of diets for barramundi/Asian seabass (*Lates calcarifer* Bloch). Aquacult. Res. 32: 415–429.

Williams, K.C. and N.P. McMeniman. 1998. Validation of sedimentation and dissection procedures for the determination of apparent digestibility. *In*: Fishmeal Replacement in Aquaculture Feeds for barramundi. K.C. Williams (ed.). Project 93/120-04. Final Report to Fisheries R&D Corporation. Canberra, Australia, pp. 17–21.

Williams, K.C., C.G. Barlow, L. Rodgers, I. Hockings, C. Agcopra and I. Ruscoe. 2003a. Asian seabass *Lates calcarifer* perform well when fed pellet diets high in protein and lipid. Aquaculture 225: 191–206.

Williams, K.C., C.G. Barlow, L. Rodgers and I. Ruscoe. 2003b. Potential of meat meal to replace fish meal in extruded dry diets for barramundi *Lates calcarifer* (Bloch). I. Growth performance. Aquacult. Res. 34: 23–32.

Williams, K.C., B.D. Patterson, C.G. Barlow, L. Rodgers, A. Ford and R. Roberts. 2003c. Potential of meat meal to replace fish meal in extruded dry diets for barramundi *Lates calcarifer* (Bloch). II. Organoleptic characteristics and fatty acid composition. Aquacult. Res. 34: 33–42.

Williams, K.C., C. Barlow, L. Rodgers and C. Agcopra. 2006. Dietary composition manipulation to enhance the performance of juvenile barramundi (*Lates calcarifer* Bloch) reared in cool water. Aquacult. Res. 37: 914–927.

Withers, P.C. 1992. Comparative Animal Physiology. Brooks/Cole, Singapore 452 pp.

Wong, F.J. and F. Chou. 1989. Dietary protein requirement for early grow-out seabass (*Lates calcarifer* Bloch) and some observations on the performance of two practical formulated feeds. *In*: Report of the Workshop on Shrimp and Finfish Feed Development, Johor Bahru, Malaysia, pp. 91–102.

9

Post-Harvest Quality in Farmed *Lates calcarifer*

Alexander G. Carton* and Ben Jones

9.1 Introduction

This chapter focuses on some current challenges and opportunities regarding post-harvest quality in farmed *Lates calcarifer*. The post-harvest quality of farmed seabass is an area that has been largely overlooked and somewhat neglected. This is in direct contrast to many farmed temperate fish species where there has been and remains a concentrated focus on optimising post-harvest quality. This focus in temperate species stems from the recognition that consumer demand and confidence is directly linked to product quality and is consequently a key driver of increased production and producer returns. As with wild fisheries, aquaculture has traditionally focused on quality assurance (undertaking of a series of planned and systematic actions that secure quality) and quality control (operational techniques and activities that are used to satisfy requirements for quality) measures (see Connell 1995). However, unlike capture fisheries, aquaculture provides an unique opportunity to manipulate, or actively control, the intrinsic qualities of the product prior to harvest, thus better meeting consumer preferences and expectations.

Centre for Sustainable Tropical Fisheries and Aquaculture, School of Marine and Tropical Biology, James Cook University, Townsville, Queensland Australia.
Email: guy.carton@jcu.edu.au

230 *Biology and Culture of Asian Seabass*

The focus of this chapter is to highlight the challenges and opportunities for improving post-harvest quality of farmed *L. calcarifer*. Due to a lack of research elsewhere, the chapter will primarily use challenges faced by the Australian barramundi industry as the illustrative example.

9.2 Harvesting

To understand product quality issues it is first important to understand how fish are commonly harvested in the Australian barramundi industry as many quality parameters stem from how the fish are initially treated at harvest. Current harvesting practices in the Australian industry involve sweeping ponds with large nets (knotless nylon ~25–50 mm is frequently used) strung from either side of the growout pond. Fish are then crowded to a single location within the pond for harvesting. A large dip net (~1–1.5 m diameter), slung from a small boom crane (or a smaller hand-held net), is lowered into the location where the fish have been crowded and a proportion of the fish are then gradually removed from the pond (Fig. 9.1). The net is then positioned over a tank containing a saline ice-water slurry. The floor of the dip net is opened and fish are euthanized by rapid chilling via immersion in

Figure 9.1. Dip net harvesting of Australian pond-reared barramundi after fish have been corralled using a seine net.

the ice-water slurry, a process that normally takes less than 1 min. Additional ice is continually added to ensure that the slurry remains close to 0°C and that fish mortality occurs quickly. The ice-water slurry containing the fish is then removed to a processing facility with further addition of ice often required to ensure that the ice-water slurry remains below 2°C until fish are packaged. Fish are packed whole into ~40L polystyrene boxes, or into plastic-lined cardboard bins ready for transportation. Following packing fish are held below 2°C during storage and transportation to maintain product quality and prevent bacterial spoilage.

Immersion in ice-water slurry is commonly used for the slaughter of many aquaculture species around the world (Benson 2004). Although a low technology method, immersion in ice-water slurries lowers production costs while rapid chilling of the fish is assumed to ensure good flesh quality (Robb and Kestin 2002, Southgate and Wall 2001). This approach also facilitates the rapid handling of large numbers of fish during harvest and ensures that fish do not remain crowded in stressful conditions for prolonged periods prior to slaughter. However, some regard this method of fish euthanasia as unacceptable with respect to animal welfare, primarily as this method does not cause immediate loss of consciousness (Ashley 2007), although it is still not clear how aversive ice-water immersions are to fish (Robb and Kestin 2002). Fish euthanised by this method often make strenuous efforts to escape, experience clonic muscle cramps (Lambooij et al. 2002, 2006a, b) and show alterations in heart rate (Lambooij et al. 2002, 2006a, b). Consequently, it is thought that fish slaughtered using this method may experience up to several minutes of distress prior to death. The effectiveness of ice-water immersion to induce loss of sensibility and death, however, is primarily thought to be a function of the ambient temperature difference between the rearing pond from which fish are removed and the ice-water slurry; the greater this difference the shorter the time to loss of brain function (Robb and Kestin 2002). For temperate cold-water species, such as rainbow trout (*Oncorhynchus mykiss*) where this temperature difference is narrow, loss of brain function can take as long as 10 min (Kestin et al. 1991). For tropical fish species harvested from outdoor ponds, such as *L. calcarifer*, this temperature difference is much greater (~16–28°C), therefore the time till loss of brain function and death could be assumed to be considerably reduced. Clearly work is needed to confirm whether ice-water immersion and the resulting thermal shock causes prolonged distress in barramundi prior to death. This work should also be accompanied by investigating the efficacy of other cost-effective slaughter techniques that result in instantaneous death.

Any discussion of fish slaughtering techniques is often coupled with the contentious and emotive issue of whether fish have the capacity to perceive pain and/or fear. It has previously been argued that fish are incapable of perceiving pain and fear, a view largely formulated on the basis that the awareness of pain in humans depends on specific areas of the cerebral cortex. Because fish lack a well-developed cerebral cortex it has been assumed that they don't possess the necessary cognitive capability to perceive pain and fear (Rose 2002). Despite this assumption, several studies have demonstrated that the perception of pain and fear in fish is similar to that of birds and mammals (Sneddon 2002, 2003, Sneddon and Gentle 2002) and there is now compelling evidence that indicates fish have the capacity for pain perception and suffering, although are they are unlikely to perceive pain in the human sense (Sneddon et al. 2003, Braithwaite and Huntingford 2004). Even in the absence of this debate fish clearly display non-conscious, neuroendocrine, and physiological stress responses to 'noxious' stimuli, thus avoiding or minimizing such stimuli or stressors during the harvest process and this is clearly an important issue regarding the welfare of farmed fish (Southgate and Wall 2001, Huntingford et al. 2006, Brännäs and Johnsson 2008).

Accordingly, the humane harvest and slaughter of fish is an important consideration for the Australian barramundi industry. Such issues are prominent topics in the minds of consumers and consumers are also increasingly becoming aware of fish welfare issues (Poli et al. 2005). However, it is interesting to note that post-harvest quality and welfare issues around humane harvest and the slaughter of fish are not only intrinsically linked, but are also complimentary (Wall 2001).

One of the primary motivations for using ice-water immersion at slaughter is the notion that rapid chilling ensures that high flesh quality is maintained post-mortem through slowing down both enzymatic and bacterial activity (Huidobro et al. 2001), removing thermal energy that is normally accessible for autolytic degradation of the muscle (Skjervold et al. 2001) and reduces muscular activity immediately prior to death (Robb and Kestin 2002). Live-chilling is effective in delaying the onset of maximum rigor (rigor mortis), however, the same effect is not observed with respect to the resolution of rigor, with no observable differences between live-chilling and other slaughter methods after 24 to 48 h post-mortem (Skjervold et al. 2001, Matos et al. 2010). However, other work has demonstrated that there is no real quality benefit to be obtained from fish being alive prior to ice-water immersion and that any such benefits could be obtained equally well if fish were slaughtered prior to rapid chilling (Pastor et al. 1998, Tejada and Huidobro 2002). Frost et al.(1999) examined post-harvest quality indices for different slaughter techniques in barramundi, comparing slaughter using ice-water immersion to other methods. Here no difference in the onset of maximum rigor or flesh texture, two important measures of post-harvest quality in fish, between different slaughter methods were found.

Most recently it has been recognized that minimizing stress immediately prior to slaughter can yield significant and positive benefits with respect to post-harvest quality. During harvesting fish are often crowded and removed from the water and these processes often elicit acute stress resulting in an increase in circulating levels of adrenaline and cortisol. This in turn elicits changes in a range of biochemical and haematological factors (Barton and Iwama 1991, Bonga 1997). Of particular relevance to post-harvest quality is that the physical activity and stress associated with pre-slaughter activities elicits an elevation of the anaerobic energy metabolism causing acidosis of the muscle and blood. This results in accelerated degradation of the muscle post-harvest and manifests as unfavourable colour changes, reduced moisture holding capability, gaping, reduced shelf-life and poor consumer acceptance (Sigholt et al. 1997, Bagni et al. 2007, Matos et al. 2010).

The recognition that physical activity and stress leads to acidosis of the blood and other flesh quality impacts has seen the evaluation of 'rested harvesting' strategies for many aquaculture species and involves sedation of the fish immediately prior to handling and slaughter via the application of an anaesthetic. The anaesthetic agent isoeugenol (2-methoxy-4-propenylphenol, trademarked as AQUI-S) is commonly used as it has a zero market withdrawal period and has been approved for use in a number of countries including New Zealand, Chile and Australia. Anaesthetising fish pre-harvest has been shown to lower physical activity and reduce the magnitude of the stress response (Robb and Kestin 2002, Iversen et al. 2003, Small 2004), resulting in improvements in post-harvest quality with fillets typically having a higher pH, greater firmness, improved colour, lower drip loss and delayed time to maximum rigor when compared to fish harvested using conventional procedures (Kiessling et al. 2004, Bosworth et al. 2007, Matos et al. 2010, Erikson et al. 2011).

Wilkinson et al. (2008) evaluated the effects on post-harvest fillet quality of barramundi subjected to simulated commercial and rested harvesting techniques. Using AQUI-S as the anaesthetic they found results similar to those observed in studies for other species, with barramundi subjected to rested harvesting exhibiting extended time to maximum rigor of ~9h and the maintenance of higher muscle pH for 12h. Although rested harvesting has the potential to improve the post-harvest quality of farmed *L. calcarifer* there needs to be direct demonstration of the economic and practical benefits to producers if it is to be adopted by the industry as a whole. Clearly rested harvesting is possible for barramundi grown in tank recirculation systems or raceways, however, it is difficult to see how rested harvesting strategies could be applied at the scale of outdoor grow-out ponds which can range in size from 0.1–1.5 ha without significant capital investment in infrastructure. Removing fish to cages or holding tanks prior to slaughter may allow rested harvest to be employed following a period of recovery. The use of fish pumps to remove fish from crowding nets may also reduce stress on fish and allow them to subsequently be anaesthetised. Further, harvest skimming is often practiced, where only a proportion of the stock is removed from the grow-out pond. In situations where rested harvest is either impractical or unachievable producers should focus on streamlining current harvesting protocols and limiting stress during the harvesting period. This includes reducing the time taken to crowd the fish, minimising the period fish spend crowded in the net, transferring fish into the ice-water slurry as quickly as practical and ensuring ice-water slurries are kept below 2°C.

9.3 Post-Harvest Quality

9.3.1 Quality deterioration and lipid oxidation

The quality of aquaculture product is greatly influenced by post-harvest handling, storage and processing. Gradual degradations in flavor quality arise due to oxidative processes and the development of off-odours that occur from enzymatic and bacterial processes (Gram 1992, Gram and Huss 1996). This may also be accompanied by changes in the texture and colour of the fillets (Olafsdóttir et al. 1997, Bonilla et al. 2007). However, high quality can be maintained for prolonged periods if post-harvest handling and storage conditions are optimal. For example, snapper stored at 0°C has a high quality life of 9 days and becomes unpleasant in sensory quality only after 15 days (Fletcher and Hodgson 1988).

Once barramundi have been removed from the grow-out pond and placed into the ice-water slurry they are transported to a packaging facility, usually a short distance from the harvesting location. Fish remain in the ice-water slurry for between 2–12 hr to ensure that the core temperature of the fish is reduced to below ~4°C. Following this period fish are removed, sorted and placed whole and undressed into 40L leak proof insulated styrene or 300–600L plastic-lined cardboard containers. Flake ice is then added and the containers sealed. The packaged fish are finally placed into refrigerated storage (<4°C) and refrigeration is maintained during road or air transport to markets. This process ensures rapid chilling and maintenance of cold temperatures throughout the supply chain. Time from harvest to the consumer can range from 24–168 hr and involve as many as 4–5 transfers in the supply chain. The high post-harvest quality of Australian farmed barramundi is generally attributed to the quality preserving effect of immediate chilling following slaughter and continuous chilling throughout the supply chain. Chilling slows the rate of undesirable biochemical and chemical reactions in the flesh and the growth and spoilage activity of microorganisms (Sikorski and Sun Pan 1994). Early and rapid chilling of fish to as low as possible, but still above freezing, is well known to extend the shelf-life of the post-harvest product (Olafsdóttir et al. 2006, Bao et al. 2007). This is a critical consideration for any aquaculture product, especially those that have protracted supply chains and long distances to major markets.

A major cause of quality deterioration in meat products results from lipid oxidation and the changes associated with this oxidative process (Ladikos and Lougovois 1990, Liu et al. 1995, Gray et al. 1996). Fish, unlike any other meat product, is acutely prone to lipid oxidation due to the high degree of polyunsaturated *n-3* fatty acids (PUFAs) in the flesh (Siu and Draper 1978, Hultin 1994, Undeland 2001), which are often promoted as having health and nutritional benefits. Post-harvest oxidation of lipid requires two substrates, molecular oxygen and lipid, unsaturated fatty acids then react with oxygen to form hydroperoxide, a primary oxidation product. Hydroperoxide then decomposes producing a variety of volatile secondary products such as aldehydes, ketones and alcohols (Hultin 1992, Undeland 2001). It is these breakdown products that are responsible for flavour deterioration or rancidity, and loss of colour, texture, moisture and nutritional value during storage.

Resistance to lipid oxidation and flavour deterioration is largely affected by the balance between pro-oxidants in the tissues, which serve to promote oxidation, and anti-oxidants, which restrain oxidation (Hultin 1992, Undeland 2001). The most well-known pro-oxidants in fish are transition metals such as iron, whereby a large part of the iron in fish muscle is bound to the oxygen carrying pigments myoglobin and haemoglobin (Hultin 1992, Undeland 2001). These pigments have catalytic properties, a result of their ability to break down hydroperoxide to volatile secondary products (Ladikos and Lougovois 1990). It is thought that haem iron is the major catalyst of lipid oxidation in mullet (Khayat and Schwall 1983).

The most widely known anti-oxidant in fish is α-tocopherol and is found in the lipid interior of membranes (Hultin 1994, Undeland 2001). This compound has the highest biological activity of the vitamin E homologues and has also been shown to be retained to a greater extent than other tocopherols in fish (Harare and Lie 1997). Other lipid soluble antioxidants have also been identified, such as ubiquinol (Petillo et al. 1998) and the carotenoid pigments important in flesh colouration in salmonids (Christophersen et al. 1992).

Slowing lipid oxidation in meat products is critical to extending shelf-life and slowing post-harvest quality deterioration and has obvious importance in terms of product acceptability at the time of consumer purchase and when supply chains are protracted due to large geographic separation between producers and major markets. Much attention has focussed on the development of post-harvest storage protocols and techniques that restrain lipid oxidation and these approaches have largely been developed for wild capture fisheries (Sikorski and Sun Pan 1994, Day 2001). Inhibiting lipid oxidation can be achieved by limiting the exposure to, or complete exclusion of, oxygen via vacuum or modified atmosphere packaging. The effect of ice storage on lipid oxidation is somewhat less clear, large losses of antioxidants in the early stages of storage and substantial increases in lipid oxidation products have been reported (Undeland et al. 1999, Haugen and Undeland 2003). Other aspects have focussed on minimising tissue disruption, such as filleting and skinning (Undeland et al. 1998). Such processing introduces excessive amounts of molecular oxygen into the tissues which has the effect of accelerating the development of rancidity (Sato and Hegarty 1971). To this end filleted fish have low stability with regard to lipid oxidation (Mendenhall 1972, Hultin 1994). The general view is that any post-harvest tissue disruption serves to accelerate lipid oxidisation and should therefore be kept to a minimum. However, there does appear to be some net benefit to fillet stability if the red muscle (slow oxidative) is removed as this tissue contains more pro-oxidative and less anti-oxidative capacity in comparison to white muscle (fast glycolitic) (Slabyj and Hultin 1983, Hultin 1992, Sohn et al. 2005).

Australian farmed barramundi would benefit from the development of an anti-oxidative strategy focussed at minimising post-harvest lipid oxidation and prolonging shelf-life. Under current practices, farmed barramundi are packed on ice whole and ungutted. This method of processing limits the exposure of tissues to molecular oxygen and therefore helps to constrain lipid oxidation. Following packing and despatch, barramundi farmers have no control over the subsequent handling and processing of fish. The fish may remain whole throughout the supply chain, or they may be filleted by the wholesaler, retailer or the end user, with time to consumption varying between hours to days. Developing techniques that fortify barramundi flesh against post-harvest lipid oxidation may help to prolong shelf life when storage and handling conditions are extremely variable. One method of fortifying fish against post-harvest lipid oxidation is through supra-nutritional dietary supplementation of α-tocopherol. This appears to be one of the most effective means of delaying lipid oxidation and prolonging shelf-life under commercial and retail storage conditions for a range of animal meat products (Liu et al. 1995). Studies on salmonids (Frigg et al. 1990, Harare et al. 1998, Chen et al. 2008), catfish (O'Keefe and Noble 1978, Bai and Gatlin III 1993, Baker and Davies 1996), red seabream (Murata and Yamauchi 1989), turbot (Ruff et al. 2003) and southern bluefin tuna (Buchanan and Thomas 2008), have clearly demonstrated the protective effects of vitamin E against lipid oxidation and post-harvest quality deterioration. These studies have shown that supplementation of α-tocopherol well in excess of normal dietary requirements can retard lipid oxidation during commercial storage or forced oxidation. It is also notable that growth efficiency and the sensory attributes of the post-harvest fillet are not adversely affected by high levels of α-tocopherol in the diet or muscle tissue.

The deposition of α-tocopherol in the muscle tissue depends on the level of supplementation, with higher doses resulting in increased deposition. This relationship has been shown to be linear, at least up to α-tocopherol levels of 1000–1500 mg kg^{-1} (Boggioa et al. 1985, Baker and Davies 1996, Harare et al. 1997, Ruff et al. 2003). However, increases in α-tocopherol in the muscle are not immediate and this is assumed to result from the order in which it is distributed through the tissues, occurring initially in the liver and later in the muscle (Frigg et al. 1990, Harare and Lie 1997, Akhtar et al. 1999, Ruff et al. 2003). Despite this most studies have shown that increases in muscle α-tocopherol occurs within the first few weeks of feeding, although deposition may continue for several more weeks at high levels (Frigg et al. 1990, Bai and Gatlin III 1993, Ruff et al. 2003). However, prolonged feeding periods (> ~10 weeks) do not appear to result in any additional deposition, with muscle tissue seeming to reach saturation (Frigg et al. 1990, Onibi et al. 1996, Bjerkeng et al. 1999). Although only a handful of species have been investigated, α-tocopherol uptake maybe highly dependent on the species, due to differences in fat content, rearing temperature and/or metabolic rates.

The oxidative deterioration of lipids is most often assessed by measuring the concentration of thiobarbituric acid reactive substances (TBARS-test) in the flesh which are secondary oxidation products. When fillets are subjected to storage or forced oxidation TBARS values (oxidative deterioration) are inversely related to the level of α-tocopherol present in the muscle and diet (Harare et al. 1998, Gatta et al. 2000). Ruff et al. (2003) have clearly demonstrated the protective properties of α-tocopherol under commercially stimulated storage conditions, showing that fillets with high levels of α-tocopherol had approximately five-fold lower TBARS values when compared to fillets with low levels of α-tocopherol, with TBARS values diverging after only 2 days of storage.

At present nothing is known about the effect of supra-nutritional dietary supplementation of α-tocopherol on tissue levels, or the efficacy of α-tocopherol in reducing lipid oxidation, for farmed barramundi. Based on that demonstrated for other species the expectation would be that supra-nutritional dietary supplementation of α-tocopherol would increase tissue levels of α-tocopherol and extend the shelf life of farmed barramundi. In fact, preliminary trials conducted by the authors whereby groups of fish were fed one of two diets, the first with the standard level of α-tocopherol (200 mg kg^{-1}) and a second with considerably elevated α-tocopherol (600 mg kg^{-1}), have shown that fish on the elevated diet showed increases in muscle levels of α-tocopherol after only 14 days. After 56 days muscle tissue appeared to reach equilibrium with the diet, with no additional deposition observed after this time (Fig. 9.2). This demonstrates that feeding elevated levels of α-tocopherol beyond 8 weeks will not increase deposition into

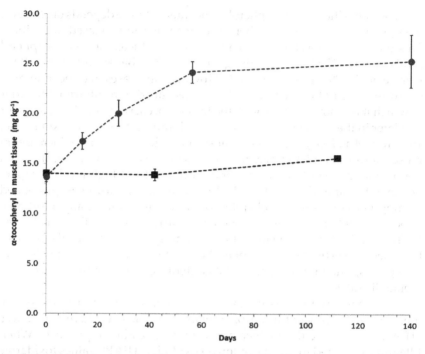

Figure 9.2. Muscle tissue levels of α-tocopherol (vitamin E) of farmed barramundi. Fish were fed a standard commercial diet containing α-tocopherol at 200 mg kg^{-1} (filled squares) and 600 mg kg^{-1} (filled circles). Fish were held in cages within earthen rearing ponds and feed via standard commercial practices.

the muscle tissue for barramundi. However, the lag time between feeding and uptake of α-tocopherol can be compressed by increasing the level of supplementation in the feed, thereby resulting in earlier deposition (Frigg et al. 1990, Chen et al. 2008). In contrast barramundi fed the standard control diet failed to show any significant increase in muscle levels of α-tocopheryl. Future experiments will examine whether elevated α-tocopherol in the muscle has the desired benefits of delaying lipid oxidation, preventing loss of colour and moisture, and extending shelf-life.

9.3.2 Visual appearance

Visual appearance is the most important characteristic of food in determining consumer selection prior to actual consumption (MacDougall 1988). Visual appeal is therefore the single most critical aspect of the purchasing decision. In the Australian market, consumers clearly show a stronger preference for lighter pigmented barramundi and white-pink firm fillets. Percival et al. (2008) have shown that skin pigmentation differs between wild and farmed barramundi, with wild fish typically being paler than farmed fish. Tank background colour also affects skin pigmentation, with barramundi reared in white tanks being considerably paler than those reared in black tanks (Howieson et al. 2012). This affect has also been identified in other aquaculture species (van der Salm et al. 2004, Doolan et al. 2008, Pavlidis et al. 2008). Although the exact mechanism(s) controlling skin pigmentation are not yet fully understood (Kawauchi and Baker 2004) adaptation to a white background is known to activate the expression of the melanin-concentrating hormone (MCH) gene in the hypothalamus of fish, with MCH expression and plasma levels of MCH being higher than fish reared on black backgrounds (van Eys and Peters 1981, Kishida et al. 1989, Gröneveld et al. 1995, Suzuki et al. 1995). Similarly, long term administration of MCH is known to reduce the melanin content of the skin and circulating levels of α-melanocyte-stimulating hormone (MSH) (Baker et al. 1986). Classically, αMSH is considered the main hormone causing dispersion of melanin granules in melanophores and the subsequent darkening of the skin (Kawauchi and Baker 2004). Barfin flounders (*Verasper moseri*) reared in white or yellow tanks for 8 months, as well as having higher levels of MCH than those reared on a black background, were also observed to have a greater body length and weight (Yamanome et al. 2005). It has been suggested that lighter backgrounds stimulate the production of MCH and that MCH signalling may also be involved in regulating food intake in teleosts.

The external coloration of fish flesh is often used as an instinctive indicator of product freshness and quality, and is considered a key driver for many consumers. It has often been noticed that in farmed barramundi following harvesting that the dorsal section of the fillet begins to take on a grey colouration. This has the effect of reducing the visual appeal of the fillet to the consumer. The pre and/or post-harvest factors responsible for fillet greying in farmed barramundi are not yet fully understood. Melanin has been identified as a possible agent responsible and is found intercellularly near fat cells within the muscle. Interestingly melanin levels are lower in wild caught barramundi, which coincidently also have lower lipid levels when compared to farm reared barramundi (Howieson et al. 2012). In an attempt to resolve the underlying cause of fillet greying Howieson et al. (2012) included substrates in the diet that would serve to promote (high tyrosine) or limit (low copper) melanin production. However, the fillet colour attributes of fish on these modified diets were similar to those fish fed the standard reference diet. Clearly further research is required to understand and develop commercially feasible and practical on-farm production techniques that influence skin pigmentation and reduce dorsal greying of fillets. Obtaining control over both of these processes will enable farmers to produce a more visually appealing whole fish and raw fillet product.

9.4 Flavour and Taste

9.4.1 Off-flavour tainting

Muddy-earthy-musty-type flavours are generally regarded as a natural characteristic of wild caught freshwater fish, although the occurrence of such flavours has also been reported for a diverse range of freshwater aquaculture species (Lovell 1983, Tucker 2000, Yamprayoon and Noonhorm 2000, Howgate 2004, Robertson et al. 2005, Petersen et al. 2011). Fish presenting with these flavour characteristics are often referred to as being 'off-flavour', or 'tainted', and are generally considered to be spoiled and of low quality. Most off-flavour tainting reports in aquaculture originate from catfish production in the USA and trout production in Europe. However, as highlighted by Howgate (2001), finfish aquaculture is dominated by freshwater production in tropical and sub-tropical localities, mostly employing earthen ponds. Consequently, episodes of off-flavour tainting are expected to be more prevalent than currently reported.

The source of muddy-earthy-musty flavours in freshwater fish is commonly attributed to the presence of two compounds, geosmin (GSM) and/or 2-methylisoborneol (2-MIB). These compounds are metabolites of certain groups of algae, actinomycetes and cyanobacteria (Tucker 2000) and are commonly found in various natural water sources such as lakes, reservoirs and streams (Jüttner and Watson 2007). Brief exposure to even low concentrations of these compounds is known to impart an intense muddy-earthy-type flavour to cultured fish (Martin et al. 1988, Persson 1980, Howgate 2001, Robertson et al. 2006, Selli et al. 2009). Uptake is an extremely rapid and passive process that primarily occurs across the gills (From and Hørlyck 1984). Tainting compounds then begin to rapidly accumulate in the muscle tissue. The rate of accumulation is primarily driven by the concentration of the compounds in the holding water relative to the concentration in the fish (Neely 1979, Clark et al. 1990, Streit 1998, Howgate 2004). The importance of uptake across the gills is linked to the octanol/water partition coefficient of a chemical (K_{OW}). Uptake across the gills dominates up to a log K_{OW} of 6, above this value uptake across the gills is considerably less important and uptake from the gut begins to dominate (Clark et al. 1990). The off-flavour tainting compounds GSM and 2-MIB have log K_{OW}'s of 3.57 and 3.31; consequently uptake of these compounds occurs across the gills.

The compounds GSM and 2-MIB are much more soluble in lipid than in water (Howgate 2004), with the lipid content of the fish becoming an important factor when considering bioconcentration of such tainting compounds. As a consequence, the level of tainting compounds in the tissues of the fish exceeds that in the surrounding water. Unfortunately, intensively farmed fish often have much higher lipid content than wild fish and this will serve to intensify the off-flavour taint and increase the susceptibility of aquacultured fish to tainting episodes. Higher lipid also makes the elimination of tainting compounds considerably more protracted as the rate at which these compounds are lost is inversely proportional to lipid content (Howgate 2004). The loss of tainting compounds from the tissue reduces exponentially with time, initially being extremely rapid, but is then followed by a much slower rate of loss which can take days or even weeks (Johnsen and Lloyd 1992, Perkins and Schlenk 1997).

At present there are no effective practices to prevent the growth of taint producing organisms or remove taint causing metabolites from the water at the scale of commercial activities. Geosmin and 2-MIB is lost from the water by volatilization and biodegradation. Volatilization of GSM and 2-MIB occurs slowly with half-lives for GSM decay ranging from 19-35 days (Li et al. 2012). Biodegradation appears to be the dominant mechanism of loss in aquatic environments, with half-life times due to biodegradation being generally less than one day, although biodegradation is considerably for 2-MIB with half-life ranging from 5–90 days (Li et al. 2012). Biodegradation has also been shown to exhibit extended lag times before significant degradation is observed (Izaguirre et al. 1982). Due to the above limitations solutions to episodes of off-flavour tainting in aquaculture have largely focused on the removal of tainting compounds from the fish prior to harvest. This process, known as depuration or purging involves transferring fish into clean (taint-free) water so that tainting compounds can be eliminated across the gills. As previously highlighted the elimination of tainting compounds from the fish, to below sensory detection thresholds, can be extremely slow taking several days and thereby represents an additional cost to production.

Detection thresholds for GSM and 2-MIB in water are extremely low approximately 0.015 and 0.035 µg l^{-1}, respectively (Howgate 2004). However, of more relevance, is the detection threshold of these compounds in fish muscle which has been shown to range from 0.25–0.5 µg kg^{-1} for GSM in channel catfish (Grimm et al. 2004) to 0.9 µg kg^{-1} for rainbow trout (Robertson et al. 2006). However, it should be recognized that detection thresholds are influenced by variations in sensory evaluation panels, the masking effect of the flavor characteristics of the species, and also the lipid content of the fish.

9.4.2 Off flavour taint in barramundi

Episodes of muddy-earthy tainting of freshwater farmed barramundi are frequently reported in Australia (Percival et al. 2008, Jones et al. 2013). Recently, this issue has been highlighted as the primary cause of an escalation in negative consumer perceptions of Australian farmed barramundi and a growing resistance to future purchases (Phillips 2010). Episodes of off-flavour tainting ultimately erode the market value of end products and returns to producers. Research is now being undertaken to address the mechanisms of off-flavour tainting and the development and implementation of practices aimed at regaining consumer confidence in the quality of Australian farmed barramundi.

Using a combination of gas chromatography-olfactometry and descriptive sensory analysis Frank et al. (2009) demonstrated that barramundi grown under different production techniques had wide differences in their flavour attributes. Barramundi grown in outdoor ponds had considerably higher muddy, dirt and musty flavor notes than either fish grown in recirculation tanks, or even wild caught fish. This affect was attributed to the presence of the off-flavour tainting compounds GSM and 2-MIB in pond grown fish.

The impact of off-flavour tainting compounds on the sensory attributes of barramundi is similar to that observed for other species (Percival et al. 2008, Jones et al. 2013). Barramundi tainted with GSM and/or 2-MIB show wide differences in several sensory attributes and descriptors use to assess flavour and aroma (Fig. 9.3). Tainted fish show elevated scores for muddy-

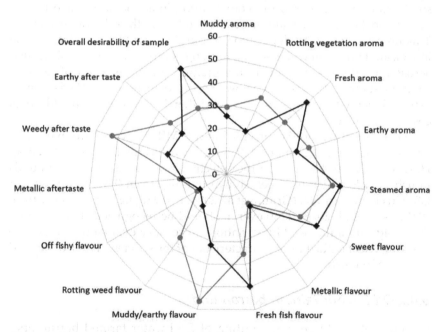

Figure 9.3. Sensory taste and aroma attributes (%) of farmed sourced barramundi with geosmin muscle tissue concentrations of 4.47 µg kg^{-1} (grey line) and 0.74 µg kg^{-1} (black line). The impact of geosmin tainting is most apparent on the attributes of weedy after taste, muddy-earthy flavour and rotting weed flavour.

earth flavour, rotting weed flavour, weedy after taste and reduced scores for fresh fish flavour, sweet flavour and fresh aroma. However, off-flavour taint appears to show regionalisation within the fillet and this is assumed to be related to the distribution of body lipid, with off-flavour taint being most perceptible in the high lipid belly cut (Percival et al. 2008). Although there has been no direct assessment of the threshold detection for off-flavour taint in barramundi some inferences can be made. Jones et al. (2013) recorded significant reductions in muddy/earthy flavour and significant increases in overall desirability at GSM flesh levels of 0.74 ug kg^{-1}. Accordingly, the threshold for detection is likely to be below this value which is in good agreement with the detection thresholds found for other species (Persson 1980, Grimm et al. 2004, Robertson et al. 2006).

Episodes of off-flavour tainting of Australian barramundi have been investigated in Lake Argyle, Western Australia (Percival et al. 2008) and from freshwater earthen ponds in Queensland (Jones et al. 2013). However, the compound responsible for tainting was found to vary between the two sites with 2-MIB identified as the primarily compound responsible for tainting in Lake Argyle (Percival et al. 2008) and GSM identified in freshwater earthen ponds (Jones et al. 2013). Geosmin levels in earthen rearing ponds were monitored over a three month period by Jones et al. (2013), whereby GSM was found to be persistent over the entire sampling period with values most often ranging between 0.2–1.75 ug l^{-1}. However, extreme levels (>14.0 ug l^{-1}) of GSM were also observed, with rapid increases taking place over several weeks. This is not overly surprising given that the prevailing climatic conditions in tropical localities, namely high solar radiation and high temperatures, would clearly favour the growth of taint producing organisms. However, more research is needed to identify the underlying mechanisms of off flavour taint production and the conditions and factors that promote or trigger episodes of tainting and periods when rearing ponds are more susceptible to tainting episodes.

As previously demonstrated with other species, the uptake of GSM by barramundi is initially exceptionally rapid and then begins to plateau. This can be seen in the example that illustrates the concentration of GSM in the muscle tissue of 1.5kg barramundi sampled at varying intervals following exposure to a water GSM concentration of 14.0 ug l^{-1} (Fig. 9.4a). The concentration of GSM in muscle tissue increases rapidly with exposure time, such that after 3 hours exposure muscle tissue concentrations are already ~60% of the water GSM concentration, although flesh concentrations were stable after 48 hours exposure and failed to reach equilibrium with the water. This represents a bioconcentration factor (BCF) of 0.6 which is somewhat low given that total body lipid levels for farmed barramundi are typically around ~10% (wet weight) (Glencross et al. 2008). Jones et al. (2013) reported similarly low BCF's in the range of 1.13–1.42 for 2.5kg

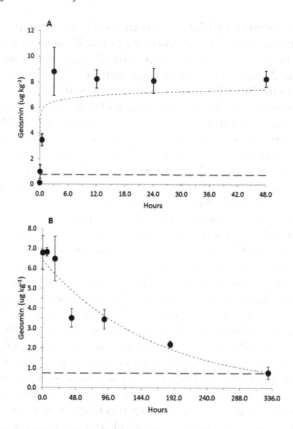

Figure 9.4. Uptake (A) and depuration (B) of geosmin. Barramundi were held in water containing 14 ug l^{-1} geosmin and muscle tissue samples were taken from the dorsal shoulder region. The grey dashed line indicates the assumed threshold detection level for geosmin in barramundi muscle.

barramundi exposed to GSM in the range of 1.2–4.0 ug l^{-1}. The finding of such low BCF's for barramundi is somewhat perplexing given that total lipid levels for farmed barramundi are comparable to other species. Such low BCF's may result from taking muscle samples from the low lipid (1.3% ww) shoulder region of the fillet, or simply represent a limitation in the bioconcentration of GSM in muscle tissue at such extreme concentrations. Most previous studies exploring bioconcentration of GSM in fish have utilised serial dilutions of synthetic GSM to explore uptake. In contrast, Jones et al. (2013) used water sourced directly from aquaculture ponds containing naturally occurring GSM. Differences in the uptake of GSM may be related to the type of GSM to which fish are exposed (synthetic or naturally occurring). Variations in uptake may also occur due to variations in the relative proportions of cellular and intracellular GSM present in the water. In aquaculture systems GSM occurs as intracellular GSM, either present in the cytosol or bound to cellular proteins, or as extracellular GSM, present in the water. The relative proportions of intracellular to extracellular GSM can vary greatly, with extracellular fractions ranging from 0.5–40% of total GSM (Wu and Jüttner 1988a, b, Li et al. 2012).

Clearly further work is required to establish the relationship between bioconcentration of off-flavour tainting compounds in muscle tissue of barramundi and concentrations of GSM in water and lipid content of fish. Further work is also required to explore differences in the uptake of synthetic versus microbial produced GSM as well as differences in uptake of intracellular and extracellular GSM and how these differences influence rates of uptake and bioconcentration. Perhaps of most importance is that even brief exposure of fish to GSM tainted water is sufficient to impart off-flavour. After only 5 min exposure to a water GSM concentration of 14.0 ug l^{-1} muscle tissue concentrations were ~1.0 ug kg^{-1}, which exceeds the proposed detection threshold for barramundi. This demonstrates that even short term exposure to GSM is sufficient to impart off-flavour taint in barramundi. Small changes in absolute values at low GSM flesh concentrations are more important than the equivalent absolute change at high concentrations (Howgate 2004), an effect that relates to the logarithmic relationship of the concentration of chemical to its perceived intensity. Depuration of off-flavour taint typically occurs in two phases, an initial phase that proceeds rapidly and a secondary phase that progresses more slowly. The prolonged nature of depuration of GSM from the muscle tissue is demonstrated by the period needed to reach the assume detection threshold for GSM in barramundi muscle tissue, which in this instance is ~14 days (Fig. 9.4b). However, lipid levels in these fish were not quantified, so it is difficult to draw any firm conclusions regarding depuration time to reach below threshold concentrations. In addition these fish were not fed during the depuration period, therefore loss of GSM from the fish is likely to be more rapid as lipid stores are accessed to meet energy requirements. As Howgate (2004) highlights the rate of elimination of tainting compounds is inversely proportional to the lipid content of the fish. Therefore starving fish during the depuration period would assist in shorting the depuration period. The protracted nature of depuration presents a real issue for farmers who must be ready to respond to market demand and volatility. It is here that mechanisms that shorten the depuration period need to be identified such as increasing metabolic rate via elevated water temperatures.

The implementation of effective and economically feasible quality assurance methods for the control of off-flavour tainting episodes requires a greater understanding of the mechanisms responsible for the production of off-flavour taint and the development of protocols to remove these compounds from the fish prior to harvest. A valuable starting point would be gaining a solid understanding of the relationship between the accumulation, concentration and elimination of off-flavour tainting compounds and lipid levels of farmed barramundi. Further to this understanding the distribution or regionalisation of lipid and the relationship to off-flavour taint compounds throughout the fillet would also be highly valuable. In addition, identifying the factors that accelerate the loss of off-flavour tainting compounds prior to harvest would also be valuable in developing responsive and effective industry wide depuration protocols.

9.5 Flavour Enhancement

It is well-known that marine and freshwater fish are divergent in their flavour characteristics (Boyle et al. 1992, Farmer et al. 2000, Frank et al. 2009). Some of this difference can be attributed to the presence or absence of certain compounds in the freshwater and marine environments (Josephson et al. 1984, 1991, Lindsay 1994). Flavour and taste attributes, alongside consistency, are perhaps the single most important factors influencing consumer acceptability and demand for fish products.

Consumers often describe freshwater fish as being bland or lacking in desirable flavour qualities, with consumers identifying the post-harvest product as being devoid of complex 'marine-like' flavours that are common in wild-caught marine fish (Ostrander and Martinsen 1976, Thomassen and Røsjø 1989, Boyle et al. 1992, Kummer 1992, Farmer et al. 2000, Grigorakis et al. 2003, Grigorakis 2007). Australian farmed barramundi is known to suffer similar criticisms, often being described as bland, flavourless, and lacking the marine-like flavour notes that characterise wild caught marine or estuarine barramundi. The difference in the flavor and taste characteristics between freshwater and marine fish is most often attributed to the presence of low concentrations of naturally occurring bromophenols. It has been suggested that the primary reason farmed fish, such as barramundi, lack these flavour and taste attributes is that although manufactured diets utilize wild sourced marine ingredients, the level of bromophenols in the rearing diet is relatively low (Boyle et al. 1992, Whitfield et al. 2002). As feed manufacturers seek to reduce levels of wild sourced marine ingredients (e.g., fish meal, krill meal, fish oil) and/or seek more sustainable terrestrial alternatives, flavour and taste issues are likely to become more prevalent in aquaculture.

Bromophenols are widely distributed in marine fish and known to impart the desirable marine, sea salt and iodine-like flavours that characterise marine seafoods (Boyle et al. 1992, 1993a, b, Lindsay 1994). These compounds are synthesized by marine primary producers and other organisms in the marine environment (Whitfield et al. 1999). Bromination occurs through secondary metabolic processes and as such bromophenols are regarded as defensive chemicals against micro-organisms and predatory marine animals (Woodin et al. 1987). These naturally occurring compounds are passed up through the food web to higher order marine organisms, eventually accumulating in fish and crustaceans (Whitfield et al. 1990, Frank et al., 1999). A comprehensive survey of marine fish by Whitfield et al. (1995, 1998) revealed that bromophenols were present in virtually all species sampled, although bromophenol content was highly variable between species. Using separate analysis of bromophenol concentration in the gut and flesh Whitfield et al. (1998) also demonstrated that high levels of bromophenols were present in the gut, therefore confirming the notion that these compounds are derived from the diet (Whitfield et al. 1998, Oliveira et al. 2009). The relationship between feeding history and the concentration of bromophenols in the flesh has also been clearly demonstrated (Boyle et al. 1992, Whitfield et al. 1992, 1998). In contrast, freshwater environments lack bromine and the flora/fauna capable of bromination (Fenical 1982) and, accordingly, freshwater fish are devoid of bromophenols (Lindsay 1994).

Aquaculture provides significant opportunities to intentionally alter the flavor and taste attributes of fish prior to harvest through the elevation of natural bromophenols in the flesh (Boyle et al. 1993b). Ma et al. (2005) evaluated the inclusion of bromophenol in the diet of aquacultured silver seabream (*Sparus sarba*), through the addition of marine algae (*Padina arborescens, Sargassum siliquastrum*), known to be relatively high in bromophenols. Inclusion levels of 30% algae in the diet resulted in total bromophenol contents of 132 to 340 ng/g, 40 times higher than that observed in the commercial diet. After 8 weeks of feeding, fish fed the algal supplemented diet contained significantly higher concentrations of three bromophenols (2-BP 2,4-DBP, 2,4,6-TBP). Uptake of bromophenols into the flesh was observed to be relatively rapid with fish on the algal diets showing elevated flesh concentrations after only 2 weeks of feeding. Subsequent sensory evaluations demonstrated that consumers could clearly detect a difference between fish fed the bromophenol enhanced diet, with the flavour of these fish being described by some consumers as 'seafood-like'.

Fuller et al. (2008) measured bromophenol concentrations in wild caught and freshwater farmed Australian barramundi. Surprisingly the concentrations of bromophenol compounds (2-BP, 4-BP, 2,6-DBP, 2,4-DBP, 2,4,6-TBP) were below detectable flavour thresholds in both groups. However, Fuller et al. (2008) highlights that the similarity in bromophenol concentrations may have resulted from the unique weather conditions that prevailed around the time of harvest, with large and prolonged freshwater input into normally estuarine areas. These conditions may have caused a dietary shift in wild fish from a marine based diet, containing bromophenol compounds, to a freshwater diet devoid of bromophenols.

Current research by the authors on flavour enhancement in farmed barramundi has followed a similar line of enquiry to that of Ma et al. (2005), where the potential of the green seaweed *Ulva pertusa*, which is known to be high in bromophenol compounds (Whitfield et al. 1999, Whitfield et al. 2002), as a dietary additive is being evaluated. Groups of barramundi were fed formulated diets with differing levels of bromophenol, a control or reference diet consisting of standard commercial barramundi feed and two bromophenol enriched feeds containing 30% and 50% *Ulva*, added in the form of dried ground power. Groups of ~1.5kg barramundi housed in indoor freshwater recirculation tanks were fed these diets for a period of 30 days. Initial observations revealed that palatability of the diet was influenced by the inclusion level of *Ulva*, with the consumption of the 50% enriched diet being very pulsatile and generally having poor acceptability when compared to the other two diets. After 30 days fish were harvested and the flesh subjected to sensory analysis by a triangle test undertaken by tasting panels using established protocols (see Johansson 2001). Barramundi fed the 30% *Ulva* enriched diet was preferred over the standard commercial barramundi diet, with sensory descriptors from panel members being descriptors that are often associated with bromophenols (Table 9.1). This trial highlights the potential to alter the flavour quality of farmed barramundi through dietary inclusion of ingredients that are high in natural bromophenols. However, clearly more information regarding the relationship between inclusion levels of bromophenols in the diet, feeding period and the resultant influence on the flavour and taste attributes of the post-harvest product are required. At high levels bromophenols are known to cause off-flavour tainting, and are responsible for unpleasant 'iodoform-like' flavours (Whitfield 1988, Anthoni et al. 1990). Considerably more understanding is required regarding the uptake and depuration of bromophenols into and out of the flesh. The uptake and persistence of a compound depends not only on the concentration of the chemical accumulated during the period of exposure, but also the rate of loss by either active or passive means. It has been previously suggested that dietary bromophenols do not bioaccumulate in the flesh, but are slowly depurated, or even metabolised (Anthoni et al. 1990). Loss of bromophenols from the flesh of barramundi appears to be somewhat rapid (Fuller et al. 2008). Barramundi fed Mantis shrimp (*Squila*

Table 9.1. Results of the sensory triangle test assessing if an overall difference exists between barramundi fed the standard commercial diet and barramundi fed the same diet with a 30% *Ulva* inclusion level for 30 days.

Sensory attribute	Correct/Incorrect	Percentage correct	P value
Aroma	19/13	58%	0.02
Taste	17/15	53%	0.05

mantis) that had been spiked with a 1.0 mg/mL solution of bromophenols showed elevated levels of bromophenols in the flesh after only 14 days. However, rapid loss of bromophenols from the flesh was observed after the enriched diet was withheld, with 2-BP, 4-BP and 2,4-DBP being completely lost after 7 days, while only <3% of 2,6-DBP and 2,4,6-TBP remained after this period. This result is not all that surprising when considering the octanol-water partition coefficients for bromophenol compounds (Boyle et al. 1992). The octanol-water partition coefficient provides an estimate of the ratio of the solubility of a compound in a non-polar solvent to its solubility in a polar solvent, such as water (Lyman et al. 1990). Numerous studies have shown that octanol/water partition coefficients are related to bioaccumulation factors for aquatic animals, compounds with values ≥ 3 are regarded as favouring bioaccumulation (Poels et al. 1988). For bromophenol compounds these values are >3, although 2,4-DBP and 2,4,6-TBP possess low values of 3 and 3.74, respectively. As a result although the uptake of bromophenols into the flesh is likely to be rapid its loss is also likely to be rapid and it would not be expected to bioaccumulate to a great degree (Fuller et al. 2008).

The above use of natural bromophenols in diets is the first attempt to modify the flavour and taste attributes of Australian farmed barramundi to better meet consumer preferences and expectations. To better satisfy consumer demands requires the development of a comprehensive finishing strategy for farmed barramundi. A similar approach is often used to modify or recover the final fatty acid composition of cultured fish that have been reared on diets containing terrestrial oils as the lipid source (Robin et al. 2006, Jobling 2004a, b, Turchini et al. 2006). However, before a comprehensive finishing strategy can be developed a greater understanding about the relationship between dietary levels of bromophenols and corresponding changes in flavour and taste attributes of the post-harvest product, and rates of uptake and loss of bromophenols in a farming situation are required. Understanding the kinetics of bromophenol loss from the flesh is also of critical importance, as it is common practice to withhold feed 24hrs prior to harvest to ensure complete gastrointestinal evacuation has occurred.

9.5 Conclusions

Research on post-harvest product quality is in its infancy in the Australian barramundi industry. There is an increasing awareness and acknowledgement that post-harvest quality attributes are critical for the consumer and can act as a point of differentiation. For the Australian industry there remain considerable opportunities to improve, enhance, or manipulate the post-harvest quality of an already highly regarded product. In this regard a great deal can be learnt from the strategies and techniques that are used to optimise post-harvest quality in other aquaculture species.

The economic endpoint of any further research on post-harvest product quality will be the creation of a premium quality product that can be produced repeatedly and with confidence and builds consumer demand.

References

Akhtar, P., J.I. Gray, T.H. Cooper, D.L. Garling and A.M. Booren. 1999. Dietary pigmentation and deposition of α-tocopherol and carotenoids in rainbow trout muscle and liver tissue. J. Food Sci. 64: 234–239.

Anthoni, U., C. Larsen, P.H. Nielsen and C. Christophersen. 1990. Off-flavor from commercial crustaceans from the North Atlantic zone. Biochem. Systematics. Ecol. 18: 377–379.

Ashley, P.J. 2007. Fish welfare: Current issues in aquaculture. Appl. Anim. Behav. Sci. 104: 199–235.

Bagni, M., C. Civitareale, A. Priori, A. Ballerini, M. Finoia, G. Brambilla and G. Marinoa. 2007. Pre-slaughter crowding stress and killing procedures affecting quality and welfare in seabass (*Dicentrarchus labrax*) and sea bream (*Sparus aurata*). Aquaculture 263: 52–60.

Bai, S.C. and D.M. Gatlin III. 1993. Dietary vitamin E concentration and duration of feeding affect tissue α-tocopherol concentrations of channel catfish (*Ictalurus punctatus*). Aquaculture 113: 129–135.

Baker, R.T.M. and S.J. Davies. 1996. Changes in tissue α-tocopherol status and degree of lipid peroxidation with varying α-tocopheryl acetate inclusion in diets for the African catfish. Aquacult. Nutri. 2: 71–79.

Baker, B.I., D.J. Bird and J.C. Buckingham. 1986. Effects of chronic administration of melanin-concentrating hormone on corticotrophin, melanotrophin, and pigmentation in the trout. Gen. Comp. Endocrin. 63: 626–629.

Bao, H.N.D., S. Arason and K.A. Þórarinsdóttir. 2007. Effects of dry ice and superchilling on quality and shelf life of Arctic Charr (*Salvelinus alpinus*) fillets. Int. J. Food Engin. 3: 1556–3758.

Barton, B.A. and J.K. Iwama. 1991. Physiological changes in fish from stress in aquaculture with emphasis on the response and effects of corticosteroids. Ann. Rev. Fish Dis. 1: 3–26.

Benson, T. 2004. Advancing aquaculture: fish welfare at slaughter.Available from http://seafood.ucdavis.edu/pubs/fishwelfare.pdf (accessed 15 March 2013).

Bjerkeng, B., B. Hatlen and E. Wathne. 1999. Deposition of astaxanthin in fillets of Atlantic salmon (*Salmo salar*) fed diets with herring, capelin, sandeel, or Peruvian high PUFA oils. Aquaculture 180: 307–319.

Boggioa, S.M., R.W. Hardy, J.K. Babbitt and E.L. Brannona. 1985. The influence of dietary lipid source and alpha-tocopheryl acetate level on product quality of rainbow trout (*Salmo gairdneri*). Aquaculture 51: 13–24.

Bonga, S.E.W. 1997. The stress response in fish. Phys. Rev. 77: 591–625.

Bonilla, A.C., K. Sveinsdottir and E. Martinsdottir. 2007. Development of Quality Index Method (QIM) scheme for fresh cod (*Gadus morhua*) fillets and application in shelf life study. Food Control 18: 352–358.

Bosworth, B.G., B.C. Small, D. Gregory, J. Kim, S. Black and A. Jerrett. 2007. Effects of rested-harvest using the anesthetic AQUI-S™ on channel catfish, *Ictalurus punctatus*, physiology and fillet quality. Aquaculture 262: 302–318.

Boyle, J.L., R.C. Lindsay and D.A. Stuiber. 1992. Bromophenol distribution in salmon and selected seafoods of fresh- and saltwater origin. J. Food Sci. 57: 918–922.

Boyle, J.L., R.C. Lindsay and D.A. Stuiber. 1993a. Occurrence and properties of flavor-related bromophenols found in the marine environment: A review. J. Aquatic Food Prod. Tech. 2: 75–112.

Boyle, J.L., R.C. Lindsay and D.A. Stuiber. 1993b. Contributions of bromophenols to marine-associated flavors of fish and seafoods. J. Aquacult. Food Prod. Tech. 1: 43–63.
Braithwaite, V.A. and F.A. Huntingford. 2004. Fish and welfare: do fish have the capacity for pain perception and suffering. Anim. Welf. 13: 587–592.
Brännäs, E. and J.I. Johnsson. 2008. Behaviour and welfare in farmed fish. *In*: C. Magnhagen, V.A. Braithwaite, E. Forsgren and B.G. Kapoor (eds.). Fish behaviour. Science Publishers, New Hampshire, pp. 593–627.
Buchanan, J.G. and P.M. Thomas. 2008. Improving the color shelf life of farmed southern bluefin tuna (*Thunnus maccoyii*) flesh with dietary supplements of vitamins E and C and selenium. J. Aquacult. Food Prod. Tech. 17: 285–302.
Chen, Y.C., J. Nguyen, K. Semmens, S. Beamer and J. Jaczynski. 2008. Effects of dietary alpha-tocopheryl acetate on lipid oxidation and alpha-tocopherol content of omega-3-enhanced farmed rainbow trout (*Oncorhynchus mykiss*) fillets. LWT-Food Sci. Tech. 41: 244–253.
Christophersen, A.G., G. Bertelsen, H.J. Andersen, P. Knuthsen and L.H. Skibsted. 1992. Storage life of frozen salmonids. Effect of light and packaging conditions on carotenoid oxidation and lipid oxidation. Z. Lebensm. Unters. For. 194: 115–119.
Clark, K.E., A.P.C. Gobas and G. Mackay. 1990. Model of organic chemical uptake and clearance by fish from food and water. Environ. Sci. Tech. 24: 1203–1213.
Connell, J.J. 1995. Quality terminology. *In*: J.J. Connell (ed.). Control of Fish Quality. Fishing New Books, Oxford. United Kingdom.
Day, B.P.F. 2001. Modified atmosphere packaging of chilled fish and seafood products. *In*: S.C. Kestin and P.D. Warriss (eds.). Farmed Fish Quality. Fishing News Books, Blackwell Science, Oxford, pp. 276–282.
Doolan, B.J., G.L. Allan, M.A. Booth and P.L. Jones. 2008. Effect of carotenoids and background colour on the skin pigmentation of Australian snapper *Pagrus auratus* (Bloch & Schneider, 1801). Aquacult. Res. 39: 1423–1433.
Erikson, U., B. Lambooij, H. Digre, H.G.M. Reimert, M. Bondø and H. van der Vis. 2011. Conditions for instant electrical stunning of farmed Atlantic cod after de-watering, maintenance of unconsciousness, effects of stress, and fillet quality—A comparison with AQUI-S™. Aquaculture 324–325: 135–144.
Farmer, L.J., J.M. McConnell and D.J. Kilpatrick. 2000. Sensory characteristics of farmed and wild Atlantic salmon. Aquaculture 187: 105–125.
Fenical, W. 1982. Natural products chemistry in the marine environment. Science 215: 923–928.
Fletcher, G.C. and J.A. Hodgson. 1988. Shelf-life of sterile snapper (*Chrysophrys auratus*). J. Food Sci. 53: 1357–1332.
Frank B. Whitfield, Melissa Drew, Fay Helidoniotis and Denice Svoronos. 1999. Distribution of Bromophenols in Species of Marine Polychaetes and Bryozoans from Eastern Australia and the Role of Such Animals in the Flavor of Edible Ocean Fish and Prawns (Shrimp) J. Agric. Food Chem. 47(11): 4756–4762.
Frank, D., S. Poole, S. Kirchhoff and C. Forde. 2009. Investigation of sensory and volatile characteristics of farmed and wild barramundi (*Lates calcarifer*) using gas chromatography–olfactometry mass spectrometry and descriptive sensory analysis. J. Agri. Food Chem. 57: 10302–10312.
Frigg, M., A.L. Prabucki and E.U. Ruhdel. 1990. Effect of dietary vitamin E levels on oxidative stability of trout fillets. Aquaculture 84: 145–158.
From, J. and V. Hørlyck. 1984. Sites of uptake of geosmin, a cause of earthy-flavor, in rainbow trout (*Salmo gairdneri*). Can. J. Fish. Aqua. Sci. 41: 1224–1226.
Frost, S., S. Poole and S. Grauf. 1999. Improving the quality of Australian aquacultured barramundi (*Lates calcarifer*) through modified harvesting, handling and processing techniques. Report to the Queensland Department of Primary Industries, Australia.
Fuller, S.C., D.C. Frank, M.J. Fitzhenry, H.E. Smyth and S.E. Poole. 2008. Improved approach for analyzing bromophenols in seafood using stable isotope dilution analysis in combination with SPME. J. Agri. Food Chem. 56: 8248–8254.

Gatta, P.P., M. Pirini, S. Testi, G. Vignola and P.G. Monetti. 2000. The influence of different levels of dietary vitamin E on seabass *Dicentrarchus labrax* flesh quality. Aquacult. Nutri. 6: 47–52.

Glencross, B., R. Michael, K. Austen and R. Hauler. 2008. Productivity, carcass composition, waste output and sensory characteristics of large barramundi *Lates calcarifer* fed high-nutrient density diets. Aquaculture 284: 167–173.

Gram, L. 1992. Reviewe valuation of bacteriological quality of seafood. Int. J. Food Microbiol. 16: 25–39.

Gram, L. and H.H. Huss. 1996. Microbiological spoilage of fish and fish products. Int. J. Food Microbiol. 33: 121–137.

Gray, J.I., E.A. Gomaa and D.J. Buckley. 1996. Oxidative quality and shelf life of meats. Meat Sci. 43, Suppl. 1: 111–123.

Grigorakis, K. 2007. Compositional and organoleptic quality of farmed and wild gilthead sea bream (*Sparus aurata*) and seabass (*Dicentrarchus labrax*) and factors affecting it: A review. Aquaculture 272: 55–75.

Grigorakis, K., A. Taylor and M.N. Alexis. 2003. Organoleptic and volatile aroma compounds comparison of wild and cultured gilthead sea bream (*Sparus aurata*): sensory differences and possible chemical basis. Aquaculture 225: 109–119.

Grimm, C.C., S.W. Lloyd and P.V. Zimba. 2004. Instrumental versus sensory detection of off-flavors in farm-raised channel catfish. Aquaculture 236: 309–319.

Gröneveld, D., P.H.M. Balm and S.E.W. Bonga. 1995. Biphasic effect of MCH on alpha-MSH release from the tilapia (*Oreochromis mossambicus*) pituitary. Peptides 16: 945–949.

Harare, K. and Ø. Lie. 1997. Retained levels of dietary α-, γ- and δ-tocopherols in tissues and body fluids of Atlantic salmon (*Salmo salar* L). Aquacult. Nutri. 3: 99–107.

Harare, K., R.K. Berge, R. Waagbø and Ø. Lie. 1997. Vitamins C and E interact in juvenile Atlantic salmon (*Salmo salar*, L.). Free Rad. Biol. Med. 22: 137–149.

Harare, K., R.K. Berge and Ø. Lie. 1998. Oxidative stability of Atlantic salmon (*Salmo salar* L.) fillet enriched in α-, γ- and δ-tocopherols through dietary supplementation. Food Chem. 62: 173–178.

Haugen, J.E. and I. Undeland. 2003. Lipidoxidation in herring fillets (*Clupea harengus*) during ice storage measured by a commercial hybrid gas-sensor array system. J. Agric. Food Chem. 51: 752–759.

Howgate, P. 2001. Tainting of aquaculture products by natural and anthropogenic contaminants. *In*: S.C. Kestin and P.D. Warris (eds.). Farmed Fish Quality. Blackwell Science, Oxford, pp. 192–201.

Howgate, P. 2004. Tainting of farmed fish by geosmin and 2-methyl-iso-borneol: a review of sensory aspects and of uptake/depuration. Aquaculture 234: 155–181.

Howieson, J., B. Glencross, S. Little, A. Aris, N. Wade, G. Partridge, N. Paton, R. Tonkin, D. Allan, R. Wilkinson and R. Smullen. 2012. Identification of factors impacting on greying and other enduser quality attributes in farmed barramundi fillets. Ridley Aqua-Feed Prawn and Barramundi Conference, Sydney, Australia.

Huidobro, A., R. Mendes and M. Nunes. 2001. Slaughtering of gilthead seabream (*Sparus aurata*) in liquid ice: influence on fish quality. Euro. Food Res. Tech. 213: 267–272.

Hultin, H.O. 1992. Lipid oxidation in fish muscle. *In*: G.J. Flick and R.E. Martin (eds.). Advances in Seafood Biochemistry: Composition and Quality. Technomic Publishing Co., Inc. Lancaster, Basel, pp. 99–122.

Hultin, H.O. 1994. Oxidation of lipids in seafood. *In*: F. Shahidi and J.R. Botta (eds.). Seafoods: Chemistry, Processing Technology and Quality. Chapman & Hall, London, pp. 49–74.

Huntingford, F.A., C. Adams, V.A. Braithwaite, S. Kadri, T.G. Pottinger, P. Sandøe and J.F. Turnbull. 2006. Current issues in fish welfare. J. Fish Biol. 68: 332–372.

Iversen, M., B. Finstad, R.S. McKinley and R.A. Eliassen. 2003. The efficacy of metomidate, clove oil, Aqui-S™ and Benzoak® as anaesthetics in Atlantic salmon (*Salmo salar* L.) smolts, and their potential stress-reducing capacity. Aquaculture 221: 549–566.

Izaguirre, G., C.J. Hwang, S.W. Krasner and M.J. McGuire. 1982. Geosmin and 2-methylisoborneol from cyanobacteria in three water systems. Appl. Environ. Microbiol. 43: 708–714.
Jobling, M. 2004a. Are modifications in tissue fatty acid profiles following a change in diet the result of dilution? Test of a simple dilution model. Aquaculture 232: 551–562.
Jobling, M. 2004b. 'Finishing' feeds for carnivorous fish and the fatty acid dilution model. Aquacult. Res. 35: 706–709.
Johansson, L. 2001. Eating quality of farmed rainbow trout (*Oncorhynchus mykiss*). *In*: S.C. Kestin and P.D. Warriss (eds.). Farmed Fish Quality. Blackwell Science, Oxford, pp. 76–88.
Johnsen, P.B. and S.W. Lloyd. 1992. Influence of fat content on uptake and depuration of the off-flavor 2-Methylisoborneol by channel catfish (*Ictalurus punctatus*). Can. J. Fish. Aqua. Sci. 49: 2406–2411.
Jones, B., S. Fuller and A.G. Carton. 2013. Earthy-muddy tainting of cultured barramundi linked to geosmin in tropical northern Australia. Aquacult. Environ. Interact. 3: 117–124.
Josephson, D.B., R.C. Lindsay and D.A. Stuiber. 1984. Variations in the occurrences of enzymatically derived aroma compounds in salt- and freshwater fish. J. Agric. Food Chem. 32: 1344–1347.
Josephson, D.B., R.C. Lindsay and D.A. Stuiber. 1991. Volatile carotenoid-related oxidation compounds contributing to cooked salmon flavor. Lebensmitt Wissensch. Tech. 24: 424–432.
Jüttner, F. and S. Watson. 2007. Minireview: biochemical and ecological control of geosmin and 2-methylisoborneol in source waters. Appl. Environ. Microbiol. 73: 4395–4406.
Kawauchi, H. and B.I. Baker. 2004. Melanin-concentrating hormone signaling systems in fish. Peptides 25: 1577–1584.
Kestin, S.C., S.B. Wootton and N.G. Gregory. 1991. Effect of slaughter by removal from water on visual evoked activity in the brain and reflex movement of rainbow trout (*Oncorhynchus mykiss*). Vet. Rec. 128: 443–446.
Khayat, A. and D. Schwall. 1983. Lipid oxidation in seafood. Food Tech. 37: 130–140.
Kiessling, A., M. Espe, K. Ruohonen and T. Morkore. 2004. Texture, gaping and colour of fresh and frozen Atlantic salmon flesh as affected by pre-slaughter iso-eugenol or CO_2 anaesthesia. Aquaculture 236: 645–657.
Kishida, M., B.I. Baker and A.N. Eberle. 1989. The measurement of melanin-concentrating hormone in trout blood. Gen. Comp. Endo. 74: 221–229.
Kummer, C. 1992. Food: Farmed fish. Atlantic 270: 88.
Ladikos, D. and D. Lougovois. 1990. Lipid oxidation in muscle foods: A review. Food Chem. 35: 295–314.
Lambooij, E., J.W. Van Der Vis, R.J. Kloosterboer and C. Pieterse. 2002. Welfare aspects of live chilling and freezing of farmed eel (*Anguilla anguilla* L.): neurological and behavioural assessment. Aquaculture 210: 159–169.
Lambooij, B., K. Kloosterboer, M.A. Gerritzen, G. Andre, M. Veldman and H. Van de Vis. 2006a. Electrical stunning followed by decapitation or chilling of African catfish (*Clarias gariepinus*): assessment of behavioural and neural parameters and product quality. Aquacult. Res. 37: 61–70.
Lambooij, E., R.J. Kloosterboer, M.A. Gerritzen and J.W. Van de Vis. 2006b. Assessment of electrical stunning in fresh water of African catfish (*Clarias gariepinus*) and chilling in ice water for loss of consciousness and sensibility. Aquaculture 254: 388–395.
Li, Z., P. Hobson, W. Ana, M.D. Burch, J. House and M. Yanga. 2012. Earthy odor compounds production and loss in three cyanobacterial cultures. Water Res. 46: 5165–5173.
Lindsay, R.C. 1994. Flavour of fish. *In*: F. Shahidi and J.R. Botta (eds.). Seafoods: Chemistry, Processing Technology and Quality. Chapman & Hall, London, pp. 75–84.
Liu, Q., M.C. Lanari and D.M. Schaefer. 1995. A review of dietary vitamin E supplementation for improvement of beef quality. J. Anim. Sci. 73: 3131–3140.
Lovell, R.T. 1983. New off-flavors in pond-cultured channel catfish. Aquaculture 30: 329–334.

Lyman, W.J., W.F. Reehl and D.H. Rosenblatt. 1990. Handbook of Chemical Property Estimation Methods: Environmental Behavior of Organic Compounds. American Chemical Society, Washington D.C.

Ma, W.C.J., H.Y. Chung, P.O. Ang Jr. and J.S. Kim. 2005. Enhancement of bromophenol levels in aquacultured silver seabream (*Sparus sarba*). J. Agric. Food Chem. 53: 2133–2139.

MacDougall, D.B. 1988. Colour vision and appearance measurement. *In*: J.R. Piggot (ed.). Sensory Analysis of Foods. Elsevier Applied Science, London, pp. 103–130.

Martin, J.F., L.W. Bennett and W.H. Graham. 1988. Off-flavor in the channel catfish (*Ictalurus punctatus*) due to 2-methylisoborneol and its dehydration products. Water Sci. Tech. 20: 99–105.

Matos, E., A. Gonçalves, M.L. Nunes, M.T. Dinis and J. Dias. 2010. Effect of harvesting stress and slaughter conditions on selected flesh quality criteria of gilthead seabream (*Sparus aurata*). Aquaculture 305: 66–72.

Mendenhall, V.T. 1972. Oxidative rancidity in raw fish fillets harvested from the Gulf of Mexico. J. Food Sci. 37: 547–550.

Murata, H. and K. Yamauchi. 1989. Relationship between the 2-thiobarbituric acid values of some tissues from cultured red sea bream and its dietary α-tocopherol levels. Nippon Suisan Gakkaishi 55: 1435–1439.

Neely, W.B. 1979. Estimating rate constants for the uptake and clearance of chemicals by fish. Environ. Sci. Tech. 13: 1506–1510.

O'Keefe, T.M. and R.L. Noble. 1978. Storage stability of channel catfish (*Ictalurus punctatus*) in relation to dietary level of α-tocopherol. J. Fish. Res. Bd. Can. 35: 457–460.

Olafsdóttir, G., E. Martinsdóttir, J. Oehlenschläger, P. Dalgaard, B. Jensen, I. Undeland, I.M. Mackie, G. Henehan, J. Nielsen and H. Nilsen. 1997. Methods to evaluate fish freshness in research and industry. Trends Food Sci. Tech. 8: 258–265.

Olafsdóttir, G., H.L. Lauzon, E. Martinsdóttir, J. Oehlenschláuger and K. Kristbergsson. 2006. Evaluation of shelf life of superchilled cod (*Gadus morhua*) fillets and the influence of temperature fluctuations during storage on microbial and chemical quality indicators. J. Food Sci. 71: S97–S109.

Oliveira, A.S., V.M. Silva, M.C.C. Veloso, G.V. Santos and J.B. de Andrade. 2009. Bromophenol concentrations in fish from Salvador, BA, Brazil. An. Acad. Bras. Ciênc. 81: 165–172.

Onibi, G.E., J.R. Scaife, T.C. Fletcher and D.F. Houlihan. 1996. Influence of α-tocopherol acetate in high lipid diets on quality of refrigerated Atlantic salmon (*Salmo salar*) fillets. Conference of IIR Commission C2: Refrigeration and aquaculture. Bordeaux, France. pp. 145–152.

Ostrander, J. and C. Martinsen. 1976. Sensory testing of pen-reared salmon and trout. J. Food Sci. 41: 886–890.

Pastor, A., A. Huidobro, C. Alvarez and M. Tejada. 1998. 28th Western European Fish Technologists' Association (WEFTA) Meeting. Tromsø, Norway.

Pavlidis, M., M. Karkana, E. Fanouraki and N. Papandroulakis. 2008. Environmental control of skin colour in the red porgy, *Pagrus pagrus*. Aquacult. Res. 39: 837–849.

Percival, S., P. Drabsch and B. Glencross. 2008. Determining factors affecting muddy-flavour taint in farmed barramundi, *Lates calcarifer*. Aquaculture 284: 136–143.

Perkins, E.J. and D. Schlenk. 1997. Comparisons of uptake and depuration of 2-methylisoborneol in male, female, juvenile, and 3MC-induced channel catfish (*Ictalurus punctatus*). J. World Aquacult. Soc. 28: 158–164.

Persson, P. 1980. Sensory properties and analysis of two muddy odour compounds, geosmin and 2 methylisoborneol, in water and fish. Water Res. 14: 1113–1118.

Petersen, M.A., G. Hyldig, B.W. Strobel, N.H. Henriksen and N.O.G. Jørgensen. 2011. Chemical and sensory quantification of geosmin and 2-methylisoborneol in rainbow trout (*Oncorhynchus mykiss*) from recirculated aquacultures in relation to concentrations in basin water. J. Agri. Food Chem. 59: 12561–12568.

Petillo, D., H.O. Hultin, J. Krzynowek and W. Autio. 1998. Kinetics of antioxidant loss in mackerel light and dark muscle. J. Agri. Food Chem. 46: 4128–4137.

Phillips, M. 2010. Addressing cheap imports. Australian Barramundi Farmers Association. Mid-Year Conference Proceedings. Cairns, Australia.
Poels, C.L.M., R. Fischer, K. Fukawa, P. Howgate, B.G. Maddock, G. Persoone, R.R. Stephenson and W.J. Bontinck. 1988. Establishment of a test guideline for the evaluation of fish tainting. Chemosphere 17: 751–765.
Poli, B.M., G. Parisi, F. Scappini and G. Zampacavallo. 2005. Fish welfare and quality as affected by pre-slaughter and slaughter management. Aquacult. Int. 13: 29–49.
Robb, D.F.H and S.C. Kestin. 2002. Methods used to kill fish: field observations and literature reviewed. Anim. Welf. 11: 269–282.
Robertson, R.F., K. Jauncey, M.C.M. Beveridge and L.A. Lawton. 2005. Depuration rates and the sensory threshold concentration of geosmin responsible for earthy-musty taint in rainbow trout, *Onchorynchus mykiss*. Aquaculture 245: 89–99.
Robertson, R.F., A. Hammond, K. Jauncey, M.C.M. Beveridge and L.A. Lawton. 2006. An investigation into the occurrence of GSM responsible for earthy-musty taints in UK farmed rainbow trout, *Onchorhynchus mykiss*. Aquaculture 259: 153–163.
Robin, J., J.P. Cravedi, A. Hillenweck, C. Deshayes and D. Vallod. 2006. Off flavor characterization and origin in French trout farming. Aquaculture 260: 128–138.
Rose, J.D. 2002. The neurobehavioural nature of fishes and the question of awareness and pain. Rev. Fish. Sci. 10: 1–38.
Ruff, N., R.D. Fitzgerald, T.F. Cross, K. Harare and J.P. Kerry. 2003. The effect of dietary vitamin E and C level on market-size turbot (*Scophthalmus maximus*) fillet quality. Aquacult. Nutr. 9: 91–103.
Sato, K. and G.R. Hegarty. 1971. Warmed-over flavor in cooked meats. J. Food Sci. 36: 1098–1102.
Selli, S., C. Prost and T. Serot. 2009. Odour-active and off-odour components in rainbow trout (*Oncorhynchus mykiss*) extracts obtained by microwave assisted distillation-solvent extraction. Food Chem. 114: 317–322.
Sigholt, T., U. Erikson, T. Rustad, S. Johansen, T.S. Nordtvedt and A. Seland. 1997. Handling stress and storage temperature affect meat quality of farmed-raised Atlantic salmon (*Salmo salar*). J. Food Sci. 62: 898–905.
Sikorski, Z.E. and B. Sun Pan. 1994. Preservation of seafood quality. In: F. Shahidi and J.R. Botta (eds.). Seafoods: Chemistry, Processing Technology and Quality. Chapman & Hall, London, pp. 168–195.
Siu, G.M and H.H. Draper. 1978. A survey of the malonaldehyde content of retail meats and fish. J. Food Sci. 43: 1147–1149.
Skjervold, P.O., A.M.B. Rørå, S.O. Fjaera, A. Synstad, Å. Vorre and O. Einen. 2001. Effects of pre-, in- or post-rigor filleting of live chilled Atlantic salmon. Aquaculture 194: 315–326.
Slabyj, B.M. and H.O. Hultin. 1983. Lipid peroxidation by microsomal fractions isolated from light and dark muscles of herring (*Clupea harengus*). J. Food Sci. 47: 1395–1398.
Small, B.C. 2004. Anesthetic efficacy of metomidate and comparison of plasma cortisol responses to tricaine methanesulfonate, quinaldine and clove oil anesthetized channel catfish *Ictalurus punctatus*. Aquaculture 218: 177–185.
Sneddon, L.U. 2002. Anatomical and electrophysiological analysis of the trigeminal nerve in the rainbow trout, *Oncorhynchus mykiss*. Neurosci. Letters 319: 167–171.
Sneddon, L.U. 2003. The evidence for pain in fish: the use of morphine as an analgesic. Appl. Anim. Behav. Sci. 83: 153–162.
Sneddon, L.U. and M.J. Gentle. 2002. Receptor types on the head of the rainbow trout: are nociceptors present? Comp. Biochem. Physiol. A. 32(Suppl. 1): S42.
Sneddon, L.U., V.A. Braithwaite and M.J. Gentle. 2003. Novel object test: examining nociception and fear in the rainbow trout. J. Pain 4: 431–440.

Sohn, J.H., Y. Taki, H. Ushio, T. Kohata, W. Shioya and T. Ohshima. 2005. Lipid oxidations in ordinary and dark muscles of fish. Influence on rancid off-odor development and color darkening of yellowtail flesh during ice storage. J. Food Sci. 70: 490–496.
Southgate, S.P. and T. Wall. 2001. Welfare of farmed fish at slaughter. In Practice 23: 277–284.
Streit, B. 1998. Bioaccumulation of contaminants in fish. *In*: T. Braunbeck, D.E. Hinton and B. Streit. (eds.). Fish Ecotoxicology. Birkhäuser Verlag, Basel, pp. 353–387.
Suzuki, M., Y.K. Narnaware, B.I. Baker and A. Levy. 1995. Influence of environmental colour and diurnal phase on MCH gene expression in the trout. J. Neuroendo. 7: 319–328.
Tejada, M. and A. Huidobro. 2002. Quality of farmed gilthead seabream (*Sparus aurata*) during ice storage related to the slaughter method and gutting. Euro. Food Res. Tech. 215: 1–7.
Thomassen, M. and C. Røsjø. 1989. Different fats in feed for salmon: Influence on sensory parameters, growth rate and fatty acids in muscle and heart. Aquaculture 79: 129–135.
Tucker, C.S. 2000. Off-flavor problems in aquaculture. Rev. Fish. Sci. 8: 45–48.
Turchini, G.M., D.S. Francis and S.S. De Silva. 2006. Modification of tissue fatty acid composition in Murray cod (*Maccullochella peelii peelii*, Mitchell) resulting from a shift from vegetable oil diets to a fish oil diet. Aquacult. Res. 37: 570–585.
Undeland, I. 2001. Lipid oxidation in fatty fish during processing and storage. *In*: S.C. Kestin and P.D. Warriss (eds.). Farmed Fish Quality. Fishing News Books, Oxford, pp. 261–275.
Undeland, I., M. Stading and H. Lingnert. 1998. Influence of skinning on lipid oxidation in different horizontal layers of herring (*Clupea harengus*) during frozen storage. J. Sci. Food Agri. 78: 441–450.
Undeland, I., G. Hall and H. Lignert. 1999. Lipid oxidation in fillets of herring (*Clupea harengus*) during ice storage. J. Agric. Food Chem. 47: 524–532.
van der Salm, A.L., M. Martínez, G. Flik and S.E.W. Bonga. 2004. Effects of husbandry conditions on the skin colour and stress response of red porgy, *Pagrus pagrus*. Aquaculture 241: 371–386.
van Eys, G.J. and P.T. Peters. 1981. Evidence for a direct role of α-MSH in morphological background adaptation of the skin in *Sarotherodon mossambicus*. Cell Tissue Res. 217: 272–361.
Wall, A.J. 2001. Ethical considerations in the handling and slaughter of farmed fish. *In*: S.C. Kestin and P.D. Warriss (eds.). Farmed Fish Quality. Fishing News Books, Oxford, pp. 108–115.
Whitfield, F.B. 1988. Chemistry of off-flavours in marine organisms. *In*: P.E. Persson, F.B. Whitfield and T. Motohiro (eds.). Water Science and Technology. IAWPRC International Symposium on Off-flavours in the Aquatic Environment. Kagoshima, Japan.
Whitfield, F.B. 1990. Flavour of prawns and lobsters. Food Rev. Int. 6: 505–519.
Whitfield, F.B., K.J. Shaw and D.I. Walker. 1992. The source of 2,6-dibromophenol: Cause of an iodoform taint in Australian prawns. *In*: P.E. Persson, F.B. Whitfield and S.W. Krasner (eds.). Water Science and Technology. IAWPRC International Symposium, Off-flavours in Drinking Water and Aquatic Organisms. Los Angeles, California.
Whitfield, F.B., F. Helidoniotis, D. Svoronos, K.J. Shaw and G.L. Ford. 1995. The source of bromophenols in some species of Australian ocean fish. Water Sci. Tech. 31: 113–120.
Whitfield, F.B., F. Helidoniotis, K.J. Shaw and D. Svoronos. 1998. Distribution of bromophenols in species of ocean fish from eastern Australia. J. Agri. Food Chem. 46: 3750–3757.
Whitfield, F.B., M. Drew, F. Helidoniotis and D. Svoronos. 1999. Distribution of bromophenols in species of marine polychaetes and bryozoans from eastern Australia and the role of such animals in the flavor of edible ocean fish and prawns (shrimp). J. Agri. Food Chem. 47: 4756–4762.
Whitfield, F.B., F. Helidoniotis and D. Smith. 2002. Role of feed ingredients in the bromophenol content of cultured prawns. Food Chem. 79: 355–365.

Wilkinson, R.J., N. Patona and M.J.R. Porter. 2008. The effects of pre-harvest stress and harvest method on the stress response, rigor onset, muscle pH and drip loss in barramundi (*Lates calcarifer*). Aquaculture 282: 26–32.

Woodin, S.A., M.D. Walla and D.E. Lincoln. 1987. Occurrence of brominated compounds in soft-bottom benthic organisms. J. Exp. Mar. Biol. Ecol. 209–217.

Wu, J.T. and F. Jüttner. 1988a. Differential partitioning of geosmin and 2-methylisoborneol between cellular constituents in *Oscillatoria tenuis*. Arch. Microbiol. 150: 580–583.

Wu, J.T. and F. Jüttner. 1988b. Effects of environmental factors on geosmin production by *Fischerella muscicola*. Water Sci. Tech. 20: 143–148.

Yamanome, T., M. Amano and A. Takahashi. 2005. White background reduces the occurrence of staining, activates melanin-concentrating hormone and promotes somatic growth in barfin flounder. Aquaculture 244: 323–329.

Yamprayoon, J. and A. Noonhorm. 2000. GSM and off-flavour in Nile tilapia (*Oreochromis niloticus*). J. Aquacult. Food Prod. Tech. 9: 29–41.

10

Farming of Barramundi/Asian Seabass: An Australian Industry Perspective

Paul Harrison,[1,]* *Chris Calogeras*[2] *and Marty Phillips*[3]

10.1 Introduction to the Australian Barramundi Industry

In Australia, Asian seabass (*Lates calcarifer*) is commonly referred to as barramundi and is widely popular for both food and recreational activities, as well as having significant totemic value to indigenous Australians. Widely distributed throughout the Southeast Asian region, barramundi are endemic to tropical northern regions of Australia between the Mary River, eastern Queensland to the Ashburton River, Western Australia, where at least 21 definable and unique genetic strains exist (Jerry and Smith-Keune 2013—Chapter 7). Throughout its Australian range a commercial net-based catch fishery exists for barramundi, which in 2011 landed 1,996 tonnes live weight. Barramundi is also successfully farmed in Australia, with commercial farms operating in all Australian mainland states and the Northern Territory. Farming methods used in Australia include land based ponds and raceways, open ocean sea cages, and recirculation aquaculture systems. In northern regions where barramundi are endemic, the majority of commercial farmed

---

[1]Mainstream Aquaculture, PO Box 2286, Werribee VIC 3030 Australia.
Email: www.mainstreamaquaculture.com
[2]Director C-AID Consultants, 38 Lake Ridge Crt, Lake Macdonald, Qld, 4563 Australia.
[3]Pejo Enterprises, PO Box 2103, Innisfail Qld 4860 Australia.
*Corresponding author

production is land-based in open fresh- and brackish-water ponds, as well as in saltwater flow-through concrete raceway systems. Barramundi is also produced in marine based sea cages (Fig. 10.1). In southern climates outside the barramundi's natural thermal range, production occurs in fully and semi-enclosed recirculating aquaculture systems, some of which utilise naturally occurring geothermal spring water to meet heating requirements (Fig. 10.1).

The Australian barramundi aquaculture industry had an annual production volume of 4,352 tonnes valued at AUS$ 35.7 million in 2011 according to the latest industry statistics (Skirtun et al. 2012). Although not yet available, these figures are likely to rise dramatically for 2012 and 2013 based on the most recent industry assessment at the half yearly Australian Barramundi Farmers Association (ABFA) meeting. This nascent industry, which emerged in the late 1980's, has seen rapid growth, with production increasing over 530% between 2000 and 2011 (898 to 4,352 tonne of farmed

Figure 10.1. Diverse farming practices are used to produce Australian barramundi. Photos are of three common approaches—that of pond-based (Photo: Pejo Enterprises), indoor recirculated (Photo: Mainstream Aquaculture) and marine sea cage(Photo: Marine Produce Australia) aquaculture systems. Barramundi are also produced in concrete raceway farming systems which also contribute a large amount to Australia's overall barramundi production (photo not shown).

product per annum) (Fig. 10.2). This represents a cumulative annualised growth of 15.5% for the Australian industry over that period. Importantly, this growth has occurred without a corresponding reduction in per unit (kg) value, reflecting the ability of the Australian market to absorb increasing amounts of farmed product without downwardly influencing price (Fig. 10.2). This relatively small, but high value, Australian industry accounts for only approximately 6.7% of global farmed seabass volume, but 14% of global value, reflecting the strong consumer preference and willingness to pay a premium for Australian product.

The Australian industry is characterised by being small, but technologically advanced compared to its Southeast Asian counterparts. Collaboration between farms is facilitated by a non-profit organisation called the Australian Barramundi Farmers Association (ABFA) which oversees identification of Research Development and Extension (RD&E) priorities for the industry and strongly advocates for increased efficiencies grounded in strong science and has a mandate to promote and develop the industry. Membership in the ABFA is voluntary, but still represents over 70% of operational barramundi farms and over 85% of the total value of the industry in Australia. Specific activities that the ABFA are involved in are detailed in a separate section below. The recent growth of the Australian farmed barramundi industry has been largely fuelled by innovation and technological advances arising out of targeted collaborative research between commercial farms, universities, private and government research organisations, such as the Fisheries Research and Development Corporation (FRDC) and the Seafood Co-operative Research Centre (SCRC) and the ABFA.

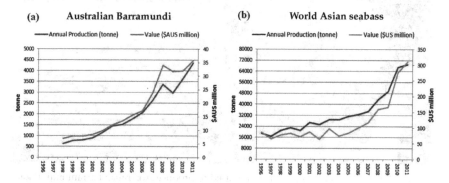

Figure 10.2. Annual volume and value of (a) barramundi produced and sold in Australia compared to that in (b) the world market. Australian value figures are in $AUS and world figures in $US. Australian Barramundi Data—Australian Bureau of Agricultural and Resource Economics and Sciences, Australian Fisheries Statistics 2011. World Asian Seabass data—©FAO —Fisheries and Aquaculture Information and Statistics Service—accessed 11/03/2013.

Hatchery technology and hatchery capacity is an example where the Australian industry is particularly advanced. Industry production statistics do not account for the commercial value of seed stock supply, but advancements in this area certainly underpin the growth that this industry has enjoyed recently and is likely to further fuel growth in the future. In Australia, numerous farms and three separate government organisations operate hatcheries and maintain improved brood stock populations. The majority of stock produced in private hatcheries linked to farms is used in-house for production, whereas the majority of stock produced at government hatcheries is used to supply regional farms and in some instances for augmentation of natural and impoundment populations. A current focus of the ABFA is to coordinate collaboration among farmers in this area to enhance the genetics of Australian barramundi hatchery stocks to improve industry performance.

The industry is diverse, with approximately 370 registered farms which include the large number (around 300) of very small decorative, or hobby, ponds in Queensland that require the property owner to be licensed to stock barramundi and other operators who have multiple species attached to their aquaculture permits/licences. However, production volume is dominated by around 12 companies, each of which produces in excess of 100 tonne per annum. The industry has seen steady year on year growth except for a production drop in 2008–2010. This was a result of severe cyclone damage that destroyed assets and/or impeded production in north Queensland, where the bulk of the country's pond and sea-cage based barramundi production occurred. Productive capacity has now been restored and it is likely that there will be a significant rebound in productive output over the next 3–5 years. In fact the barramundi market price has seen positive growth over the last 18 months and the ABFA estimates the value of the industry in 2013 to be approximately AU$60 million (www.abfa.org.au).

A similar trend is seen in global production of Asian seabass with a substantial increase since 2000. Between 2000 and 2011, global production of Asian seabass increased 330% from approximately 20,986 to 69,116 tonnes (FAO2013a). Over this period, the Australian barramundi aquaculture industry increased 530% from 814 to 4,352 tonnes. Despite this rapid growth neither the Australian or global markets experienced a decline in per unit value.

10.2 Markets for Barramundi in Australia

The Australian barramundi aquaculture industry is fuelled predominantly by markets in the domestic food service and gourmet retail sector. Coupled with product from the wild fishery, the industry contributes approximately 33% of the domestic product, with the majority of other Asian seabass product consumed in Australia imported, primarily from Thailand and Taiwan (Fig. 10.3).

The Australian barramundi seafood market is therefore dominated by imported product. This trend holds for the majority of seafood products in Australia, a country that is a net importer of fish despite having abundant natural resources that are conducive to aquaculture and fishing. In Australia, the wholesale fish market for barramundi is currently estimated to be AU$120 million in annual sales from total food fish sales of AU$3 billion. This equates to an estimated 16,000 tonnes of whole fish per annum and implied wholesale price of AU$7.50 per kilogram regardless of origin. The data indicates that barramundi account for about 3% of Australia's food fish consumption by volume and 4% by value. The market for barramundi product, and seafood more generally, can be broadly defined in two channels, retail and food service, with an approximate 65% to 35% market share, respectively (Fig. 10.4).

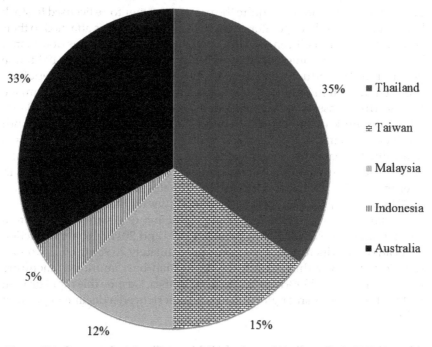

Figure 10.3. Country of origin of *Lates calcarifer* product sold in Australia in 2011.

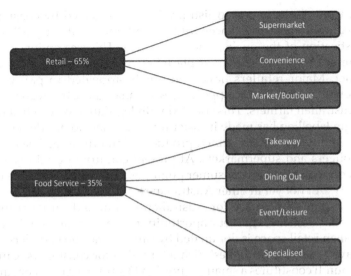

Figure 10.4. End use of barramundi product in the Australian marketplace.

10.3 The Australian Barramundi Marketplace: Trends, Opportunities and Threats

10.3.1 Trends

To date the Australian industry has been able to increase productive output significantly without any longer term down-ward impact on per unit price. This is largely because barramundi are sold in Australia as a premium product in the food service sector. Cheaper imported product is more typically directed toward the retail sector.

Market trends and research suggest that:

- a premium is obtained for fresh local product
- a premium is obtained for environmentally sustainable product
- health and product availability are important purchase considerations

10.3.2 Opportunities

In Australian markets major psycho-social factors determining the purchase of seafood include product origin and sustainability. The perception of consumers within Australia is that, compared to imports, Australian-grown products are of superior quality and grown in more sustainable ways. Conversely, imported fish product is perceived by consumers as originating from vast distances with limited traceability, regardless of the sophistication of the quality control process. This perception influences consumer choice in the market place between Australian and imported product. Major retailers have vigorously appealed to patriotism by promoting the notion of supporting Australian industry, Australian jobs and Australian farmers. This coupled with legislation on product country of origin labelling has made it easier for consumers to decide on whether they will buy Australian farmed product or imports at retail sites such as fishmongers and supermarkets. At present country of origin labelling is also required at the end consumer point (i.e., restaurants) in the Northern Territory, but not yet in other Australian states.

Consumer preference for Australian barramundi product provides substantial domestic market opportunity to the Australian industry. The Australian retail space is dominated by salmon and tuna retailers who do not have a reliable white flesh fish alternative for consumers. Currently, barramundi constitutes a small proportion (3%) of per capita consumption of food fish products, but is the third most popular fish consumed in Australia in a highly fragmented market. Indicators suggest that domestic demand is artificially low because consumers do not have access to a consistent and reliable high quality, fresh, Australian-grown product and therefore are not habitually purchasing barramundi as it is not familiar to them like salmon and tuna. If additional high-quality Australian product was present at point of sale in the marketplace, research suggests that the domestic market could further absorb higher volumes without negatively influencing price. Therefore opportunities exist in further up-scaling of the Australian barramundi industry. Expansion of the Australian industry will occur through increased scale which will reduce production costs and be further driven by macro-economic conditions as consumers become more health conscious, more affluent, and consume more fish.

A positive trend is seen in per capita fish consumption in Australia, which has increased steadily for the past 50 years along with increased population size. Of particular significance for the Australian market is the trend in China, where per capita fish consumption has spiked dramatically upward over the past twenty years (Fig. 10.5). This trend in China is likely to continue with increasing affluence of China's middle class. These strengthening global demands will likely present opportunity for the Australian industry in two ways. First, by creating direct export opportunities for Australian companies; second, by decreasing the volume of imported product as food security issues are likely to see more product retained in Southeast Asia in the future.

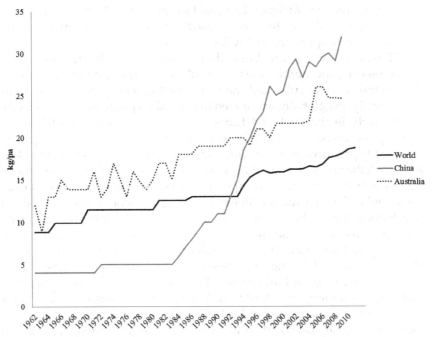

Figure 10.5. Per capita fish consumption globally, in China and Australia, between 1962–2010. (Source: FAO State of the Worlds Fisheries and Aquaculture 2012).

Capitalising on this opportunity will be largely centred on the Australian industry's ability to ensure continuity of supply, and to continue the provision of a premium product. These opportunities will be further facilitated by innovative improvement to product and production technologies. This will reduce the production cost base of Australian product to align more closely with that in Southeast Asia. It is likely that this will be further driven by consolidation and industrialisation of the Australian industry.

10.3.3 Threats

The major threat to the Australian industry is price sensitivity through increased volume of imported product. Market trends outlined above indicate that there is some protection afforded to the industry in its current state. But technological advances and industrialisation of the broader seabass industry is resulting in perceivably better quality imports and this trend is likely to continue. The ability for the Australian industry to sustain its quality-based position will depend largely on internal factors, such as its ability to assure a high quality domestic product and to ensure there is ongoing production efficiency. External factors such as the extent of import pressure and the degree to which Australian consumer sentiment can be shifted toward import product will be significant.

"Barramundi" is an Australian indigenous aboriginal name. Consequently, the name "barramundi" is strongly embedded within the Australian vernacular and the name itself is widely recognised, both domestically and globally, and linked to Australian product. Internationally (particularly in the USA and Europe), the use of "barramundi" in the marketplace has been shown to be perceived by consumers as a high-quality, sustainably produced product. This is also the case in the Australian market. As a result the use of the name barramundi for non-Australian Asian seabass is a threat, both from a direct competitive market perspective and from risks associated with image diminishment. Labelling legislation exists in Australia which doesn't help this situation. In an effort to better inform consumers, imported product now has to be clearly labeled with its country of origin until the point of first sale (i.e., wholesaler, fishmongers and supermarkets). However, separate labeling legislation ensures that all Asian seabass sold in Australia are labeled as barramundi. This can be misleading to consumers that link barramundi with Australia. A recent case demonstrates this point and highlights the sentiment of Australian consumers. A large domestic fast food chain recently launched an advertising campaign around 'barramundi burgers'. This product was immediately popular, but quickly removed after the public became aware that the barramundi were Asian seabass imported from Thailand. From a public awareness perspective, Australian consumers link barramundi with Australia and consider marketing of imported fish under the name barramundi to be misleading.

There is growing interest within Australia to protect the name barramundi as a geographic indication. To enable product identification and prevent fraudulent labeling, there is a genetic test available to determine if "barramundi" product originates from Australian stocks (Jerry and Smith-Keune 2010). It follows that Australian barramundi is unique in being genetically distinct from non-Australian Asian seabass. Currently the European Union is leading the way with agricultural policy identifying geographic indications as intellectual property. Geographical indications such as Bordeaux or Tequila are usually linked with specific locations (O'Connor and Company 2005). However policy also extends to cover traditional names where the product is unique. An example is Feta, which is reserved for a unique variety of "white cheese" produced only in some areas of Greece.

Might this threat turn to opportunity? With global expansion of Asian seabass markets including barramundi, both originating in Australia and elsewhere, this threat may indeed turn into opportunity for the Australian industry. Consumer rights and truth in labeling has received increased attention in Europe recently, after it was found that horse meat was being used as a substitute for beef. Cases like this will continue to heighten regulation by governments and consumer groups and may prove to be an interesting area of development in the near future for the barramundi industry. 'Barramundi' markets established on non-Australian Asian seabass may provide additional opportunity for the Australian industry should labeling legislation firm worldwide.

On a broader horizon, food security is a macro-economic issue that is likely to shape market trends and provide opportunity for the Australian industry. Australia trails behind its Asian counterparts and other regions in the world with focus and concern on food security. With Southeast Asia becoming increasingly more affluent, particularly reflected in the increasing demands of China's middle class, the demand will increase on retaining food products locally. This would create a two part opportunity for the Australian industry. Increasing demand in Asia has the potential to reduce the availability of imports to Australia, thus decreasing competition in the domestic marketplace. Furthermore, this increasing demand would directly increase export opportunity.

10.4 ABFA—Coordinating Innovation

The ABFA is a non-profit voluntary membership organisation for the Australian barramundi aquaculture industry (www.abfa.org.au). The ABFA fosters links between the industry, farmers, suppliers, wholesalers and retailers and as highlighted previously plays a central role in coordinating key priority areas of this industry and in promoting and facilitating innovation necessary to drive growth.

The structure of ABFA with respect to how it can best serve the Australian barramundi industry is currently under review. Current members pay a prescribed membership fee that is directed toward industry relevant RD&E funding. As this industry continues to grow and mature it is likely that structural change will see both a RD&E and a marketing levy tied to production to provide pooled funds to focus on research and marketing of Australian barramundi. Throughout Asia, Asian seabass marketing is not well developed and the fish is typically sold as a low value product. The Australian industry to date has been able to protect the iconic barramundi image and enjoy increased growth along with stable price per unit. This will be an increasing challenge as the industry expands. Quality and reliability are hallmarks of the Australian barramundi aquaculture industry. Examples of current critical focus areas for the ABFA to drive these are:

1) Market development and positioning of Australian barramundi. ABFA coordinates a collaborative project with a view to achieving a marketing and quality assurance program. This includes a proposal to work with the supply chain to develop a simple to use and effective quality scheme. The ABFA has recently supported further development of a branding option for Australian barramundi as part of a more structured marketing program.
2) Quality standards for farmed barramundi. This is a multi-faceted program aimed at ensuring best practice throughout the entire production and market cycles of barramundi product. It is a highly collaborative project involving farms, universities and government research institutes. It is delivering on methods for optimising flavour, colour and texture in farmed barramundi, coordinating nation-wide residue testing to meet EU standards and development of best practice farming guidelines and compliance monitoring systems.
3) Selective breeding and genetic improvement of barramundi seedstock. The Australian industry has a very high representation of quality barramundi hatcheries. The bulk of these are privately owned and linked to farms, however, ABFA has recently started a concept company called "Barratek" that aims to consolidate intellectual property in this area for the benefit of the Australian industry. As part of this program, the ABFA coordinated an unprecedented genetic audit on currently held Australian barramundi hatchery brood-stock in order to as certain the genetic diversity/integrity of Australian commercial brood-stock. This has demonstrated a high level of quality genetic stock with the sophistication in this level of the industry demonstrated by the fact that all hatcheries were found to be maintaining brood-stock with appropriate levels of diversity to avoid inbreeding. The industry is now working together to develop a large-scale commercial breeding program and the necessary funding required for Australian farmed barramundi. Assessments of the benefits that can accrue to industry through a well maintained and operated breeding program have demonstrated that even a basic program can have highly favourable benefit-cost ratios in the range 11:1 to 16:1 after 10 years with internal rate of return between 25% and 65%.
4) Enhanced Return On Feed. The ABFA works closely with major feed companies in Australia, who in turn provide sponsor support. Barramundi diet formulation is under constant review with feed companies and government organisations striving to maximise production from innovative formulated feeds and investigate suitable protein substitutions in barramundi diets to maximise the environmentally sustainable credentials of farmed barramundi (see also Glencross et al. 2013—Chapter 8).

10.5 The Australian Industry—the Next 15 Years?

Asian seabass is now well-established as an important aquaculture species, but how far can the industry grow? Worldwide, seabass aquaculture is experiencing upward growth, a trend sustained by the industry for the past 20 years. Opportunities to farm this species are increasing throughout the world given the seabass's adaptability in a wide range of geographic locations, and growth potential well outside its natural range. The Australian barramundi industry has outperformed the global trend with respect to growth and its ability to maintain per unit price in the face of increased supply. It is a given that the industry will continue to grow, but is this industry capable of realising the expansion necessary to establish Asian seabass as a major global finfish product? With the answer almost certainly yes, we consider below the likely position that the Australian barramundi industry will fill.

Barramundi display numerous attributes necessary for a successful aquaculture species and has some unique traits that make it particularly compelling for commercial production. Commercial feeds have been developed and the life cycle has been closed enabling controlled timing of seed-stock inputs. Challenges remain in these two important areas, but development is sufficient to fuel a rapidly expanding industry. From a biological perspective, production capabilities are greatly facilitated by high fecundity and rapid and efficient growth. Growth rates are not impeded by sexual differentiation due to the species protandrous hermaphroditic life-strategy, whereby the majority of sexually maturemales remain as males through the time-span of commercial culture. Furthermore, growth efficiency is achieved by virtue of the fact that individual barramundi mature in the 3–7 kg size range and can grow beyond 60kg, but commercial size (500g–3kg) is reached during juvenile stages where high appetite and food conversion efficiencies levels are sustained. A coordinated selective breeding program will further enhance these traits.

A further important attribute is that barramundi is a true euryhaline species. As such it is highly suitable to a wide range of salinities, with commercial culture successfully demonstrated in pure freshwater, brackish and marine salinities. With food security challenges linked to limited land and water resources, barramundi offers an economically feasible option across a wide range of environments.

To consider the possible future of the Australian barramundi industry, and more broadly the global seabass industry, it is useful to draw on examples of other like industries. The salmon aquaculture industry is the dominant and most successful global aquaculture industry. Salmon farming practices have been established for hundreds of years, but the industry began to commercialize in the 1960's. With abundant sites for sea pens and less regulation than seen today, the industry grew to approximately 200,000 tonnes of annual production in its first 20 years. It then expanded nearly 10-fold in the next 25 years to a record 1.83 million tonnes in 2008 (ISA 2013). This rapid expansion occurred despite numerous challenges with disease outbreaks and in the face of increasing regulation. It was facilitated by a close link between industry, universities and scientific institutes that led to innovative development of improved domesticated seedstock, improved and industrialized production technology, particularly on net pen construction, optimisation of feed monitoring and distribution, and bulk-handling logistics for fish. Typical of a maturing industry, the salmon industry experienced a period of rapid proliferation followed by consolidation. Today it is a highly professional, highly regulated industry that in 2007 produced 1,433,708 tonnes of Atlantic salmon with a value of US$7.578 billion (FAO 2013b). In the same year Norwegian-owned PanFish merged with Marine Harvest to create the largest salmon aquaculture company in the world. Consolidation has been a feature of growth in the modern, mature salmon industry.

The Asian seabass industry clearly has high growth potential. Whether it can emulate the growth of the salmon industry remains to be determined, but opportunities for growth are evident given the species' adaptability to a wide range of environments and provided that growth can occur in an efficient and sustainable way. The world seafood market including fresh, canned and frozen products is continuing to expand, with growth driven by increasing population size and increased affluence, particularly in developing countries. High demand is anticipated in Latin America and the Asia-Pacific region, with the overall market for aquaculture and fisheries products predicted to exceed 135 million tons worth US$370 billion by 2015 (Global Industry Analysts 2012). In this global marketplace the seabass industry currently contributes less than 0.04% of the total product, but is well positioned for growth.

What role will the Australian industry play as the global market expands? The Australian industry is advanced with respect to best practice quality assurance, environmental sustainability and technological development. This has been fuelled by necessity as the affluent Australian society is highly focused on environmental issues and imparts high operational costs, primarily through strict regulatory compliance and high wages, on its primary industries. For this reason, and despite Australia's abundance of natural resources, it is unlikely that Australia will emerge as a major volume producer of barramundi or Asian seabass for worldwide markets. It is more likely that the industry will continue to expand to meet domestic needs for premium quality finfish products and continue to attract a premium for products with high levels of quality assurance. The extent to which the Australian industry can compete with imports from Asian neighbours will depend largely on Asian-based companies providing product with high levels of quality assurance. Regardless of this, a domestic niche for Australian sourced product will remain given social market trends. The extent to which the Australian industry can capitalise on increased export demand over the next 15 years depends largely on the level of demand for premium quality product from increasingly affluent Southeast Asian markets and over a broader horizon on regional pressure on food security which would both limit imports and increase export opportunities.

The Australian industry is poised to play a major role in the rapidly expanding worldwide industry through the provision of core technology. Genetic based stock improvement, feeds, and production technology are three core areas. Australian hatcheries are leading the development of highly domesticated and optimised stock. Feed companies are actively working with the industry, along with government research institutes, universities and private research providers to optimise diets and match diets to improved stock. It is likely that Australian technology providers will play an integral role driving innovation in this industry globally, as well as on domestic production.

Over the next 15 years, the Australian aquaculture industry is likely to experience substantial further growth. The overall market forecast for the global aquaculture and fisheries market is favourable and core technology is in place. Emerging technology that drives higher profits and enables companies to become more sustainable, coupled with industrialisation and consolidation of farming and technology players, will likely be a feature of the industry in this time and are essential elements to realise this growth opportunity.

References

FAO State of the Worlds Fisheries and Aquaculture. 2012. Food and Agriculture Organization of the United Nations. Rome.

FAO. 2013a. Species Fact Sheets: *Lates calcarifer* (Bloch 1790). Fisheries and Aquaculture Information and Statistics Service http://www.fao.org/fishery/species/3068/en – accessed 11/03/2013.

FAO. 2013b. Species Fact Sheets: *Salmosalar* (Linnaeus 1758). Fisheries and Aquaculture Information and Statistics Service http://www.fao.org/fishery/species/2929/en accessed 06/03/2013.

Glencross, B., N. Wade and K. Morton. 2013. *Lates calcarifer* nutrition and feeding practices. *In:* D.R. Jerry (ed.). Ecology and Culture of Asian Seabass *Lates calcarifer*. CRC Press.

[ISA] International Salmon Farming Association website. 2013. http://www.salmonfarming.org accessed 06/03/2013.

Global Industry Analysts. 2012. Aquaculture and Fisheries: A global business strategic report. http://www.strategyr.com/Aquaculture_And_Fisheries_Market_Report.asp.

Jerry, D.R. and C. Smith-Keune. 2010. Molecular discrimination of imported product from Australian barramundi *Lates calcarifer*. Project 2008/758 Final Report to the SEAFOOD CRC, Flinders University, Adelaide.

Jerry, D.R. and C. Smith-Keune. 2013. The Genetics of Asian Seabass *Lates calcarifer*. *In:* D.R. Jerry (ed.). Ecology and Culture of Asian Seabass *Lates calcarifer*. CRC Press.

O'Connor and Company. 2005. Geographical Indications and the challenges for ACP countries —A discussion paper. Website http://agritrade.cta.int/accessed 06/03/2013.

Skirtun, M., P. Sahlqvist, R. Curtotti and P. Hobsbawn. ABARES 2012. Australian Fisheries Statistics 2011, Canberra, December.

11

Nursery and Grow-out Culture of Asian Seabass, *Lates calcarifer*, in Selected Countries in Southeast Asia

Felix G. Ayson,[1,]* Ketut Sugama,[2] Renu Yashiro[3] and Evelyn Grace de Jesus-Ayson[1]

11.1 Introduction

The popularity of Asian seabass as a food commodity and as an attractive species for aquaculture production differs dependent on country. In Australia, where it is locally known as barramundi and where acceptability of the species as a food item and its affordability to the people are high, this species is highly popular and therefore a very active aquaculture industry is developing. In many Southeast Asian (SEA) countries, the popularity of seabass is largely influenced by its affordability. Although it is generally

---

[1]Southeast Asian Fisheries Development Center, Aquaculture Department (SEAFDEC/AQD), Tigbauan, Iloilo, Philippines.
Email: fgayson@seafdec.org.ph and edjayson@seafdec.org.ph
[2]Center for Research and Development of Aquaculture, Agency for Marine and Fisheries Research of Indonesia, Jl Ragunan No 20 Jatipadang, Pasar Minggu, Jakarta, Indonesia.
Email: sugama@indosat.net.id
[3]86 Mu 3, Tumbol Parkred, Parkred District, Nothaburi Province 11120, Thailand.
Email: renuyashiro@gmail.com
*Corresponding author: fgayson@seafdec.org.ph

accepted that seabass is a nice tasting fish, its high market price relative to the purchasing power or incomes of people in SEA countries is the major factor that determines the level of seabass aquaculture in these countries. Consumption of seabass is popular in Singapore and Thailand, where the purchasing power of the people is relatively higher than in other SEA countries. Interestingly, despite being an important endemic fish in the Philippines, seabass is not known in many places in the country, even if it is present in the market, and as a result market demand for this fish product remains relatively low.

In SEA, seabass culture exists in Brunei Darussalam, Indonesia, the Philippines, Malaysia, Singapore, Thailand, Taiwan and Vietnam. In these countries, seabass aquaculture production is largely intended for the export market. Because of the differences in the demand of this species in these countries, various levels of production also exist.

The Asian seabass is a hardy species that is relatively easy to culture in cages or in brackish and freshwater ponds and is a species that can tolerate a wide range of salinity levels from freshwater (FW) to full-strength seawater (SW). It is one of the priority species among the 10 selected species for aquaculture development in Indonesia. Research into the culture of seabass in the country began in 1980 and initially focused on developing techniques for seed production and grow-out culture in floating net cages. However, although hatchery and grow-out technologies have been developed, commercial adoption by the private sector has been slow for the culture of this species in Indonesia. There are currently only three registered seabass farms and two governmental hatcheries in Indonesia carrying out research and development (R&D) for fry production and grow-out culture. Seabass culture in floating net cages is presently conducted only in the coastal waters of Riau Island Province, Lampung Bay Province and Seribu Island of Jakarta Bay, the production of which is mostly destined for the export market which is usually in the form of fillet from 2–3 kg size fish. Conversely, seabass for domestic consumption mostly originates from wild capture with only small amounts of fish product coming from the net cage farms, the later which are sold mainly in seafood restaurants at sizes between 400–500 g. In 2010, the seabass aquaculture production in Indonesia was 5.2 tonnes (Directorate General of Aquaculture 2011).

In the Philippines, seabass culture is highly localized. Fish are produced largely in sea cages, or in brackishwater ponds. Seabass fry are produced only in selected fish hatcheries like that of the SEAFDEC Aquaculture Department (SEAFDEC/AQD), and a few commercial hatcheries. The local market is not that developed except in supermarkets in big cities, or in mid- or high-end restaurants where the fish are sold live.

Compared to Indonesia and the Philippines, however, Asian seabass culture is becoming a mature aquaculture industry in Thailand. The species was first successfully bred in 1973 using wild broodstock at Songkhla Fisheries Station, Department of Fisheries (DOF). The use of hatchery-bred broodstock for seed production started in 1985 and led to the expansion of seabass culture in the country. Government and private hatcheries produced seabass fingerlings to support the fry requirement of the grow-out culture farms which are mostly cages and ponds in coastal areas. Excess fingerlings to local requirements are also exported to other countries. Thailand produces about 30–50 million seabass fingerlings per year, 60–70% of which are destined for restocking seabass populations in coastal areas where populations have become depleted due to overfishing and habitat degradation. The remainder of fingerlings are sold for aquaculture production.

In other Asian countries such as Taiwan and Vietnam, seabass aquaculture is developing very rapidly, primarily in the form of marine sea cage culture. However, specific information on the scale and culture technologies within these countries is unknown to the authors and therefore will not be discussed further.

In this chapter, the practices of growing Asian seabass in nursery and grow-out culture systems in selected Southeast Asian countries like the Philippines, Thailand and Indonesia are described.

11.2 Nursery Culture

In the nursery phase, seabass fry are grown from an initial size of 1.5 cm until 10–12 cm total length (TL). This is considered the most difficult phase of seabass culture because cannibalism is usually more intense during this period than during later stages of culture (Parazo et al. 1991). Grading or sorting is the single most effective method of reducing cannibalism during the nursery phase. Grading of the fry is undertaken every 3–4 days during the early nursery phase; larger fish need to be graded at least once a week. It is also during the nursery phase that the fry are weaned to artificial feeds. Maintaining optimal feeding is also effective in reducing cannibalism. In some farms, the fry are transferred to grow-out farms when they reach 5–6 cm in length.

11.2.1 Philippines

In the Philippines, 3 week old seabass fry produced in the hatchery (0.3–0.5 cm TL) are brought to the nursery culture systems for growing until they reach 10–12 cm size or 20–50 g body weight (BW), the appropriate size for stocking in grow-out culture systems. Like the other carnivorous fish species that are highly cannibalistic during the fry/juvenile stage, seabass nursery rearing is divided into two phases (Phase 1 and Phase 2) to have better control of the culture operation, especially related to maintaining uniform size of stocks in order to minimize cannibalism. In Phase 1, the fry from the hatchery (0.3–0.5 cm TL) are grown until they reach 2.5–5.0 cm TL. The fry produced from Phase 1 are grown further until the size of 10–12 cm TL or 20–50 g BW in Phase 2. In both phases, culture duration takes about 45–60 days to reach the desired size.

In Phase 1, nursery rearing is usually performed in land-based tanks or in hapa net cages set inside earthen ponds. Land-based tanks are usually made of concrete, with tank water capacity ranging from 5–20 tons. Hapa net cages should be of manageable size to facilitate sorting of stocks. Hapa nets with dimensions of 2 x 3 x 1 m (L x W x H) and initially of 1–2 mm mesh size and later 0.5 cm mesh size (B-nets) are usually used. The 21-day old fry can be stocked at a density of 1000–2000/m^3. The density is gradually decreased as the fry grows bigger. The fry in tanks are fed brine shrimp (*Artemia* sp.) of increasing sizes. Mysids may be added to supplement or replace brine shrimp, depending on availability. In net cages set in the ponds, natural populations of mixed zooplankton exist. Various species of copepods (e.g., *Acartia, Pseudodiaptomus*) and mysids are nutritionally excellent food and are therefore highly recommended. Copepods and mysids are easily collected from the pond water. Illumination of the cages at night may be also provided, as many zooplankton are attracted by light and the abundance of zooplankton in the cages will encourage feeding on zooplankton during the night (Fermin et al. 1996, Fermin and Seronay 1997). In both the tanks and net cages, weaning the fry to feed on artificial feeds is initiated. In the weaning process, the amount of natural food (zooplankton) is gradually decreased at the same time gradually increasing the amount of artificial feed until the fish is completely weaned to the artificial diet. At this stage, feeding is *ad libitum* and feeding frequency is 4–5 times daily.

During the early stage of Phase 1 nursery rearing, water quality is not critical since the stocks feed mainly on live food organisms. Daily water exchanges of 30–50% of the tank volume is enough to maintain good water quality since daily siphoning to remove wastes and dead fry in the tanks are also performed. During the latter part of culture when minced trash fish or artificial feeds are provided, water quality deteriorates rapidly because of the decomposing uneaten feed inside the culture tanks. In this case, a flow-through or a re-circulating water system is usually recommended. In the ponds, however, water exchanges are undertaken for successive 3–4 days twice a month to coincide with the tidal exchange. For more modern ponds, water pumps are also available so that water exchange can be performed more frequently and is not limited only during the normal tidal cycle.

Phase 2 nursery culture can also be done in land-based tanks, or in B-net cages set in ponds (Madrones-Ladja and Catacutan 2012, Madrones-Ladja et al. 2012). For land-based tanks, tank volume can be 10–20 tons or larger. For cages, the cage dimension is usually 2 x 5 x 1.5 m and initially use B-net with 0.5 cm mesh size. This mesh size is increased to 1 cm later on as the fry grow bigger. Bigger mesh size of the nets allows for a better water circulation and thus prevents fouling of the nets and deterioration of water quality inside the net cage. The initial stocking density is 500–1000 fish m^3 and is gradually reduced as the fish grows bigger. During this phase of nursery rearing, it is advantageous if the fingerlings have already been weaned to feed on artificial feeds. Feeding rate is 4–12% of average BW per day and divided into five rations. Feeding is usually set at 6 AM, 9 AM, 12 PM, 3 PM and 6 PM, with bigger rations given in the early morning and late afternoon. The artificial feed contains about 46–48% crude protein (CP), and 10% lipid. In the absence of artificial feeds, stocks are usually fed with chopped small fish given *ad libitum* two to three times a day.

11.2.2 Thailand

Nursery culture in Thailand is undertaken in concrete tanks, earthen ponds and net cages. Earthen pond nursery culture is practiced on most farms because of the low production cost and the faster growth of fry due to the presence of various kinds of natural food in the pond environment (at the same stocking density, it takes 50–60 days to reach a size of 5 cm in tank culture systems, whereas it takes only about 30 days in the pond nursery system). The disadvantage of a pond nursery system, however, is the heterogeneous size of fry at harvest and low survival rate of only 10–20%. Since seabass can thrive in freshwater as well as seawater, the pond nurseries are located not only in coastal areas, but also in inland areas in the central provinces of Thailand.

Pond nurseries have an area of not more than 2 rai (3,200 m^2) for one pond. The ponds are prepared in the traditional way as in shrimp culture. The ponds are drained and dried to eradicate pests and predators. The bottom is leveled and the area is divided into two parts of 1 rai (1,600 m^2) each for better management before refilling with water. Seabass fry (1–2 cm) from the hatchery are first stocked at a density of 80,000–100,000 fry/rai (50–60 fry/m^2) in a small area in the pond (about 30% of the pond area), using mosquito nylon nets (Fig. 11.1). The area usually measures 9 x 10 m or 10 x 10 m with water depth of 30 cm. The fry are acclimatized to the new environment and trained to accept artificial feeds in these enclosures. Feeding is conducted in this area at the same point twice a day.

Nursery rearing is also undertaken in rows of floating net cages with sizes of 2 x 2 x 2 m or 3 x 3 x 2 m set along the coastal line in brackish water or marine environments. Cages can be set in canals, or in ponds, with moderate water movement. The fry are usually fed minced fresh fish because of the ease in obtaining small fishes from fishermen in the area

278　*Biology and Culture of Asian Seabass*

Figure 11.1. Mosquito net enclosure for nursery culture of sea bass in Thailand.

(sometimes the cages are also operated by the fishermen themselves). The minced fish is placed on a square plastic nylon tray and given twice daily at 10–15% of BW for 1 week to 10 days. Mixed food of chopped fresh fish and pellet (initial ratio of 10:1; fresh fish and pellet) is provided afterwards to train the fry to accept an artificial diet (Fig. 11.2). The amount of pellet is gradually increased in the next 10–15 days, while at the same time reducing the amount of minced fresh fish in the feed. Feeding solely with artificial feeds is done when the fry totally accepts pellet already. Different sizes of pellets are used depending on the size of the fry. Grading is done every 5 days to minimize cannibalism.

11.2.3 Indonesia

In Indonesia, the nursery culture tanks are usually rectangular (1 x 2 x 1 m) or circular (diameter 2.5 m, 1.2 m depth) in shape and the inside of the tank is painted light blue or yellow. The fry (1.2–2.5 cm TL) are stocked at an initial density of 1000–2000 fish/m^3 and are reared in the nursery tanks for 30–35 days until a size of 5–8 cm TL. The density is reduced during grading every 10 days to ensure uniformity in size of the fry and in order to reduce mortality by cannibalism. The fish are fed commercial pellets from feed companies that produce pellet feed specifically for Asian seabass. During nursery rearing, fish under 8 cm TL are fed 3–4 times a day usually at 4–6 h intervals.

Culture of Asian Seabass in Southeast Asia 279

Figure 11.2. Mixed chopped fish and pellets as feeds for nursery culture of seabass in Thailand.

A summary of nursery culture practices in Indonesia, Philippines and Thailand is shown in Table 11.1. A suggested feeding schedule and optimum physico-chemical parameters of water during nursery culture of seabass in the Philippines are shown in Tables 11.2 and 11.3.

Table 11.1. Nursery culture practices of Asian seabass in three Southeast Asian countries.

	Indonesia	Philippines	Thailand
Culture System	Tanks	Tanks or in hapa nets in ponds	Tanks, ponds, cages set inside ponds
Type of operation (straight run or modular)	Straight run	Modular	Straight run
Stocking density	1000–2000 fish/m^3 in tanks	Phase 1: 1000–2000 fish/m^3 in tanks Phase 2: 500–1000/m^3 in tanks	50–62 fish/m^3 in ponds
Size at stocking	1.2–2.4 cm	Phase 1: 0.3–0.5 cm Phase 2: 2.5–5 cm	1–2 cm
Size at harvest	5–8 cm	Phase 1: 2.5–5 cm Phase 2: 10–12 cm	10–15 cm
Feeding management	3–4 times daily using formulated pellets	Mixed feeding with live food (rotifers, copepods, *Artemia*, mysids, minced fish and formulated pellets)	Formulated pellets two times a day in ponds, trash fish two times a day in net cages
Water management	Flow through exchange	Flow through exchange in tanks, tidal flow in ponds (sometimes pumped)	Flow through exchange in tanks, tidal exchange in ponds
Stock management	Size grading every: 10 days	Phase 1: 2–3 times a week Phase 2: 1–2 times a week	Size grading every 5 days

Table 11.2. Suggested feeding schedule for Asian seabass during nursery culture (from Madrones-Ladja et al. 2012).

Days of culture	Feeding rate (% average body weight)	Feeding frequency (times in a day)
0–15	12	10
16–30	10	8
31–45	8	6
46–60	6	5
61–75	5	4
76–90	4	-

11.3 Managing Juvenile Cannibalism

Asian seabass are highly cannibalistic during the nursery stage. To ensure high survival during this stage, cannibalism needs to be minimized. Cannibalism in seabass fry can be observed even at the hatchery phase

Culture of Asian Seabass in Southeast Asia   281

Table 11.3. Optimum levels of physico-chemical parameters of pond water for nursery and grow-out culture of Asian seabass in the Philippines (from Madrones-Ladja et al. 2012, Jamerlan and Coloso 2010).

Water Parameters	Nursery Culture	Grow-out Culture
Salinity (ppt)	20–32	10–30
Temperature (°C)	25–32	26–32
Dissolved oxygen (ppm)	4–8	4–8
pH	7.0–8.5	7.5–8.5
Nitrite-nitrogen (ppm)	0.05	< 1
Unionized ammonia (ppm)	0.02	< 1
Transparency (cm)	30–50	30–50

at 10–12 days of age when unusually big size fry or shooters are present. Cannibalism becomes intense when variation in size is very pronounced. A 30% difference in body size of the fry is enough for cannibalism to occur (Parazo et al. 1991). Stocking fry of uniform sizes and maintaining the uniformity in size during culture is therefore necessary. To achieve this, uniform-sized fry are stocked during the start of culture and these are sorted and size-graded regularly (e.g., 2–3 times a week during Phase 1 nursery rearing and 1–2 times during Phase 2) since the cannibalistic behavior decreases as the fry grows bigger (Fig. 11.3). Grading involves collecting all the fish in the tank and placing them into a grading box or mechanical grader. The smaller fish are able to swim out of the box through narrow openings and the larger fish that are left behind are transferred to another tank. Care must be taken during grading to minimize stress. This can be achieved by minimizing the time the fish spend out of the water,

Figure 11.3. Example of size graders used in Thailand to grade juvenile Lates calcarifer.

not collecting too many fish in the scoop net at one time and performing the whole operation as quickly and efficiently as possible. Frequent and regular feeding of the stocks to make sure that they do not become hungry also minimizes cannibalism. In some cases, providing artificial shelters like bamboo or PVC cuttings that measures 7.5–9 cm diameter and 15–30 cm long also helps since these serve as hiding places for the smaller fry (Qin et al. 2004).

11.4 Grow-out Culture

Growing seabass in earthen brackish water ponds was the traditional method used in much of SEA. Recently, cage culture is the preferred method for growing seabass in Thailand, Hong Kong, Malaysia, Vietnam, Singapore, the Philippines and Indonesia because this operation approach is simpler and more profitable than pond culture. Traditional Asian cage farming uses either simple fixed or floating cages. Fixed cages are made by fastening a cage net to four wooden poles or stakes installed in its four corners. Fixed cages are usually set in shallow bays with narrow tidal fluctuation. In SEA, floating cages are made from timber, bamboo or steel walkways mounted from some sort of floatation from which the small cage net is hung. Floating cages are normally used in relatively deep, sheltered areas that are subjected to moderate tidal flushing.

11.4.1 Philippines

In the Philippines, rearing of seabass from fingerlings to a marketable size is conducted either in ponds or in cages. For pond culture, the water level is a little deeper, about 1–1.5 m deep, than the usual milkfish pond culture (Fig. 11.4). A manageable pond area of about a hectare is mostly used. Water salinity is not a problem since seabass are euryhaline and can thrive in FW, BW or SW. Seabass are grown either as a single species (monoculture), or stocked together with other species (polyculture). In monoculture, the stocks are fed either minced trash fish or artificial feeds. In polyculture, the other species that are stocked together with seabass are forage fish. The forage fish will be preyed upon by seabass thereby reducing the amount of artificial feeds that will be given to the stocks. The forage fish makes use of the natural food in the ponds and continue to reproduce in sufficient quantities to sustain the growth of seabass. Tilapia is an example of a commonly used forage species in these polyculture systems.

Pond preparation is performed before stocking. Pond bottom drying, plowing, flushing and liming are normally undertaken during pond preparation. Plowing breaks the soil, loosens and exposes the organic matter content and reduces or eliminates hydrogen sulfide content. Liming can

Culture of Asian Seabass in Southeast Asia 283

Figure 11.4. Earthen ponds for sea bass culture in the Philippines.

neutralize soil acidity. A small net enclosure, the size of which is dependent on the volume of fry to stock, is set up inside the pond. This serves as the temporary enclosure for the stocks to get used to pellet feeding in the ponds.

Depending on the degree of sophistication of the culture system, stocking density in ponds may range from 5,000–20,000 fry/ha. The higher the stocking density, the more is the demand for feeds, frequent water change and provision of life support systems like paddle wheel aerators to ensure sufficient dissolved oxygen in the water when the biomass increases. In sea cages, the density can range from 30–50 fry/m^3, assuming the water environment is ideal for good growth and survival.

Trash fish (fish by catch), formulated feeds, or their combination, are given during the grow-out stages. Using trash fish over formulated feeds, however, has many disadvantages including high seasonality and variability of the fish species composition, polluting the water, acting as possible vectors of pathogens, results in high feed conversion ratios (FCR > 8), has short storage life, and often are also food for the people in some areas. Formulated feeds are easily available in the market. Using these feeds, however, requires that the stocks are already weaned to feed on artificial feeds when they are brought to the farms. The advantages of formulated feeds over that of trash fish are that they provide a nutritionally complete diet, are less polluting because they are more stable in the water column compared to trash fish, contain less moisture (roughly 10%) than trash fish (roughly 75%), and result in a low feed conversion ratio (FCR of 1–2). With formulated feeds however, proper feeding management is very important. Proper feeding frequency which may be 3–4 times a day with the right feeding ration is very important in order to reduce wastage of feeds and prevent pollution due to overfeeding.

11.4.2 Thailand

In Thailand, grow out culture of seabass is undertaken in both coastal and inland areas where environmental conditions can vary from a salinity of 0–31 ppt, dissolved oxygen level of 3–5 ppm, pH of 7.0–8.5 and alkalinity of more than 100 ppm. Typically, the culture period is about 4–6 months and is usually conducted in three types of environmental conditions such as in FW rice fields, BW ponds and marine cage culture. In rice fields, seabass are grown to increase the income of the farmers. Rice fields are prepared for the culture of fish by digging canals (50–60 cm depth and 1 m wide) around the field beside the dike. Fry already trained to accept live food and acclimatized to FW are used. Stocking rate is low at 1–2 fry/m^2 and culture duration is typically only for 3–5 months before the rice fields dry up. In most cases, the harvested fish are sold for further rearing in ponds or in cage culture. Survival rate is 50–60% and total production values of 150–200 kg per farm are attainable with harvest size of mostly 200–400 g per fish. Only about 5–10% of fish grown this way reach marketable size.

Due to the collapse of shrimp culture, some farmers in central Thailand have turned to seabass culture (without feeding) as an alternative livelihood activity. Shrimp ponds are prepared by cleaning the pond bottom and exposing the top soil to the sun for 2 weeks. Water is allowed inside ponds for 2 weeks to allow small fishes to come to the ponds before stocking fry (2–5 cm TL) at a density of 0.5–1 fry/m^2. Small fishes like *Tilapia* sp. fry are added if the natural food becomes limiting. Culture period is for 6 months and survival is about 60% with production of 360–600 kg/rai (1,600 m^2). There are also some ponds that are developed specifically for seabass culture. The area ranges from 800–3,200 m^2 and 1.5–2.0 m water depth. Pond preparation is the same as in shrimp culture. The juveniles used for stocking are already trained to feed on pellets because artificial feeds are normally the feeds given. Because of the large size of the pond, a feeding area is set using nylon mosquito net and set at a depth of 30 cm. The feeding area is set for 20 days to train the fish for the location of the feeds during feeding time. If this is not done, the juveniles will spread out around the pond and will not know the feeding area. As a consequence, size range of the stocks will be highly variable, survival rate is low, growth is also slow, and disease problems may arise.

The top three provinces in Thailand that produced seabass from cage culture are Chachoensao, Pattani and Songkhla. In 2009, these three provinces produced 8,697 tons of seabass which is equivalent to 58.7% of the total seabass production of the country (DOF 2009). The majority of the production (86%) came from cage culture along coastal areas and fresh water canals. Square-shaped floating net cages are the most popular type. Fixed cage or small pen culture types are operated in Songkhla and Pattani provinces in southern Thailand and areas along the east coast as in Chantaburi province. Big round cages are used for research and demonstration under the Thai-Norway aquaculture project in Phuket, southern Thailand.

Traditional floating net cage cultures are operated along the coast and in canals in mangrove areas. Most of these are small scale and operated/owned by the fishermen themselves, hence fresh small fish that comes from their own catches are used as feeds. In Songkhla Lake, the traditional fish cages are fixed net cages which may be more difficult to manage, especially when problems of diseases occur.

Commercial floating net cages are operated in the Andaman Sea, Pangna and Phuket areas (Fig. 11.5). Square-shaped nylon net cages (4 x 4 x 2.5 m) with mesh size of 1–2 cm are set in rows and fixed together in a group of about 40 cages. A distance of at least 1–2 m in between groups is allowed.

Figure 11.5. Commercial floating cages for sea bass culture in Andaman Sea in Thailand.

Feeds are composed of minced fresh fish mixed with 1–2% vitamin mineral pre- mixed per 1 kg of feed and fed every other day in the morning or late afternoon.

The juveniles that are stocked in floating cages are produced in the hatchery. These are ensured to be healthy and free from any kinds of diseases. For juveniles that come from abroad, the government requires that the juveniles are certified free of diseases. The stocking density depends on the location of the farm and the carrying capacity of the farm site. In general, the stocking density in floating net cages for juveniles sized between 5–8 cm TL is 80–100 fish/m^3 for the first 3–4 months and thereafter the density is reduced gradually up to 10–15 fish/m^3 at harvest size of 2.0 kg.

The preferred market size of seabass in Thailand is mostly around 0.4–0.7 kg. Because of the low market price of oversized seabass ($\geq$ 0.7 kg) and the large meat volume of seabass, some big farm operators continue to culture the fish until 3–4 kg. At an initial size of $\geq$ 0.7 kg, stocking density ranges from 1–2 fish/m^2. The fish will grow to 3–4 kg within 12–15 months with FCR of 1.7–2.0 and survival rate of more than 90%. Oversized fish are usually processed into various kinds of products like fish fillet, half cooked pieces, and other forms of value adding.

11.4.3 Indonesia

Under normal conditions with good water quality, Asian seabass are able to grow up to 1.5–2.5 kg within 12–18 months of culture. This size is mainly for export in fillet form. Various domestic markets require different sizes between 400–500 g and the fish are sold live or fresh (not frozen) to seafood restaurants and chilled for supermarket and local buyers. The price ranges from US$3.5–4.7/kg in the domestic marketplace.

Because most of the coastal earthen ponds are utilized for shrimp and milkfish culture, seabass grow-out culture in Indonesia is mostly performed in floating net cages (Fig. 11.6). Generally, the floating net cages comprise several units of net cages suspended from a floating raft and anchored to the bottom of the sea. The raft frame is made of wood or galvanized iron. To improve on the construction, some farmers use modern frames made of high-density polyethylene (HDPE) pipes. The shape of cages are usually square or circular. For square-shaped cages, the sizes of the frames are 4 x 4 x 4 m, 5 x 5 x 5 m, 8 x 8 x 5 m and 10 x 10 x 5 m (L x W x H), while for circular-shaped cages they are generally 10–12 m in diameter and 5–8 m deep. The catwalk along the cages allows the farmers to walk and move easily around the cages for feeding, checking the fish and changing the nets (Sugama et al. 2008). Two local manufacturers have successfully produced rectangular or circular frame HDPE cages which have been well accepted by farmers.

Culture of Asian Seabass in Southeast Asia 287

Figure 11.6. Commercial floating cages for seabass culture in Indonesia.

The cages are located in calm sea areas with minimum water depths of 10 m, water current flow of ~15–30 cm/s, water temperature of 27–32°C, water transparency of > 5 m, salinity of 30–33 ppt, dissolved oxygen of > 5 ppm, pH of 7.5 to 8.1 and nil nitrite and ammonia levels (0 ppm).

Stocking densities depends on the location of the farm and the carying capacity of the farm site. In general, the stocking density in floating net cages for juveniles with sizes between 5–8 cm TL is 80–100 fish/m^3 for the first 3–4 months and thereafter the density is reduced gradually down to 10–15 fish/m^3 until harvest size of 2.0 kg. In Indonesia, some companies produce pellet feeds specifically for Asian seabass. The stocks require different size feeds as they grow. Fish size of 50–100 g are fed 3–4 times a day at a feeding rate of 5–7% BW, while larger fish are fed twice a day at 3–4% BW. Feed conversion ratio (FCR) ranges from 1.2 to 1.5 in these systems (Mariculture Development Center Lampung, pers. comm. 2012).

A summary of grow-out culture practices in Indonesia, Philippines and Thailand is shown in Table 11.4. A suggested feeding schedule during grow-out culture of seabass in the Philippines is shown in Table 11.5.

Table 11.4. Grow-out culture practices for Asian seabass in Indonesia, Philippines and Thailand.

	Indonesia	Philippines	Thailand
Culture System	Cages	Ponds or cages	Ponds, cages, pens, rice and fish culture
Stocking density	80–100/m^3 later 10–15/m^3	0.5–2/m^2 in ponds 30–50/m^3 in cages later reduced to 10–15/m^3	1–2/m^2 in rice fields 80–100/ m^3 later 10–15/ m^3 in cages
Size at stocking	5–8 cm	10–12 cm	10–15 cm in ponds 5–10 cm in rice fields
Size at harvest (g)	400–500	400–500	400–700 in ponds
Feeding management	Commercial pellets 3–4 times daily	Trash fish, pelleted feeds 3–4 times daily	Trash fish, pelleted feeds 3–4 times daily
Water management	Tidal flow	Tidal exchange in ponds, pumps	Tidal exchange in ponds, pumps

Table 11.5. Suggested feeding schedule for Asian seabass during grow-out culture (modified from Jamerlan and Coloso 2010).

Average body weight (ABW, g)	Feeding rate (% ABW)	Type of feed	Feeding frequency (times daily)
20–50	7	Starter	3
50–100	6	Starter	3
100–200	5	Grower	2–3
200–300	4	Grower	2–3
300–400	3	Finisher	2–3
400–500	2.5	Finisher	2–3

11.5 Health Management

A spectrum of diseases and pathogens are known to affect Asian seabass and there is agreement that as intensification of seabass culture continues it is likely that further occurrences of known or previously undescribed diseases will be experienced (Schipp et al. 2007). Prevention is better than cure hence, regular monitoring for the presence of diseases in the stocks is necessary to ensure production success. Basic health management for seabass during nursery and grow-out phases includes cleaning of tanks and cages, including all used equipment, using either formalin (200 ppm) or chlorine (100 ppm) and then subsequent sun-drying. Regular cleaning of tanks and cages are usually performed after transfer of stock. A close observation of the swimming and feeding activity of fry to check signs of abnormal behavior will help. For new stocks, examination of the presence of disease-causing organisms should be done before stocking.

The major diseases in Southeast Asia that frequently occur during the grow-out culture in marine net cages are bacterial infections caused by *Vibrio* and *Streptococcus*, mainly *Streptococcus iniae, Streptococcus agalactiae* and *Vibriosis* sp. Vaccination of fingerlings before entry to the farm appears to be effective in preventing infection. The vaccine for preventing infection of *S. iniae, S. agalactiae,* and *Vibriosis* are commercially available. Vaccination may be conducted using immersion techniques whereby fish are swam in water containing the vaccine. However, immunity may not persist for long using this approach in comparison with vaccination via intramuscular injection. Another cause of mass mortality of Asian seabass juveniles during the nursery phase is the bacteria *Pasteurella piscicida*, the causative agent of pseudotuberclosis in marine fish (Koesharyani et al. 2001, Zafran et al. 2005). Clinical signs of moribund fish which this disease classically is a slightly swollen abdomen.

Parasitic infections due to protozoans (e.g., cryptocaryonosis caused by *Cryptocaryon irritans; Amyloodimium* sp.), monogeneans (*Dactylogyrus, Pseudorhabdosynochus*), digeneans (*Pseudometadena, Transversatrema, Prosorhynchus, Lecithochirium*), fish lice like *Caligus*, isopods and marine leech are common in seabass culture (see Cruz-Lacierda 2010 and Hutson 2013—Chapter 6). These parasites cause skin ulcers, tail rot, loss of scales and mucus, respiratory difficulties and swollen muscles. Affected fish usually rub their body against objects, exhibit body discoloration, reduced food intake and retarded growth and mortality. Parasitic diseases like skin fluke are caused by infection with *Neobenedenia girellae* and *Benedenia epinepheli* in Lampung Buy-Sumatra Island (Rakerts et al. 2008). *Neobenedenia girelae* showed higher pathogenicity than *B. epinepheli*. Affected fish lost appetite, showed lethargic swimming behavior and skin lesions with secondary bacterial infection. *Neobenedenia girellae* infected seabass have opaque eyes which caused blindness and infected fish finally die due to their inability to find food (Koesharyani et al. 2001, Zafran et al. 2005). The parasite is transparent so that it is difficult to confirm infection. To prevent parasitic infections, the stocks may be treated simply by bath treatment with freshwater for about 1 h for three successive days. The culture water should also be changed, or the treated fish should be transferred to another tank or pond. Alternatively, bath treatment for 30–60 min with 200 ppm of formalin or for 1–2 days with 25–30 ppm formalin can be used. During treatment, strong aeration is recommended to prevent stress due to oxygen depletion as a result of the formalin.

Viral Nervous Necrosis (VNN) infection in larval and juvenile marine fish including seabass is now regarded as the most dangerous disease affecting marine fish during seed production and also during the nursery and grow-out stages (Lio-Po 2010). Nodavirus, the causative virus has a wide range of hosts and is known to occur in more than 40 known species of marine fish, including Asian seabass. The virus infects the central nervous system of the fish resulting in vacuolation and degeneration. The virus was first identified in farmed barramundi in Australia in the 1980s (Munday et al. 1992). VNN infection can cause 100% mortality and in Indonesia, it initially occured in hatchery reared Asian seabass in 1997 (Koesharyani et al. 2001). Incidence of mass mortalities in hatchery-reared seabass was also reported in the Philippines (Maeno et al. 2004, Maeno et al. 2007). Recent work has shown that vaccination of seabass juveniles with formalin-deactivated betanodavirus conferred protective immunity to seabass lasting for several months (Pakingking et al. 2009). As long as the farmers use only VNN-free juveniles during stocking, VNN infection usually does not become a serious limiter to production. In fact, the presence of the virus in fish does not necessarily result in mortalities. Fish may carry the virus without showing disease symptoms and it is often only as a result of external stressors that manifestation of the disease in fish carrying the virus is often initiated (de la Peña et al. 2011).

Finally, an unknown disease was observed to have infected Asian seabass culture in marine net cages in the coastal area of Batam Island close to Singapore during May–July 2011. Clinical signs of infected fish included abnormal swimming behavior, loss of appetite, and removal or dropping off of scales. Hence the disease is temporarily known as *"scale drop disease"* (Fig. 11.7). The causative agent of this disease is yet unknown and so far has not reoccurred after this initial infection in Singapore.

Figure 11.7. Seabass suffering from "scale drop" disease in Indonesia.

References

Cruz-Lacierda, E.R. 2010. Parasitic diseases and pests. *In:* G.D. Lio-Po and Y. Inui (eds.). Health Management in Aquaculture (2nd edition). Southeast Asian Fisheries Development Center, Aquaculture Department, Tigbauan, Iloilo, Philippines.

de la Peña, L.D., V.S. Suarnava, G.C. Capulos and M.N.M. Satos. 2011. Prevalence of viral nervous necrosis (VNN) virus in wild caught and trash fish in the Philippines. Bull. Eur. Assoc. Fish Pathol. 31: 129–138.

[DOF] Department of Fisheries. 2009. Fisheries Statistics of Thailand 2009. No. 9 / 2011. Information Technology Center, Department of Fisheries. 91 pp.

Fermin, A.C. and G.A. Seronay. 1997. Effects of different illumination levels on zooplankton abundance, feeding periodicity, growth and survival of the Asian seabass, *Lates calcarifer* (Bloch), fry in illuminated floating nursery cages. Aquaculture 157: 227–237.

Fermin, A.C., E.C. Bolivar and A. Gaitan. 1996. Nursery rearing of the Asian seabass, *Lates calcarifer*, fry in illuminated floating net cages with different feeding regimes and stocking densities. Aquat. Living Resour. 9: 43–49.

Hutson, K. 2013. Infectious diseases of asian seabass and health management, *In:* D.R. Jerry (ed.). Ecology and Culture of Asian Seabass *Lates calcarifer*. CRC Press.

Jamerlan, G.S. and R.M. Coloso. 2010. Intensive culture of seabass *Lates calcarifer* Bloch, in brackishwater earthen ponds. Southeast Asian Fisheries Development Center, Aquaculture Department, Tigbauan, Iloilo, Philippines. AEM#46.

Koesharyani, I., D. Rosa, K. Mahardika, F. Johni, R. Zafran and K. Yuasa. 2001. *In:* K. Sugama, K. Hatai and T. Takai (eds.). Manual for Fish Diseases Diagnoses II. Marine fish and crustacean diseases in Indonesia. Gondol Research Institute for Mariculture, Central Research Institute for Aquaculture and Japan Cooperation Agency.

Lio-Po, G.D. 2010. Viral diseases. *In:* G.D. Lio-Po and Y. Inui (eds.). Health Management in Aquaculture (2nd edition). Southeast Asian Fisheries Development Center, Aquaculture Department, Tigbauan, Iloilo, Philippines, pp. 77–146.

Madrones-Ladja, J., N. Opina, M. Catacutan, E. Vallejo and V. Cercado. 2012. Cage nursery of high-value fishes in brackishwater ponds: Seabass, grouper, snapper, pompano. Southeast Asian Fisheries Development Center, Aquaculture Department, Tigbauan, Iloilo, Philippines. pp. 42. AEM#54.

Maeno, Y., L.D. de la Peña and E.R. Cruz-Lacierda. 2004. Mass mortalities associated with viral nervous necrosis in hatchery-reared seabass *Lates calcarifer* in the Philippines. Japan Agricultural Research Quarterly 38: 69–73.

Maeno Y., L.D. de la Peña and E.R. Cruz-Lacierda. 2007. Susceptibility of fish species in mangrove brackish area to piscine nodavirus. Japan Agricultural Research Quarterly 41: 95–99.

Madrones-Ladja, J.A and M.R. Catacutan. 2012. Netcage Rearing of the Asian Seabass *Lates calcarifer* (Bloch) in Brackishwater Pond: The technical and economic efficiency of using high protein diets in fingerling production. Philipp. Agricult. Scientist 95: 79–86.

Munday, B.L., J.S. Langdon, A. Kyatt and J.D. Humphrey. 1992. Mass mortality associated with a viral-induced vacuolating encephalopathy and retinopathy of larvae and juvenile barramundi, *Lates calcarifer* Bloch. Aquaculture 103: 197–211.

Pakingking, R. Jr., R. Seron, L.D. de la Peña, K. Mori, H. Yamashita and T. Nakai. 2009. Immune responses of Asian seabass (*Lates calcarifer*) against the inactivated betanodavirus vaccine. J. Fish Dis. 32: 457–463.

Parazo, M.M., E.M. Avila and D.M. Reyes Jr. 1991. Size-dependent and weight-dependent cannibalism in hatchery-bred seabass (*Lates calcarifer* Bloch). J. Appl. Ichthyol. 7: 1–7.

Qin, J.G., L. Mittiga and F. Ottolenghi. 2004. Cannibalism reduction in juvenile barramundi *Lates calcarifer* by providing refuges and low light. J. World Aquacult. Soc. 35: 113–118.

Rakerts, S., H.W. Palm and S. Kumpel. 2008. Parasite fauna of seabass under mariculture condition in Lampung Bay Indonesia. J. Appl. Ichthyol. 24: 321–327.

Schipp, G., J. Bosmans and J. Humphrey. 2007. Northern Territory Barramundi Farming Handbook. Department of Primary Industry, Fisheries and Mines, Darwin Aquaculture Center, Darwin, Australia 80 pp.

Statistical Data of Aquaculture Production. 2010. Directorate General of Aquaculture, Ministry of Marine Affairs and Fisheries of Indonesia. 56 pp.

Statistical Data Aquaculture Production. 2011. Directorate General of Aquaculture, Ministry of Marine Affairs and Fisheries of Indonesia 58 pp.

Sugama, K.I., I. Innsan, I. Koesharyani and K. Suwirya. 2008. Hatchery and grow-out technology of grouper in Indonesia. *In*: I.C. Liao and E.M. Leano (eds.). The Aquaculture of Groupers. World Aquaculture Society, pp. 61–78.

Zafran, R., D. Koesharyani, F. Johni and K. Yuasa. 2005. Marine fish and crustacean diseases in Indonesia. *In*: K. Sugama, K. Hatai and T. Takai (eds.). Manual for Fish Diseases, Diagnoses and Treatment. Gondol Research Institute for Mariculture, Central Research Institute for Aquaculture and Japan Cooperation Agency.

12

Muscle Proteins in Asian Seabass—Parvalbumin's Role as a Physiological Protein and Fish Allergen

Michael Sharp and *Andreas L. Lopata**

12.1 Introduction

Asian seabass (*Lates calcarifer*) is commonly consumed throughout India, South-east Asia and Australia, as well as increasingly being marketed into North America and Europe. Paralleling this growing global consumption of Asian seabass, and seafood in general, is the increasing prevalence of human allergic reactions to fish products. Allergic reactions to Asian seabass, and other bony fish, are primarily caused by the small Ca^{2+} binding muscle protein parvalbumin. This chapter will discuss the physiologic and allergenic roles of fish parvalbumin and what we currently know about this protein in Asian seabass and how it differs from other commonly consumed fish species.

Centre for Sustainable Tropical Fisheries and Aquaculture and School of Pharmacy and Molecular Sciences, James Cook University, Townsville, QLD, 4811, Australia.
Email: michael.sharp2@my.jcu.edu.au
*Corresponding author: andreas.lopata@jcu.edu.au

12.2 Physiological Function of Parvalbumin

Calcium binding proteins are a diverse group of proteins which are involved in multiple cellular processes such as Ca^{2+} homeostasis and Ca^{2+} signaling pathways. Calcium binding proteins can be divided into two different categories: membrane-intrinsic proteins which regulate and transport Ca^{2+} across the cell membrane and calcium-modulating proteins which modulate Ca^{2+} concentration and signaling in and between cells. Parvalbumin proteins belong to the latter group and due to their predominance in fish white muscle are one of the most abundant muscle proteins in Asian seabass.

Parvalbumins are small, acidic, heat stable proteins often found in the sarcoplasmic reticulum of vertebrate muscle myofibrils. Parvalbumin molecular weights range in fish between 10–15 kDa and have an isoelectric point between 4.1 and 6.0 (Fig. 12.1, Table 12.1). There is substantial diversity in the number and molecular weight of parvalbumin isoforms in fish, with Asian seabass (Fig. 12.1—lane 1) exhibiting two prominent parvalbumin isoforms in the 10-13 kDa range, black bream (*Acanthopagrus butcheri*) and coral trout (*Plectropomus leopardus*) demonstrate one parvalbumin protein isoform, while other fish such as Atlantic salmon (*Salmo salar*) or blue eye trevella (*Hyperoglyphe antarctica*) exhibit weak or no parvalbumin isoforms. Parvalbumin proteins play an important role in calcium regulation, as each protein contains two Ca^{2+} binding ligands. They are a part of the EF hand

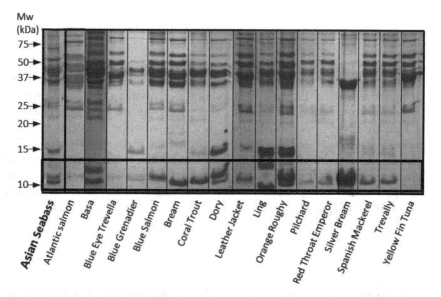

Figure 12.1. Fish muscle protein extracts separated on a sodium dodecyl sulfate polyacrylamide gel (SDS) subjected to Coomassie blue staining from a variety of fish species, including Asian seabass (*Lates calcarifer*). Bands boxed are parvalbumin isoforms.

Table 12.1. Physical characteristics of allergenic parvalbumin isoforms from Asian seabass and five other commonly aquacultured fish species.

Common Name	Species	Isoform	Isoelectric point	Molecular weight (kDa)
Asian seabass	*Lates calcarifer*	β1	4.50	11.6
		β2	4.48	11.7
Atlantic salmon	*Salmo salar*	β1	4.95	11.9
		β2	4.41	11.4
Common carp	*Cyprinus carpio*	β1	4.51	11.5
		β2	4.67	11.6
Mozambique tilapia	*Oreochromis mozambicus*	β	4.56	11.5
Channel catfish	*Ictalurus punctatus*	β	4.75	11.6
Silver carp	*Hypophthalmichthy molitrix*	β	4.46	11.6

protein family, which is the largest group of proteins involved in binding charged metals such as Ca^{2+} and Mg^{2+}. Parvalbumins form a globular shape containing 6 α-helices (A, B, C, D, E, F) which pair up and form three helix-loop-helix EF hand motifs (Fig. 12.2a). EF hand motifs are common in most Ca^{2+} binding proteins. The CD and EF helix pairs are near identical and it

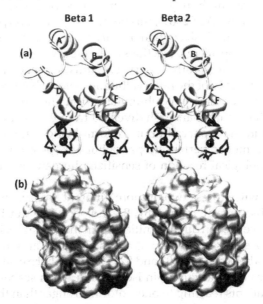

Figure 12.2. Ribbon (a) and space-fill (b) molecular structures of both β1 (left) and β2 (right) Asian seabass (*Lates calcarifer*) parvalbumin isoforms. Helices are labeled A, B, C, D, E and F. These proteins bind two Ca^{2+} ions, indicated in black. EF hand motifs are common in most Ca^{2+} binding proteins and are found in all vertebrates. The space-filling model (Fig. 12.2b) of parvalbumin highlights the Ca^{2+}-binding sites of the β1 and β2 isoform (black shaded regions).

is thought they were the result of gene duplication, this also means that these are the two hands that bind to the two Ca^{2+} ions (Arif 2009). However, AB helical pairs control the conformational structure of the protein, as seen in the space-filling model of the Asian seabass parvalbumin protein (Fig. 12.2b). Differences are commonly observed among AB domains of parvalbumin from temperature and tropical fishes, with evolution studies showing that there have been specific mutations in the AB domain which enable parvalbumin to more efficiently bind calcium in species inhabiting cooler aquatic environments (Whittington and Moerland 2012). Chelated calcium is known to change the conformational structure of parvalbumin, making it heat stable and less susceptible to protease or chemical digestion (Bugajska-Schretter et al. 2000). The allergenicity of parvalbumin is also higher in the presence of Ca^{2+}, as the major parvalbumin IgE antibody binding epitope is believed to be conformational (Bugajska-Schretter et al. 1998, Untersmayr et al. 2006). Parvalbumin also has the ability to bind to other parvalbumins to form oligomers which also may be allergenic (Rosmilah et al. 2005).

Genetically, parvalbumin is expressed as one of two distinct isoform lineages; α and β. β-parvalbumins have an isoelectric point less than 5.0, while the α-parvalbumin usually has an isoelectric point greater than 5.0 (Tanokura et al. 1987, Arif 2009), this being the greatest physical difference between the isoforms. Fish contain both α and β parvalbumin, but humans seem to only develop allergic reactions to the β-parvalbumin isoform, which is the most abundant isoform observed in muscle tissue of Asian seabass. Increased allergenicity of the β-parvalbumin isoform has been suggested to be a consequence of the β-parvalbumin isoform lineage evolving after the point at which mammals and fish diverged from their common ancestors. This resulted in the β-parvalbumin isoform being significantly different in structure to higher vertebrate parvalbumins, possibly explaining the strong immunological reaction of sensitized human individuals (Jenkins et al. 2007).

Most fish will express two or more different β-parvalbumin isoforms, which are subsequently named β1, β2 and so forth (Van Do et al. 2003). β-isoforms, however, can differ significantly in their amino acid composition. For example, Asian seabass parvalbumins have an amino acid sequence identity of 67% between their β1 and β2 isoforms (Sharp et al. unpubl. data). The differences in β-parvalbumin isoforms in Asian seabass can result in fish allergic patients reacting to one isoform stronger than the other, which adds to the complexity of diagnosing fish allergy and detecting allergenic parvalbumin (Perez-Gordo et al. 2011, Sharp and Lopata 2013). Despite differing greatly in primary structure, Asian seabass parvalbumin differ only slightly in their isoelectric point (Table 12.1, Fig. 12.3). Subsequently when anion exchange chromatograpy is applied both isoforms elute at

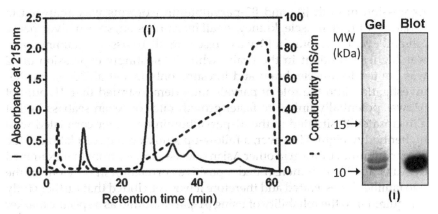

Figure 12.3. Purification of parvalbumin isoforms from Asian seabass (*Lates calcarifer*) muscle extracts. Identification and purity are demonstrated on sodium dodecyl sulfate Coomassie stained polyacrylamide gels and western blotting with anti-parvalbumin specific antibodies.

the same time demonstrating their similar negative charge. Parvalbumins also have the ability to form oligomers. For example, a 24 kDa oligomeric β-parvalbumin was detected by the anti-parvalbumin monoclonal antibody (MoAb) PARV-19 in Atlantic cod (Das Dores et al. 2002). It is unknown how often these oligomers occur in fish, or whether they form in Asian seabass parvalbumin, as many of these oligomers could be destroyed in the process of protein extraction and identification.

12.3 Other roles of Parvalbumin

Physiologically, parvalbumin is a key protein in fast muscle contraction/relaxation through its Ca^{2+} handling and signaling abilities. Yet it is not the sole muscle contraction/relaxation protein, as parvalbumin works alongside muscular proteins such as myosin, troponin and other myofibrillar structures (Arif 2009). Parvalbumin is also not restricted to muscle tissues of vertebrates. In higher vertebrates it seems to regulate electric and enzymatic functions in neural tissue as it is found in most cells containing the neurotransmitter inhibitor gamma-aminobutyric acid (GABA) (Celio 1986). Parvalbumin has also been implicated in regulation of endocrine products such as testosterone, renal transport of NaCl, and as a chemostatic ligand in the detection of prey by vertebrates containing a chemosensory vomeronasal organ (e.g., as present in snakes) (Arif 2009).

Due to the parvalbumin protein being intrinsically associated with muscle fibre contraction and relaxation, and the fact that growth is largely a result of muscle mass, this protein has been suggested in Asian seabass as a potential marker linked to growth performance (Xu et al. 2006).

Expression of both β1- and β2-parvalbumin isoforms was analyzed in *L. calcarifer* brain, muscle, kidney, small intestine, skin, spleen, liver, heart, gill and eye. Parvalbumin β1 was present in all 10 tissues examined and was highly abundant in the brain, while interestingly expression of β2 was limited to muscle, brain and intestine only (Xu et al. 2006). Further investigation into the role of parvalbumin demonstrated that β1, but not β2, was potentially linked to faster growth rates of Asian seabass in that a microsatellite situated in the β1-parvalbumin promoter correlated with higher bodyweight. However, a follow-up study on the correlation of the β1-parvalbumin isoform in other Asian seabass families from that examined by Xu et al. (2006) demonstrated a positive correlation in only one of the two families investigated and therefore it was concluded that further study is required into the reliability of using β1-parvalbumin as a specific marker protein for growth for Asian seabass (Wang et al. 2008).

Parvalbumin expression can also differ in skeletal muscle, as it is believed allergenic β-parvalbumins are abundant in fast twitching white muscle of fish, whereas parvalbumin is absent in slow twitching dark muscle. White muscle is abundant in most bony fish, including the Asian seabass, as the fast twitching properties aid their quick agile movements. However many migratory bony fish such as tunas possess more dark than white muscle and consequently appear less allergenic due to the lack of high levels of β-parvalbumin in their muscle tissues. Cartilaginous fishes, in particular elasmobranchs, are often large fish which also contain less allergenic fast muscle twitching β-parvalbumin and the more non-allergenic α-parvalbumin isoform (Kobayashi et al. 2006, Griesmeier et al. 2010).

12.4 Parvalbumin Evolution

Fish species can be divided into two main groups; the bony fish and cartilaginous fish. Most edible fish belong to the bony fish (Osteichthyes), whereas sharks and rays are cartilaginous and belong to the Class Chondrichthyes. To date, studies on fish allergens have mainly focused on allergens expressed in fish species where there is high human consumption, particularly by Europeans (e.g., Atlantic cod, carp, and Atlantic salmon) (Lindstrom et al. 1996, Van Do et al. 2005, Untersmayr et al. 2006, Perez-Gordo et al. 2011). These fish, however, belong to ancestral lineages that diverged early in the evolution of bony fishes and therefore may not provide a good representation of allergenicity of bony fishes from the Super Order Acanthopterygii.

Bony fish can be divided into 45 orders. The most commonly consumed bony fish belong to the Orders Clupeiformes (herrings and sardines), Salmoniformes (salmons and trouts), Cypriniformes (carps), Gadiformes (cods, hakes, and whiting), Siluriformes (catfish), and Perciformes (perches, mackerels, and tunas). Among taxonomic Orders, large differences in amino acid sequence of parvalbumin isoforms are observed, as are differences between isoforms within the same fish species (Fig 12.4). For example, the Asian seabass β1 isoform is more closely related to the β1 isoforms of large-mouth bass (*Micropterus salmonides*) and channel catfish (*Ictalurus furcatus*) than it is to its β2 isoform. From an evolutionary perspective, phylogenetic examination of the AB domain suggests that parvalbumin genes appear to have initially evolved over millions of years in bony fish that were denizen to a warm climate. During the cooling of the Southern Ocean, ecological niches were explored by fish species which led to evolution of parvalbumins with conformational structures which increased their efficiency of Ca^{2+} binding in these cooler aquatic environments. These proteins evolved to be as functionally efficient at 2°C as counterpart proteins in temperate and tropical species at 30°C. This physiological adaptation to cold water is reflected in specific amino acid substitutions, making this group of parvalbumins in cold-water fish species very different at the molecular level compared to other bony fish. Their molecular differences are highlighted in the six Antarctic perch fish species forming the bottom-most clade in the phylogenetic tree in Fig. 12.4.

12.5 Parvalbumin as the Major Fish Allergen

The first and most comprehensive study on any allergen derived from a food source was the analysis of parvalbumin—the major allergen in codfish, Gad c 1 (Elsayed and Aas 1970). Although most heat-resistant food allergens contain linear IgE binding protein regions (epitopes), parvalbumin has in addition conformational epitopes (Untersmayr et al. 2006), stabilized by the interaction of the metal-binding protein domains (Fig. 12.2). This means that this protein still retains its allergenic IgE epitope sites, even after cooking, as recently demonstrated with 12 fish allergic patients who tested positive for immunological reactions to cooked Asian seabass (Sharp et al. unpub. data).

This major allergen has been identified in several bony fish and in addition was recently also confirmed as an allergen in four different tropical fish species (Lim et al. 2008, Beale et al. 2009). High-molecular-weight polymers of parvalbumin have also been documented in snapper (Rosmilah et al. 2005), which appear to impact on the allergenicity of this protein. However, human clinical cross-sensitivity to parvalbumins differs among fish species and may in part be related to the large amount of diversity seen for this protein, whereby species-specific isoforms share only limited degrees of amino acid sequence identity (ranging from 60 to 80%). For example, Asian seabass parvalbumin shares 88% amino acid sequence identity with yellowfin tuna (*Thunnus albacares*) and only 66%

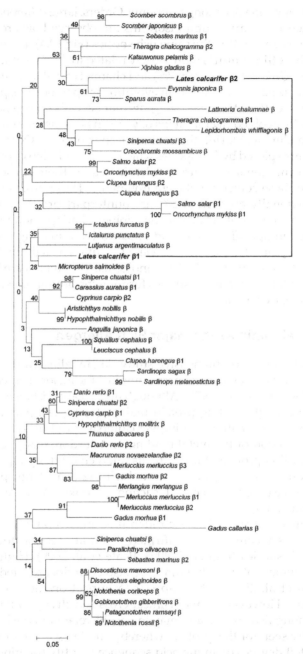

Figure 12.4. Bootstrap consensus tree (10,000 replicates) showing molecular phylogenetic relationships of β-parvalbumins from commonly consumed fish species. Analyses are based on amino-acid sequences of proteins.

with Baltic cod (*Gadus morhua*), which could reflect high or low similarity of allergic reactions, respectively. However, further research into comparative allergenicities need to be conducted to substantiate the hypothesis that molecular similarity correlates with cross-sensitivity.

In addition to parvalbumin, other allergens have been characterized in fish such as collagen (Sakaguchi et al. 2000), from the skin and muscle tissue, as well as the hormone, vitellogenin (Escudero et al. 2007, Perez-Gordo et al. 2008), found particularly in fish roe (caviar) (Gonzalez-De-Olano et al. 2011). However, these two proteins seem to be rather rarely involved in allergic reactions. It is to note that crustacean and mollusk allergens do not cross-react with fish allergens as they are very different proteins.

12.6 Mechanism and Detection of Fish Allergy

During the sensitization of individuals to fish proteins, IgE antibodies are developed by the immune system specifically directed to the allergenic protein. These IgE antibodies bind onto specific receptors present on the surface of many human cells including mast- and basophil cells. Upon subsequent exposure of the immune system to the allergenic proteins, these IgE antibodies on the cell surface mediate the allergic reactions. The pattern of allergic symptoms after ingestion of fish appears similar to those that occur during allergic reactions to other foods. Most reactions are immediate and reported within 2 h and patients may have a single symptom, but often there is a multi-organ involvement.

To determine if a person is allergic to a particular fish species and specific proteins, a range of diagnostic tests have been developed including immunoblotting, ELISA tests, protein microarray and ImmunoCAP systems, which are all based on the detection and quantification of allergen specific antibodies (Fæste and Plassen 2008). For the protection of sensitized consumers of accidental exposure to food allergens, the presence of allergic proteins has to be determined in various food products, as stipulated by the FAST-guidelines for Europe and other countries (Zuidmeer-Jongejan et al. 2012). However, due to the large amino acid divergence of allergenic parvalbumins from bony fish, no antibody based commercial assays have currently been developed to detect this protein in food products.

12.7 Prevalence of Allergies to Fish Parvalbumin

Seafood, including fish, play an important role in human nutrition and health, but can provoke serious IgE antibody-mediated adverse reactions in susceptible individuals (Lopata and Lehrer 2009, Jeebhay and Lopata 2012). A marked increase in allergic diseases is occurring in most major industrialized countries. The World Allergy Organization reports that in

2008, 20–30% of the world population was affected by allergy of some type. Seafood allergy and resulting anaphylaxis is particularly serious. Seafood allergy, including shellfish and fish, is typically life-long affecting up to 5% of all children and 2% of all adults. Prevalence rates of allergic reactions to fish vary considerably between regions and among children and adults.

Most of the population based prevalence studies come from Spain, Portugal, and Scandinavia. In Norway, allergic reactions to fish were reported in nearly 3% of children by the age of 2 years (Eggesbø et al. 1999). In the USA, allergy to fish was ~0.2% among 15,000 individuals of the general population (Sicherer et al. 2004). The major fish species reported causing allergic reactions were salmon, tuna, catfish, and cod followed by flounder, halibut, trout, and bass. As evidence for cross-sensitivity, the majority of allergic subjects reacted to multiple fish species (67%) (Sicherer et al. 2004).

Allergy to fish is common, not only in Western countries, but also in Asia, where allergic reactions to fish are significant among children and adults (Chiang et al. 2007). For example, a study from Singapore involving 227 children with food hypersensitivity confirmed that fish were significant sensitizers in approximately 13% of children. A subsequent prevalence study throughout Southeast Asia among 11,434 Filipino, 6,498 Singaporean, and 2,034 Thai test patients established that 2.3, 0.3 and 0.3% of the children suffered from allergic sensitization to fish, respectively (Connett et al. 2012). In Australia, a retrospective study among 2,999 children with food allergy demonstrated the prevalence of fish allergy being high at 5.6% of the tested population, with white fish, tuna, salmon and Asian seabass being the most implicated fish species (Turner et al. 2011).

12.8 Fish Allergy to Asian Seabass and other Fish

Recent research findings by our group identified two novel Asian seabass parvalbumin allergens, β1 and β2, and has characterized these proteins at both the molecular and immunological levels. While these two isoforms share only 67% of amino acid homology, their biochemical properties in charge and molecular weight are very similar. Interestingly though, when levels of IgE allergenicity of fish allergic patients are compared, Asian seabass parvalbumin isoforms demonstrate different immunological reactivity's (Sharp et al. unpub. data). Patients demonstrated significantly higher sensitivity to the β2-isoform suggesting that differences in protein structure of these isoforms influences allergenicity. This data lets us speculate that patients with low allergenic responses to the β1 isoform of Asian seabass will also exhibit similar low sensitivities to fish species like snapper which possess phylogenetically similar β1 isoforms. However, clinical studies will have to confirm these molecular investigations.

12.9 Conclusion

The muscle protein parvalbumin plays a major role in Asian seabass in the physiological functioning of muscle and nerve tissue. In addition parvalbumin has been identified as a possible marker for growth, however, current data are not conclusive and has to be further investigated. The protein parvalbumin from Asian seabass has also been identified as a food allergen, and has molecular and immunological similarities to other allergenic parvalbumins from a variety of bony fish, in particular large-mouth bass and channel catfish. Further detailed comparative studies on this important muscle protein will benefit the aquaculture industry by conceivably applying parvalbumin as a marker for increased growth rate, or if different isoforms are found to have different allergenicity, selecting for low allergenic Asian seabass families potentially allowing people allergic to parvalbumin to consume fish product.

References

Arif, S.H. 2009. A Ca^{2+}-binding protein with numerous roles and uses: Parvalbumin in molecular biology and physiology. BioEssays 31(4): 410–421.
Beale, J.E., M.F. Jeebhay and A.L. Lopata. 2009. Characterisation of purified parvalbumin from five fish species and nucleotide sequencing of this major allergen from Pacific pilchard, *Sardinops sagax*. Mol. Immunol. 46(15): 2985–2993.
Bugajska-Schretter, A., L. Elfman, T. Fuchs, S. Kapiotis, H. Rumpold, R. Valenta and S. Spitzauer. 1998. Parvalbumin, a cross-reactive fish allergen, contains IgE-binding epitopes sensitive to periodate treatment and Ca^{2+} depletion. J. Allergy Clin. Immunol. 101(1): 67–74.
Bugajska-Schretter, A., M. Grote, L. Vangelista, P. Valent, W. R. Sperr, H. Rumpold, A. Pastore, R. Reichelt, R. Valenta and S. Spitzauer. 2000. Purification, biochemical, and immunological characterisation of a major food allergen: different immunoglobulin E recognition of the apo- and calcium-bound forms of carp parvalbumin. Gut 46(5): 661–669.
Celio, M.R. 1986. Parvalbumin in most gamma-aminobutyric acid-containing neurons of the rat cerebral cortex. Science 231(4741): 995–997.
Chiang, W.C., M.I. Kidon, W.K. Liew, A. Goh, J.P.L. Tang and O.M. Chay. 2007. The changing face of food hypersensitivity in an Asian community. Clin. Exp. Allergy 37(7): 1055–1061.
Connett, G.J., I. Gerez, E.A. Cabrera-Morales, A. Yuenyongviwat, J. Ngamphaiboon, P. Chatchatee, P. Sangsupawanich, S.E. Soh, G.C. Yap, L.P.C. Shek and B.W. Lee. 2012. A population-based study of fish allergy in the Philippines, Singapore and Thailand. Int. Arch. Allergy Immunol. 159(4): 384–390.
Das Dores, S., C. Chopin, C. Villaume, J. Fleurence and J.L. Guéant. 2002. A new oligomeric parvalbumin allergen of Atlantic cod (Gad m 1) encoded by a gene distinct from that of Gad c 1. Allergy 57: 79–83.
Eggesbø, M., R. Halvorsen, K. Tambs and G. Botten. 1999. Prevalence of parentally perceived adverse reactions to food in young children. Pediatric Allergy and Immunology 10(2): 122–132.
Elsayed, S.M. and K. Aas. 1970. Characterization of a major allergen (cod.) chemical composition and immunological properties. Int. Arch. Allergy Appl. Immunol. 38(5): 536–48.
Escudero, R., P.M. Gamboa, J. Anton and M.L. Sanz. 2007. Food allergy due to trout roe. J. Invest. Allergol. Clin. Immunol. 17(5): 346–347.

Fæste, C.K. and C. Plassen. 2008. Quantitative sandwich ELISA for the determination of fish in foods. J. Immunol. Methods 329(1-2): 45–55.
Gonzalez-De-Olano, D., A. Rodriguez-Marco, E. Gonzalez-Mancebo, M. Gandolfo-Cano, A. Melendez-Baltanas and B. Bartolome. 2011. Allergy to red caviar. J. Invest. Allergol. Clin. Immunol. 21(6): 493–494.
Griesmeier, U., S. Vázquez-Cortés, M. Bublin, C. Radauer, Y. Ma, P. Briza, M. Fernández-Rivas and H. Breiteneder. 2010. Expression levels of parvalbumins determine allergenicity of fish species. Allergy 65(2): 191-198.
Jeebhay, M.F. and A.L. Lopata. 2012. Chapter 2—Occupational allergies in seafood-processing workers. Advances in Food and Nutrition Research. H. Jeyakumar, Academic Press. Volume 66: 47-73.
Jenkins, J.A., H. Breiteneder and E.N.C. Mills. 2007. Evolutionary distance from human homologs reflects allergenicity of animal food proteins. J. Allergy Clin. Immunol. 120(6): 1399-1405.
Kobayashi, A., H. Tanaka, Y. Hamada, S. Ishizaki, Y. Nagashima and K. Shiomi. 2006. Comparison of allergenicity and allergens between fish white and dark muscles. Allergy 61(3): 357-363.
Lim, D.L.-C., N. Keng Hwee, Y. Fong Cheng, C. Kaw Yan, G. Denise Li-Meng, S. Lynette Pei-Chi, G. Yoke Chin, P.S.V.B. Hugo and L. Bee Wah. 2008. Parvalbumin—the major tropical fish allergen. Pediatric Allergy and Immunology 19(5): 399–407.
Lindstrom, C.D., T. van Do, I. Hordvik, C. Endresen and S. Elsayed. 1996. Cloning of two distinct cDNAs encoding parvalbumin, the major allergen of Atlantic salmon (*Salmo salar*). Scand. J. Immunol. 44(4): 335–344.
Lopata, A.L. and S.B. Lehrer. 2009. New insights into seafood allergy. Curr. Opin. Allergy Clin. Immunol. 9(3): 270–277.
Perez-Gordo, M., J. Lin, L. Bardina, C. Pastor-Vargas, B. Cases, F. Vivanco, J. Cuesta-Herranz and H.A. Sampson. 2011. Epitope mapping of Atlantic Salmon major allergen by peptide microarray immunoassay. Int. Arch. Allergy Immunol. 157(1): 31–40.
Perez-Gordo, M., S. Sanchez-Garcia, B. Cases, C. Pastor, F. Vivanco and J. Cuesta-Herranz. 2008. Identification of vitellogenin as an allergen in Beluga caviar allergy. Allergy 63(4): 479–480.
Rosmilah, M., M. Shahnaz, A. Masita, A. Noormalin and M. Jamaludin. 2005. Identification of major allergens of two species of local snappers: *Lutjanus argentimaculatus* (merah/red snapper) and *Lutjanus johnii* (jenahak/ golden snapper). Tropical Biomedicine 22(2): 171–177.
Sakaguchi, M., M. Toda, T. Ebihara, S. Irie, H. Hori, A. Imai, M. Yanagida, H. Miyazawa, H. Ohsuna, Z. Ikezawa and S. Inouye. 2000. IgE antibody to fish gelatin (type I collagen) in patients with fish allergy. J. Allergy Clin. Immunol. 106(3): 579–584.
Sharp, M.F. and A.L. Lopata. 2013. Fish Allergy: In Review. Clin. Rev. Allergy Immunol.
Sicherer, S.H., A. Muñoz-Furlong and H.A. Sampson. 2004. Prevalence of seafood allergy in the United States determined by a random telephone survey. J. Allergy Clin. Immunol. 114(1): 159–165.
Tanokura, M., K. Goto, Y. Toyomori and K. Yamada. 1987. Preparation and characterization of the major isotype of parvalbumin from skeletal muscle of the toad (*Bufo bufo japonicus*). J. Biochem. 102(5): 1133–1139.
Turner, P., I. Ng, A.S. Kemp and D.E. Campbell. 2011. Seafood allergy in children—A descriptive study. Clin. Transl. Allergy 106(6): 494–501.
Untersmayr, E., K. Szalai, A.B. Riemer, W. Hemmer, I. Swoboda, B. Hantusch, I. Schöll, S. Spitzauer, O. Scheiner, R. Jarisch, G. Boltz-Nitulescu and E. Jensen-Jarolim. 2006. Mimotopes identify conformational epitopes on parvalbumin, the major fish allergen. Mol. Immunol. 43(9): 1454–1461.
Van Do, T., S. Elsayed, E. Florvaag, I. Hordvik and C. Endresen. 2005. Allergy to fish parvalbumins: Studies on the cross-reactivity of allergens from nine commonly consumed fish. J. Allergy Clin. Immunol. 116(6): 1314–1320.

Van Do, T., I. Hordvik, C. Endresen and S. Elsayed. 2003. The major allergen (parvalbumin) of codfish is encoded by at least two isotypic genes: cDNA cloning, expression and antibody binding of the recombinant allergens. Mol. Immunol. 39(10): 595–602.
Wang, C.M., L.C. Lo, F. Feng, Z.Y. Zhu and G.H. Yue. 2008. Identification and verification of QTL associated with growth traits in two genetic backgrounds of Barramundi (*Lates calcarifer*). Anim. Genet. 39(1): 34–39.
Whittington, A.C. and T.S. Moerland. 2012. Resurrecting prehistoric parvalbumins to explore the evolution of thermal compensation in extant Antarctic fish parvalbumins. The Journal of Experimental Biology 215(18): 3281–3292.
Xu, Y.X., Z.Y. Zhu, L.C. Lo, C.M. Wang, G. Lin, F. Feng and G.H. Yue. 2006. Characterization of two parvalbumin genes and their association with growth traits in Asian seabass (*Lates calcarifer*). Anim. Genet. 37(3): 266–268.
Zuidmeer-Jongejan, L., M. Fernandez-Rivas, L.K. Poulsen, A. Neubauer, J. Asturias, L. Blom, J Boye, C. Bindslev-Jensen, M. Clausen, R. Ferrara, P. Garosi, H. Huber, B. M. Jensen, S. Koppelman, M.L. Kowalski, A. Lewandowska-Polak, B. Linhart, B. Maillere, A. Mari, A. Martinez, C.E. Mills, C. Nicoletti, D.J. Opstelten, N.G. Papadopoulos, A. Portoles, N. Rigby, E. Scala, H.J. Schnoor, S. Sigursdottir, G. Stavroulakis, F. Stolz, I. Swoboda, R. Valenta, R. van den Hout, S.A. Versteeg, M. Witten and R. van Ree. 2012. FAST: Towards safe and effective subcutaneous immunotherapy of persistent life-threatening food allergies. Clin. Transl. Allergy 2(1): 5.

Index

2-methylisoborneol 240
2-MIB 240–243

A

α-melanocyte-stimulating hormone 239
α-parvalbumin 296, 298
α-tocopherol 235–238
acidification 58, 59
allergen 293–303
amino Acid Requirements 182–184, 221
Amyloodimium 289
Australian barramundi aquaculture industry 259, 261, 267
Australian Barramundi Farmers Association (ABFA) 259–261, 267, 268

B

β-parvalbumins 296, 298, 300
bacterial Disease 116, 118
Barramundi 258–271
Benedenia 108, 124, 289
Betanodavirus 115, 116
biosecurity measures 114
blood flukes 125
breeding 70, 73, 74
bromophenols 246–249
broodstock health management 74

C

Caligus 289
Caligus chiastos 128
Calreticulin 165, 170
cannibalism 24, 26, 275, 276, 278, 280–282
carbohydrates 186, 190, 192, 194, 209, 210
catch adjusted for effort (CAE) 38–43, 54, 56
catch data 33–35, 37, 38, 41, 54, 56
cell lines 162, 163

Chilodonella 104, 122
Chlamydia 118
choline 192, 193
copepods 127
cryopreservation 160, 161, 171
Cryptocarion irritans 104
Cryptocaryon 289
Cymothoa indica 113, 129

D

Dactylogyrus 289
deformities 26, 27
development of bony structures 22
digestible protein to digestible energy ratio 181
digestive system 23
Diplectanum 109, 125
Diseases 285, 286, 288, 289

E

embryonic development 17
Energy demand 182, 184, 194–196, 198, 200–203
energy requirements 178, 184, 194, 221
Epinepheli 108, 110, 124
Epitheliocystis 104, 118
Epizootic Ulcerative Syndrome 119
essential fatty acids 184, 187
essential nutrient requirements 179

F

fecundity 70, 81, 82
feed Management 201, 218
feed ration 203, 204, 219
fillet quality 233
flavor 251
Flexibacter 104, 116, 118
floating net cages 274, 277, 285–287
F_{ST} 140, 141, 143–145

G

genetic correlations 154, 156
genetic linkage 167, 168
genome 137, 138, 162, 164, 167–169, 171
genome Mapping 164
genomics 164, 167
geosmin 240–244
grading 275, 278, 280, 281
grow-out 273–275, 281–283, 286–289
GSM 240–245

H

H. macropterus 2
haploid C-value 138
harvesting 230, 233, 234, 239
hatchery 147, 150, 152, 158, 170
health 288
heritability 154–156, 169
Hypopterus 2–5

I

ice-water immersion 231, 232
Ich 122
Ichthyophthirius multifiliis 122
IgE epitope 299
impoundments 78, 79, 91
Indonesia 273–275, 278–280, 282, 286–288, 290
inositol 192, 193
insulin-like growth factor-I 166
Irritans 289
isopods 129

L

L. japonicus 5–7, 9, 10
L. lakdiva 6, 7, 9, 12
L. uwisara 6–9, 12
Labelling 264, 266
Lates calcarifer 1, 5–7, 10, 12
Lates japonicus 7, 8, 10
Lates lakdiva 8–10
Lates uwisara 9, 10
Laticola latesi 109, 125
LCCWet 51–54
Lecithochirium 289
Leeches 119, 120, 124, 127
Lernanthropus latis 112, 128, 129
Lipid and essential fatty acids 184
lipid oxidation 234–238
Lipid sources 218
Lipids 184–186, 194

luteinizing hormone releasing hormone-analog A 70
Lymphocystis 103, 116

M

Madden–Julian Oscillation 41, 56
male maturation 68
markets 261, 263, 264, 266, 267, 270, 271
maturation 68–72, 74
melanin 239
Melleni 109, 124
metamorphosis 18, 24
methyltestosterone 70
microsatellite 298
minerals 193, 222
Monogeneans 124
movements 82, 86–93
Myostatin 165

N

Neobenedenia 108, 109, 124, 289
neurocranium 22, 23
niacin 193
Nodaviruses 115
nursery culture 275–281
nursery habitats 80, 82–85, 87, 89, 90
Nybelinia indica 105, 126

O

off-odours 234
Oodinium 120
Oomycetes 119
osteological development 22
ovaries 68, 70, 71
oxidative deterioration 237

P

P. waigiensis 2
pantothenic acid 192, 193
Parasanguinicola vastispina 106, 125
Parvalbumin 165, 293–303
parvalbumin isoforms 294–299, 302
pathogens 283, 288
Philippines 273–275, 279–283, 287, 288, 290
phylogenetic 299, 300
pond nurseries 277
pond preparation 282, 284
population genetics 139, 164
Prosorhynchus 289
protandry 68

Protein 179–187, 190–192, 194, 198, 200–203, 208, 209–217, 219, 221–223
Protein and amino acid utilisation 183
Protein and amino acids 179, 181
Psammoperca 2–4
Pseudometadena 289
Pseudorhabdosynochus 289
pyridoxine 192, 193

Q

Quantitative trait loci (QTL) 167–170

R

rainfall 34, 35, 37–43, 47–52, 54–58
recruitment 80, 82, 83, 96
regional population genetic structure 139
reproduction 80, 87
reproductive season 70, 72
riboflavin 192, 193

S

scale drop disease 290
Scolex pleuronectis 105, 126
seabass culture 274, 275, 284, 287–290
SEAFDEC 273, 274
seasonality 80
sex change 81, 82
sex chromosomes 139
Sex inversion 67, 68, 72, 73
size- and age-at-maturity 82
skeletal deformities 26
spawning grounds 81, 82, 84, 87, 88
stocked fishery 78
stocking densities 287
Streptococcossis 117
Streptococcus 103, 116, 117, 289
Streptococcus iniae 117

T

tagging programs 86
Taiwan 274, 275
TBARS 237
testis 68, 69, 72
thermal tolerance 158, 159
thiamine 192, 193
thiobarbituric acid 237
Transversatrema 289
trash fish 276, 280, 282–284, 288
Trichodina 104, 121, 122
Trypanosoma 104, 119, 120
type locality 5, 6, 10, 11

U

ubiquinol 235

V

velvet disease 120
Vibrio 104, 116–118, 127, 289
Vibriosis 117, 289
Vietnam 274, 275, 282
viral disease 114, 115
viral encephalopathy and retinopathy (VER) 115
viral nervous necrosis 27
viral nervous necrosis (VNN) 115, 116
virus 114–116
vitamin 192, 193, 222, 235, 236, 238
vitamin C 193, 222
vitamin E 192, 193

Z

Zeylanicobdella arugamensis 111, 127

Color Plate Section

Chapter 1

Figure 1.2 A) *Psammoperca waigiensis*, AMS IA.1456, 175 mm SL, Queensland, Australia; and **B)** *Hypopterus macropterus*, WAM P.30162-008, 90 mm SL, Shark Bay, Western Australia. Arrows indicate characters that distinguish the genus from *Lates* (see Fig. 1.3). 1, nostrils set widely apart; 2, lower jaw not in advance of upper jaw; 3, maxilla falls under the eye, not behind it. [Figure 1.2B courtesy Barry Hutchins, Western Australian Museum.]

Color Plate Section 313

Figure 1.3. *Lates calcarifer*, AMS I.40053-001, 183 mm SL, Northern Territory, Australia.

Figure 1.4. *Lates japonicus*, AMS I.25742-001, 132 mm SL, Kyushu, Japan.

314 *Biology and Culture of Asian Seabass*

Figure 1.5. *Lates lakdiva*, AMS I.37516-001, 220 mm SL, western Sri Lanka.

Figure 1.6. *Lates uwisara*, ANFC H.6316-10, 353 mm SL, Myanmar.

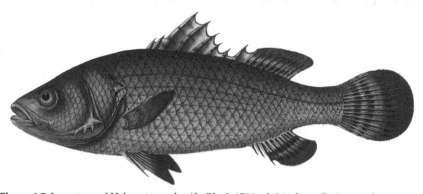

Figure 1.7. Iconotype of *Holocentrus calcarifer* Bloch 1790; pl. 244, laterally inverted.

Chapter 7

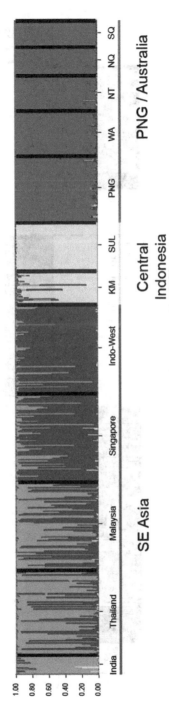

Figure 7.2. STRUCTURE ancestry membership plot for 558 individual Asian seabass sampled from India, SE Asia, the Indo-Pacific and Australasian regions and genotyped for 7 hypervariable microsatellite markers. The plot was generated using sampling locations (indicated on x-axis) as population priors and assuming a correlated allele frequency and admixture model. Some sampling location names are abbreviated (KM = Kalimantan, SUL = Sulawesi, PNG = Papua New Guinea, WA = Western Australia, NT = Northern Territory (Australia), NQ = rortheast Queensland, SQ = southeast Queensland). Ten independent runs of each value of K from 1–10 were run, the most likely K (4) following the methods of Evanno et al. (2005) was selected and the proportion of ancestry to each of the four clusters is indicated by different colors.

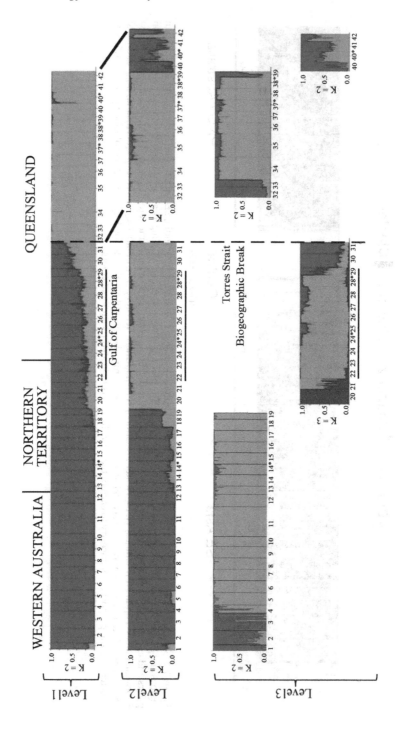

Figure 7.4. Hierarchical STRUCTURE ancestry membership bar plot for 1205 individual Asian seabass sampled from 43 locations around Australia and genotyped for 16 microsatellite loci. The plots were generated using sampling locations as population priors and assuming a correlated allele frequency and admixture model. Ten independent runs of each value of K from 1–10 were performed, the most likely K (number of clusters) for each level of the hierarchical analysis was determined by the methods of Evanno et al. (2005) and is indicated on the Y-axis. The proportion of ancestry to each of the identified clusters is indicated for each individual fish by different colors within each bar. The 43 sampling locations are indicated by differing numbers along the x-axis and are displayed from left to right in geographical order from Broome in Western Australia along the northern and eastern coasts of Australia to the Mary River in south-east Queensland. Five populations (14, 24, 28, 37, 38 and 40) have been sampled twice at different times (up to 20 years apart) and the earliest temporal sample is indicated by *. State and Territory boundaries (corresponding to different fisheries management jurisdictions) are indicated above the level 1 plot, while populations within the Gulf of Carpentaria are joined by a solid bar underneath each applicable plot. The most likely position of the Torres Strait biogeographic break is indicated by the vertical dashed line. Population codes are as follows. 1 Broome, 2 St George Basin, 3 Admiralty Gulf, 4 Swift Bay, 5 Drysdale River, 6 Salmon Bay, 7 King George, 8 Berkley River, 9 Helby River, 10 Nulla Nulla Creek, 11 Ord River, 12 Bonaparte Gulf, 13 Moyle River, 14 Daly River – 2008, 14* Daly River - 1990, 15 Bathurst Island, 16 Darwin Harbour, 17 Shoal Bay, 18 Mary River, 19 Alligator River, 20 Liverpool River, 21 Arnhem Bay, 22 Roper River, 23 McArthur river, 24 Albert River – 2011, 24* Albert/Leichhardt River – 1990/91, 25 Gilbert River, 26 Mitchell River, 27 Holroyd River, 28 Archer River – 2011, 28* Archer River – 1993, 29 Jardine River, 30 Jacky Jacky Creek - Kennedy Inlet, 31 Escape River, 32 Princess Charlotte Bay, 33 Bizant River, 34 Johnstone River, 35 Hinchinbrook Channel, 36 Cleveland Bay, 37 Bowling Green Bay – 2008, 37* Bowling Green Bay – 1988, 38 Burdekin River – 2008, 38* Burdekin River – 1989/90, 39 Broad Sound, 40 Fitzroy River – 2008, 40* Fitzroy River – 1988/90, 41 Port Alma, 42 Mary River.